Five Hundred
Mathematical Challenges

© 1995 by
The Mathematical Association of America (Incorporated)
Library of Congress Catalog Card Number 95-78644

ISBN 0-88385-519-4

Printed in the United States of America

Current Printing (last digit):
10   9   8   7   6   5   4   3   2   1

Five Hundred
Mathematical Challenges

Edward J. Barbeau
University of Toronto

Murray S. Klamkin
University of Alberta

William O. J. Moser
McGill University

Published by
THE MATHEMATICAL ASSOCIATION OF AMERICA

SPECTRUM SERIES

The Spectrum Series of the Mathematical Association of America was so named to reflect its purpose: to publish a broad range of books including biographies, accessible expositions of old or new mathematical ideas, reprints and revisions of excellent out-of-print books, popular works, and other monographs of high interest that will appeal to a broad range of readers, including students and teachers of mathematics, mathematical amateurs, and researchers.

All the Math That's Fit to Print, by Keith Devlin
Circles: A Mathematical View, by Dan Pedoe
Complex Numbers and Geometry, by Liang-shin Hahn
Cryptology, by Albrecht Beutelspacher
Five Hundred Mathematical Challenges, Edward J. Barbeau, Murray S. Klamkin, and William O. J. Moser
From Zero to Infinity, by Constance Reid
I Want to be a Mathematician, by Paul R. Halmos
Journey into Geometries, by Marta Sved
The Last Problem, by E. T. Bell (revised and updated by Underwood Dudley)
The Lighter Side of Mathematics: Proceedings of the Eugène Strens Memorial Conference on Recreational Mathematics & its History, edited by Richard K. Guy and Robert E. Woodrow
Lure of the Integers, by Joe Roberts
Mathematical Carnival, by Martin Gardner
Mathematical Circus, by Martin Gardner
Mathematical Cranks, by Underwood Dudley
Mathematical Magic Show, by Martin Gardner
Mathematics: Queen and Servant of Science, by E. T. Bell
Memorabilia Mathematica, by Robert Edouard Moritz
New Mathematical Diversions, by Martin Gardner
Numerical Methods that Work, by Forman Acton
Out of the Mouths of Mathematicians, by Rosemary Schmalz
Polyominoes, by George Martin
The Search for E. T. Bell, also known as John Taine, by Constance Reid
Shaping Space, edited by Marjorie Senechal and George Fleck
Student Research Projects in Calculus, by Marcus Cohen, Edward D. Gaughan, Arthur Knoebel, Douglas S. Kurtz, and David Pengelley
The Trisectors, by Underwood Dudley
The Words of Mathematics, by Steven Schwartzman

Mathematical Association of America
1529 Eighteenth Street, NW
Washington, DC 20036
800-331-1MAA   FAX 202-265-2384

PREFACE

This collection of problems is directed to students in high school, college and university. Some of the problems are easy, needing no more than common sense and clear reasoning to solve. Others may require some of the results and techniques which we have included in the Tool Chest. None of the problems require calculus, so the collection could be described as "problems in pre-calculus mathematics". However, they are definitely not the routine nor "drill" problems found in textbooks. They could be described as challenging, interesting, thought-provoking, fascinating. Many have the "stuff" of real mathematics; indeed quite a few are the simplest cases of research-level problems. Hence they should provide some insight into what mathematical research is about.

The collection is dedicated to students who find pleasure in wrestling with, and finally overcoming, a problem whose solution is not apparent at the outset. It is also dedicated to teachers who encourage their students to rise above the security offered by prefabricated exercises and thus experience the creative side of mathematics. Teachers will find here problems to challenge mathematically oriented students, such as may be found in mathematics clubs or training sessions for mathematical competitions. Seeking solutions could be a collective experience, for collaboration in research often succeeds when a lonely effort might not.

Pay no attention to the solutions until your battle with a problem has resulted in a resounding victory or disappointing defeat. The solutions we have given are not to be regarded as definitive, although they may suggest possibilities for exploring similar situations. A particular problem may be resolved in several distinct ways, embodying different approaches and revealing various facets. Some solutions may be straightforward while others may be elegant and sophisticated. For this reason we have often included more than one solution. Perhaps you may discover others.

While many of these problems should be new to you, we make no claim for the originality of most of the problems. We acknowledge our debt to the unsung creators of the problems, recognizing how hard it is to create a problem which is interesting, challenging, instructive, and solvable without being impossible or tedious. With few exceptions, the problems appeared in a series of five booklets which were available from the Canadian Mathematical Society. Indeed, the first of these appeared in 1973. Since then they have received a steady distribution, and now we feel that an edited, revised version of all five together is desirable. The problems are arranged in no particular order of difficulty or subject matter. We welcome communications from the readers, comments, corrections, alternative solutions, and suggested problems.

Collecting and creating the problems, and editing them, has been a rewarding learning experience for us. We will feel fully rewarded if teachers and students find this collection useful and entertaining.

E. Barbeau,   M. Klamkin,   W. Moser

Contents

PROBLEMS

Problem 1. The length of the sides of a right triangle are three consecutive terms of an arithmetic progression. Prove that the lengths are in the ratio $3 : 4 : 5$.

Problem 2. Consider all line segments of length 4 with one endpoint on the line $y = x$ and the other endpoint on the line $y = 2x$. Find the equation of the locus of the midpoints of these line segments.

Problem 3. A rectangle is dissected as shown in Figure 1, with some of the lengths indicated. If the pieces are rearranged to form a square, what is the perimeter of the square?

Problem 4. Observe that

$$3^2 + 4^2 = 5^2,$$

$$5^2 + 12^2 = 13^2,$$

$$7^2 + 24^2 = 25^2,$$

$$9^2 + 40^2 = 41^2.$$

State a general law suggested by these examples, and prove it.

Problem 5. Calculate the sum

$$6 + 66 + 666 + \cdots + \underbrace{666\ldots6}_{n\ 6's} \qquad (n \geq 1).$$

Problem 6. Alice, Betty, and Carol took the same series of examinations. For each examination there was one mark of x, one mark of y, and one mark of z, where x, y, z are distinct positive integers. The total of the marks obtained by each of the girls was: Alice—20; Betty—10; Carol—9. If Betty placed first in the algebra examination, who placed second in the geometry examination?

Problem 7. Let

$$f_0(x) = \frac{1}{1-x}, \quad \text{and} \quad f_n(x) = f_0(f_{n-1}(x)),$$

$n = 1, 2, 3, 4, \ldots$. Evaluate $f_{1976}(1976)$.

Problem 8. Show that from any five integers, not necessarily distinct, one can always choose three of these integers whose sum is divisible by 3.

Problem 9. Mr. Smith commutes to the city regularly and invariably takes the same train home which arrives at his home station at 5 PM. At this time, his chauffeur always just arrives, promptly picks him up, and drives him home. One fine day, Mr. Smith takes an earlier train and arrives at his home station at 4 PM. Instead of calling or waiting for his chauffeur until 5 PM, he starts walking home. On his way he meets the chauffeur who picks him up promptly and returns home arriving 20 minutes earlier than usual. Some weeks

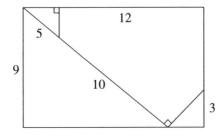

FIGURE 1

later, on another fine day, Mr. Smith takes an earlier train and arrives at his home station at 4:30 PM. Again instead of waiting for his chauffeur, he starts walking home. On his way he meets the chauffeur who picks him up promptly and returns home. How many minutes earlier than usual did he arrive home this time?

Problem 10. Suppose that the center of gravity of a water jug is above the inside bottom of the jug, and that water is poured into the jug until the center of gravity of the combination of jug and water is as low as possible. Explain why the center of gravity of this "extreme" combination must lie at the surface of the water.

Problem 11. A father, mother, and son decide to hold a family tournament, playing a particular two-person board game which must end with one of the players winning (i.e., no "tie" is possible). After each game the winner then plays the person who did not play in the game just completed. The first player to win two games (not necessarily consecutive) wins the tournament. It is agreed that, because he is the oldest, the father may choose to play in the first game or to sit out the first game. Advise the father what to do: play or not to play in the first game. (USAMO 1974)

Problem 12. $EFGH$ is a square inscribed in the quadrilateral $ABCD$ as in Figure 2. If $\overline{EB} =$

$\overline{FC} = \overline{GD} = \overline{HA}$, prove that $ABCD$ is also a square.

Problem 13. Show that among any seven distinct positive integers not greater than 126, one can find two of them, say x and y, satisfying the inequalities $1 < \frac{y}{x} \le 2$.

Problem 14. Show that if 5 points are all in, or on, a square of side 1, then some pair of them will be no further than $\frac{\sqrt{2}}{2}$ apart.

Problem 15. During an election campaign n different kinds of promises are made by the various political parties, $n > 0$. No two parties have exactly the same set of promises. While several parties may make the same promise, every pair of parties have at least one promise in common. Prove that there can be as many as 2^{n-1} parties, but no more.

Problem 16. Given a $(2m + 1) \times (2n + 1)$ checkerboard in which the four corners are black squares, show that if one removes any one red square and any two black squares, the remaining board is coverable with dominoes (i.e., 1×2 rectangles).

Problem 17. The digital sum $D(n)$ of a positive integer n is defined recursively as follows:

$$D(n) =$$
$$\begin{cases} n & \text{if } 1 \le n \le 9, \\ D(a_0 + a_1 + a_2 + \cdots + a_m) & \text{if } n > 9, \end{cases}$$

where $a_0, a_1, \ldots, a_m$ are all the digits of n expressed in base 10, i.e.,

$$n = a_m 10^m + a_{m-1} 10^{m-1} + \cdots + a_1 10 + a_0.$$

For example $D(989) = D(26) = D(8) = 8$. Prove that

$$D((1234)n) = D(n) \quad \text{for } n = 1, 2, 3, \ldots.$$

Problem 18. Given three points A, B, C construct a square with a center A such that two adjoining sides (or their extensions) pass through B and C respectively.

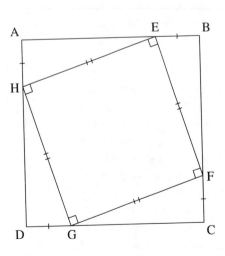

FIGURE 2

Problem 19.   Give an elementary proof that

$$\sqrt{n}^{\sqrt{n+1}} > \sqrt{n+1}^{\sqrt{n}}, \quad n = 7, 8, 9, \ldots.$$

Problem 20.   If, in a circle with center O, OXY is perpendicular to chord AB (as shown in Figure 3), prove that $\overline{DX} \le \overline{CY}$ (see Figure 4). (P. Erdős and M. Klamkin)

Problem 21.   Let $a_1, a_2, \ldots, a_n$ be n positive integers. Show that for some i and k ($1 \le i \le i + k \le n$)

$$a_i + \cdots + a_{i+k} \text{ is divisible by } n.$$

Problem 22.   Given a finite number of points in the plane with distances (between pairs) distinct, join each point by a straight line segment to the point nearest to it. Show that the resulting configuration contains no triangle.

Problem 23.   Show that if m is a positive rational number then $m + \frac{1}{m}$ is an integer only if $m = 1$.

Problem 24.   Let P be the center of the square constructed on the hypotenuse AC of the right-angled triangle ABC. Prove that BP bisects $\angle ABC$.

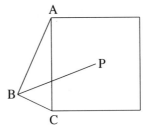

FIGURE 5

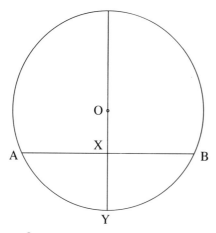

FIGURE 3

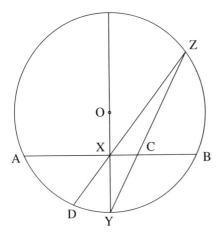

FIGURE 4

Problem 25.   Four distinct lines L_1, L_2, L_3, L_4 are given in the plane, with L_1 and L_2 respectively parallel to L_3 and L_4. Find the locus of a point moving so that the sum of its perpendicular distances from the four lines is constant.

Problem 26.   Suppose 5 points are given in the plane, not all on a line, and no 4 on a circle. Prove that there exists a circle through three of them such that one of the remaining 2 points is inside the circle while the other is outside the circle.

Problem 27.   Let ABC be an arbitrary triangle, and P any point inside. Let d_1, d_2, and d_3 denote the perpendicular distance from P to side BC, CA, and AB respectively. Let h_1, h_2, and h_3 denote respectively the length of the altitude from A, B, C to the opposite side of the triangle. Prove that

$$\frac{d_1}{h_1} + \frac{d_2}{h_2} + \frac{d_3}{h_3} = 1.$$

Problem 28. A boy lives in each of n houses on a straight line. At what point should the n boys meet so that the sum of the distances that they walk from their houses is as small as possible?

Problem 29. Let P be one of the two points of intersection of two intersecting circles. Construct the line l through P, not containing the common chord, such that the two circles cut off equal segments on l.

Problem 30. On each side of an arbitrary triangle ABC, an equilateral triangle is constructed (outwards), as in Figure 6. Show that $\overline{AP} = \overline{BQ} = \overline{CR}$.

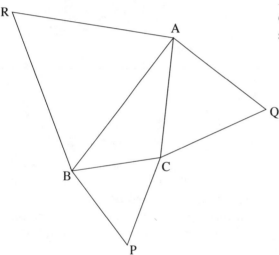

FIGURE 6

Problem 31. Show that if n is a positive integer greater than 1, then

$$1 + \frac{1}{2} + \frac{1}{3} + \cdots + \frac{1}{n}$$

is not an integer.

Problem 32. Two points on a sphere of radius 1 are joined by an arc of length less than 2, lying inside the sphere. Prove that the arc must lie in some hemisphere of the given sphere. (USAMO 1974)

Problem 33. Prove that for any positive integer n,

$$\left[\frac{n}{3}\right] + \left[\frac{n+2}{6}\right] + \left[\frac{n+4}{6}\right] = \left[\frac{n}{2}\right] + \left[\frac{n+3}{6}\right].$$

Problem 34. Prove that the sum of all the n-digit integers $(n > 2)$ is

$$494 \underbrace{99\ldots9}_{n-3 \ 9's} 55 \underbrace{00\ldots0}_{n-2 \ 0's}.$$

Problem 35. Let ABC be the right-angled isosceles triangle whose equal sides have length 1. P is a point on the hypotenuse, and the feet of the perpendiculars from P to the other sides are Q and R. Consider the areas of the triangles APQ and PBR, and the area of the rectangle $QCRP$. Prove that regardless of how P is chosen, the largest of these three areas is at least $2/9$.

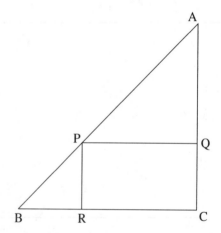

FIGURE 7 $\overline{BC} = \overline{CA} = 1$

Problem 36. Prove that a triangle with sides of lengths 5, 5, 6 has the same area as the triangle with sides of lengths 5, 5, 8. Find other pairs of incongruent isosceles triangles, with integer sides, having equal areas.

Problem 37. A quadrilateral has one vertex on each side of a square of side-length 1. Show that the lengths a, b, c, and d of the sides of the quadrilateral satisfy the inequalities

$$2 \le a^2 + b^2 + c^2 + d^2 \le 4.$$

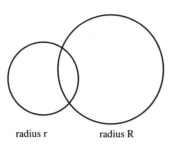

radius r   radius R

FIGURE 8

Problem 38. A circle of radius r intersects another circle, of radius R ($R > r$). (See Figure 8.) Find an expression for the difference in the areas of the nonoverlapping parts.

Problem 39. The number 3 can be expressed as an ordered sum of one or more positive integers in four ways, namely as

$$3, \qquad 1+2, \qquad 2+1, \qquad 1+1+1.$$

Show that the positive integer n can be so expressed in 2^{n-1} ways.

Problem 40. Teams $T_1, T_2, \ldots, T_n$ take part in a tournament in which every team plays every other team just once. One point is awarded for each win, and it is assumed that there are no draws. Let $s_1, s_2, \ldots, s_n$ denote the (total) scores of $T_1, T_2, \ldots, T_n$ respectively. Show that, for $1 < k < n$,

$$s_1 + s_2 + \cdots + s_k \le nk - \frac{1}{2}k(k+1).$$

Problem 41. Observe that

$$1^2 = \frac{1 \cdot 2 \cdot 3}{6}$$

$$1^2 + 3^2 = \frac{3 \cdot 4 \cdot 5}{6}$$

$$1^2 + 3^2 + 5^2 = \frac{5 \cdot 6 \cdot 7}{6}.$$

Guess a general law suggested by these examples, and prove it.

Problem 42. In the following problem no "aids" such as tables, calculators, etc. should be used.

(a) Prove that the values of x for which $x = \frac{x^2+1}{198}$ lie between $\frac{1}{198}$ and $197.99494949\ldots$.

(b) Use the result of (a) to prove that $\sqrt{2} < 1.41421356421356421356\ldots$.

(c) Is it true that $\sqrt{2} < 1.41421356$?

Problem 43. Prove that if 5 pins are stuck onto a piece of cardboard in the shape of an equilateral triangle of side length 2, then some pair of pins must be within distance 1 of each other.

Problem 44. Given an even number of points in the plane, does there exist a straight line having half of the points on each side of the line?

Problem 45. Two circles intersect in points A, B. PQ is a line segment through A and terminating on the two circles. Prove that $\overline{BP}/\overline{BQ}$ is constant for all allowable configurations of PQ.

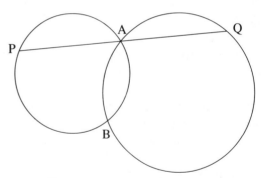

FIGURE 9

Problem 46. Let $f(n)$ be the sum of the first n terms of the sequence

$$0, 1, 1, 2, 2, 3, 3, 4, 4, \ldots, r, r, r+1, r+1, \ldots.$$

(a) Deduce a formula for $f(n)$.

(b) Prove that $f(s+t) - f(s-t) = st$ where s and t are positive integers and $s > t$.

Problem 47. Three noncollinear points P, Q, R are given. Find the triangle for which P, Q, R are the midpoints of the edges.

Problem 48. Prove that $1^{99} + 2^{99} + 3^{99} + 4^{99} + 5^{99}$ is divisible by 5.

Problem 49. Show that there are no integers a, b, c for which $a^2 + b^2 - 8c = 6$.

Problem 50. If a, b, c, d are four distinct numbers then we can form six sums of two at a time, namely $a + b$, $a + c$, $a + d$, $b + c$, $b + d$, $c + d$. Split the integers 1, 2, 3, 4, 5, 6, 7, 8 into two sets, four in each set, so that the six sums of two at a time for one of the sets is the same as that of the other set (not necessarily in the same order). List all possible ways in which this can be done.

Problem 51. If $2 \log(x - 2y) = \log x + \log y$, find $\frac{x}{y}$.

Problem 52. Let f be a function with the following properties:
(1) $f(n)$ is defined for every positive integer n;
(2) $f(n)$ is an integer;
(3) $f(2) = 2$;
(4) $f(mn) = f(m)f(n)$ for all m and n;
(5) $f(m) > f(n)$ whenever $m > n$.
Prove that $f(n) = n$ for $n = 1, 2, 3, \ldots$.

Problem 53. If a line l in space makes equal angles with three given lines in a plane π, show that l is perpendicular to π.

Problem 54. Let a be the integer

$$a = \underbrace{111 \ldots 1}_{m\ 1's}$$

(where the number of 1's is m); let

$$b = 1\underbrace{00 \ldots 00}_{m-1\ 0's}5$$

(where the number of 0's between digits 1 and 5 is $m - 1$). Prove that $ab + 1$ is a square integer. Express the square root of $ab + 1$ in the same form as a and b are expressed.

Problem 55. Two flag poles of heights h and k are situated $2a$ units apart on a level surface. Find the set of all points on the surface which are so situated that the angles of elevation, at each point, of the tops of the poles are equal.

Problem 56. Prove that, for $n = 1, 2, 3, \ldots$,

$$1 + \frac{1}{1!} + \frac{1}{2!} + \frac{1}{3!} + \cdots + \frac{1}{n!} < 3.$$

Problem 57. Let X be any point between B and C on the side BC of the convex quadrilateral $ABCD$ (as in Figure 10). A line is drawn through B parallel to AX and another line is drawn through C parallel to DX. These two lines intersect at P. Prove that the area of the triangle APD is equal to the area of the quadrilateral $ABCD$.

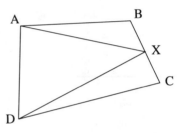

FIGURE 10

Problem 58. Let

$$s_n = 1 + \frac{1}{\sqrt{2}} + \frac{1}{\sqrt{3}} + \cdots + \frac{1}{\sqrt{n}}.$$

Show that $2\sqrt{n+1} - 2 < s_n < 2\sqrt{n} - 1$.

Problem 59. Show that for any quadrilateral inscribed in a circle of radius 1, the length of the shortest side is not more than $\sqrt{2}$.

Problem 60. Prove that if a convex polygon has four of its angles equal to $90°$ then it must be a rectangle.

Problem 61. You are given 6 congruent balls, two each of colors red, white, and blue, and informed that one ball of each color weighs 15 grams while the other weighs 16 grams. Using an equal arm balance only twice, determine which three are the 16-gram balls.

Problem 62. A plane flies from A to B and back again with a constant engine speed. Turn-around time may be neglected. Will the travel time be more with a wind of constant speed blowing in the direction from A to B than in still air? (Does your intuition agree?)

Problem 63. Tetrahedron $OABC$ is such that lines OA, OB, and OC are mutually perpendicular. Prove that triangle ABC is not a right-angled triangle.

Problem 64. Find all number triples (x, y, z) such that when any one of these numbers is added to the product of the other two, the result is 2.

Problem 65. Let nine points be given in the interior of the unit square. Prove that there exists a triangle of area at most $\frac{1}{8}$ whose vertices are three of the nine points. (See also problem 14 or 43.)

Problem 66. Let a, b, and c be the lengths of the sides of a triangle. Show that if $a^2 + b^2 + c^2 = bc + ca + ab$ then the triangle is equilateral.

Problem 67. A triangle has sides of lengths a, b, c and respective altitudes of lengths h_a, h_b, h_c. If $a \geq b \geq c$ show that $a + h_a \geq b + h_b \geq c + h_c$.

Problem 68. Let n be a five-digit number (whose first digit is nonzero) and let m be the four-digit number formed from n by deleting its middle digit. Determine all n such that $\frac{n}{m}$ is an integer.

Problem 69. Prove that for nonzero numbers x, y, z the expressions

$$x^n + y^n + z^n, \qquad (x + y + z)^n$$

are equal for any odd integer n provided this is so when $n = -1$.

Problem 70. An army captain wishes to station an observer equally distant from two specified points and a straight road. Can this always be done? Locate any possible stations. In other words, how many points are there in the Euclidean plane which are equidistant from two given points and a given line? Find them with straight-edge and compasses if possible.

Problem 71. Prove that for $n = 1, 2, 3, \ldots$,

$$\left[\frac{n+1}{2}\right] + \left[\frac{n+2}{4}\right] + \left[\frac{n+4}{8}\right] + \left[\frac{n+8}{16}\right] + \cdots = n.$$

Problem 72. Given three noncollinear points A, B, C construct a circle with center C such that the tangents from A and B to the circle are parallel.

Problem 73. Let

$$f(x) = x^4 + x^3 + x^2 + x + 1.$$

Find the remainder when $f(x^5)$ is divided by $f(x)$.

Problem 74. Let the polynomial

$$f(x) = x^n + a_1 x^{n-1} + a_2 x^{n-2} + \cdots + a_{n-1} x + a_n$$

have integral coefficients $a_1, a_2, \ldots, a_n$. If there exist four distinct integers a, b, c, and d such that $f(a) = f(b) = f(c) = f(d) = 5$, show that there is no integer k such that $f(k) = 8$.

Problem 75. Given an $n \times n$ array of positive numbers

$$\begin{matrix} a_{11} & a_{12} & \cdots & a_{1n} \\ a_{21} & a_{22} & \cdots & a_{2n} \\ \vdots & & & \vdots \\ a_{n1} & a_{n2} & \cdots & a_{nn}, \end{matrix}$$

let m_j denote the smallest number in the jth column, and m the largest of the m_j's. Let M_i denote the largest number in the ith row, and M the smallest of the M_i's. Prove that $m \leq M$.

Problem 76. What is the maximum number of terms in a geometric progression with common ratio greater than 1 whose entries all come from the set of integers between 100 and 1000 inclusive?

Problem 77. Prove that for all positive integers n, $1^n + 8^n - 3^n - 6^n$ is divisible by 10.

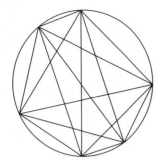

FIGURE 11

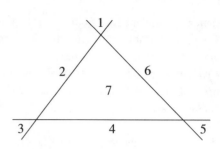

FIGURE 12

Problem 78. n points are given on the circumference of a circle, and the chords determined by them are drawn. If no three chords have a common point, how many triangles are there all of whose vertices lie inside the circle. (Figure 11 shows 6 points and one such triangle.)

Problem 79. A sequence $a_1, a_2, \ldots, a_n, \ldots$ of integers is defined successively by $a_{n+1} = a_n^2 - a_n + 1$ and $a_1 = 2$. The first few terms are $a_1 = 2$, $a_2 = 3$, $a_3 = 7$, $a_4 = 43$, $a_5 = 1807, \ldots$. Show that the integers $a_1, a_2, a_3, \ldots$ are pairwise relatively prime.

Problem 80. Show that the integer N can be taken so large that $1 + \frac{1}{2} + \frac{1}{3} + \cdots + \frac{1}{N}$ is larger than 100.

Problem 81. Let $a_1, a_2, \ldots, a_n$, $b_1, b_2, \ldots, b_n$ be $2n$ positive real numbers. Show that either

$$\frac{a_1}{b_1} + \frac{a_2}{b_2} + \cdots + \frac{a_n}{b_n} \geq n$$

or

$$\frac{b_1}{a_1} + \frac{b_2}{a_2} + \cdots + \frac{b_n}{a_n} \geq n.$$

Problem 82. Let $f(x) = a_n x^n + a_{n-1} x^{n-1} + \cdots + a_1 x + a_0$ be a polynomial of degree $n \geq 1$ with integer coefficients. Show that there are infinitely many positive integers m for which

$$f(m) = a_n m^n + a_{n-1} m^{n-1} + \cdots + a_1 m + a_0$$

is not prime.

Problem 83. Figure 12 shows three lines dividing the plane into seven regions. Find the maximum number of regions into which the plane can be divided by n lines.

Problem 84. In a certain town, the blocks are rectangular, with the streets (of zero width) running E–W, the avenues N–S. A man wishes to go from one corner to another m blocks east and n blocks north. The shortest path can be achieved in many ways. How many?

Problem 85. Given six numbers which satisfy the relations

    (1) $y^2 + yz + z^2 = a^2$
    (2) $z^2 + zx + x^2 = b^2$
    (3) $x^2 + xy + y^2 = c^2$,

determine the sum $x + y + z$ in terms of a, b, c. Give a geometric interpretation if the numbers are all positive.

Problem 86. Six points in space are such that no three are in a line. The fifteen line segments joining them in pairs are drawn and then painted, some segments red, some blue. Prove that some triangle formed by the segments has all its edges the same color.

Problem 87. Represent the number 1 as the sum of reciprocals of finitely many distinct integers larger than or equal to 2. Can this be done in more than one way? If so, how many?

Problem 88. Show how to divide a circle into 9 regions of equal area, using a straight-edge and compasses.

Problem 89. Given n points in the plane, any listing (permutation) $p_1, p_2, \ldots, p_n$ of them determines the path, along straight segments, from p_1 to p_2, then from p_2 to $p_3, \ldots$, ending with the segment from p_{n-1} to p_n. Show that the shortest such broken-line path does not cross itself.

Problem 90. Let $P(x, y)$ be a polynomial in x and y such that:

i) $P(x, y)$ is symmetric, i.e.,
$$P(x, y) \equiv P(y, x);$$

ii) $x - y$ is factor of $P(x, y)$, i.e.,
$$P(x, y) \equiv (x - y)Q(x, y).$$

Prove that $(x - y)^2$ is a factor of $P(x, y)$.

Problem 91. Figure 13 shows a (convex) polygon with nine vertices. The six diagonals which have been drawn dissect the polygon into seven triangles: $P_0P_1P_3$, $P_0P_3P_6$, $P_0P_6P_7$, $P_0P_7P_8$, $P_1P_2P_3$, $P_3P_4P_6$, $P_4P_5P_6$. In how many ways can these triangles be labelled with the names $\triangle_1$, $\triangle_2, \triangle_3, \triangle_4, \triangle_5, \triangle_6, \triangle_7$ so that P_i is a vertex of triangle $\triangle_i$ for $i = 1, 2, 3, 4, 5, 6, 7$? Justify your answer.

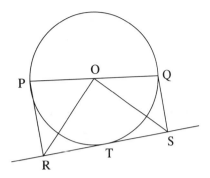

FIGURE 14

Problem 93. Let n be a positive integer and let $a_1, a_2, \ldots, a_n$ be any real numbers ≥ 1. Show that

$$(1 + a_1) \cdot (1 + a_2) \cdots (1 + a_n)$$
$$\geq \frac{2^n}{n + 1}(1 + a_1 + a_2 + \cdots + a_n).$$

Problem 94. If A and B are fixed points on a given circle not collinear with center O of the circle, and if XY is a variable diameter, find the locus of P (the intersection of the line through A and X and the line through B and Y).

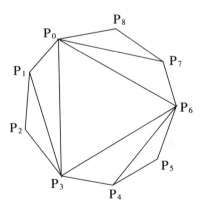

FIGURE 13

Problem 92. In Figure 14, the point O is the center of the circle and the line POQ is a diameter. The point R is the foot of the perpendicular from P to the tangent at T and the point S is the foot of the perpendicular from Q to this same tangent. Prove that $\overline{OR} = \overline{OS}$.

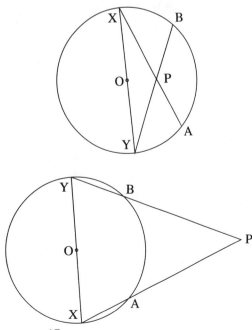

FIGURE 15

Problem 95. Observe that:

$$\frac{1}{1} = \frac{1}{2} + \frac{1}{2}; \quad \frac{1}{2} = \frac{1}{3} + \frac{1}{6};$$

$$\frac{1}{3} = \frac{1}{4} + \frac{1}{12}; \quad \frac{1}{4} = \frac{1}{5} + \frac{1}{20}.$$

State a general law suggested by these examples, and prove it. Prove that for any integer n greater than 1 there exist positive integers i and j such that

$$\frac{1}{n} = \frac{1}{i(i+1)} + \frac{1}{(i+1)(i+2)}$$

$$+ \frac{1}{(i+2)(i+3)} + \cdots + \frac{1}{j(j+1)}.$$

(*American Mathematical Monthly* 55 (1948), 427, problem E827)

Problem 96. Let $ABCD$ be a rectangle with $\overline{BC} = 3\overline{AB}$. Show that if P, Q are the points on side BC with

$$\overline{BP} = \overline{PQ} = \overline{QC},$$

then

$$\angle DBC + \angle DPC = \angle DQC.$$

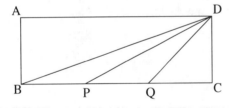

FIGURE 16

Problem 97. Let n be a fixed positive integer. For any choice of n real numbers satisfying $0 \le x_i \le 1, i = 1, 2, \ldots, n$, there corresponds the sum across the bottom of this page. Let $S(n)$ denote the largest possible value of this sum. Find $S(n)$.

Problem 98. Observe that

$$\frac{1}{1} + \frac{1}{3} = \frac{4}{3}, \qquad 4^2 + 3^2 = 5^2,$$

$$\frac{1}{3} + \frac{1}{5} = \frac{8}{15}, \qquad 8^2 + 15^2 = 17^2,$$

$$\frac{1}{5} + \frac{1}{7} = \frac{12}{35}, \qquad 12^2 + 35^2 = 37^2.$$

State and prove a generalization suggested by these examples.

Problem 99. John tosses 6 fair coins, and Mary tosses 5 fair coins. What is the probability that John gets more "heads" than Mary?

Problem 100. A hexagon inscribed in a circle has three consecutive sides of length a and three consecutive sides of length b. Determine the radius of the circle.

Problem 101. (a) Prove that 10201 is composite in any base.

(b) Prove that 10101 is composite in any base.

(c) Prove that 100011 is composite in any base.

Problem 102. Suppose that each of n people knows exactly one piece of information, and all n pieces are different. Every time person "A" phones person "B", "A" tells "B" everything he knows, while "B" tells "A" nothing. What is the minimum number of phone calls between pairs of people needed for everyone to know everything?

Problem 103. Show that, for each integer $n \ge 6$, a square can be subdivided (dissected) into n nonoverlapping squares.

$$\sum_{1 \le i \le j \le n} |x_i - x_j| = |x_1 - x_2| + |x_1 - x_3| + |x_1 - x_4| + \cdots + |x_1 - x_{n-1}| + |x_1 - x_n|$$

$$+ |x_2 - x_3| + |x_2 - x_4| + \cdots + |x_2 - x_{n-1}| + |x_2 - x_n|$$

$$+ |x_3 - x_4| + \cdots + |x_3 - x_{n-1}| + |x_3 - x_n|$$

$$+ \cdots$$

$$+ |x_{n-2} - x_{n-1}| + |x_{n-2} - x_n|$$

$$+ |x_{n-1} - x_n|$$

Problem 104. Let $ABCD$ be a nondegenerate quadrilateral, not necessarily planar (vertices named in cyclic order) and such that

$$\overline{AC}^2 + \overline{BD}^2 = \overline{AB}^2 + \overline{BC}^2 + \overline{CD}^2 + \overline{DA}^2.$$

Show that $ABCD$ is a parallelogram.

Problem 105. Show that every simple polyhedron has at least two faces with the same number of edges.

Problem 106. $ABCDEF$ is a regular hexagon with center P, and PQR is an equilateral triangle, as shown in Figure 17. If $\overline{AB} = 3$, $\overline{SB} = 1$, and $\overline{PQ} = 6$, determine the area common to both figures.

Problem 107. Prove that, for each positive integer n,

$$1 - \frac{1}{2} + \frac{1}{3} - \cdots + \frac{1}{2n-1}$$
$$= \frac{1}{n} + \frac{1}{n+1} + \cdots + \frac{1}{2n-1}.$$

Problem 108. For every positive integer n, let

$$h(n) = 1 + \frac{1}{2} + \frac{1}{3} + \cdots + \frac{1}{n}.$$

For example,

$$h(1) = 1, \ h(2) = 1 + \frac{1}{2}, \ h(3) = 1 + \frac{1}{2} + \frac{1}{3}.$$

Prove that

$$n + h(1) + h(2) + h(3) + \cdots + h(n-1) = nh(n),$$

$$n = 2, 3, 4, \ldots .$$

Problem 109. For which nonnegative integers n and k is

$$(k+1)^n + (k+2)^n + (k+3)^n + (k+4)^n + (k+5)^n$$

divisible by 5?

Problem 110. Describe a method for the construction (with straight-edge and compasses) of a triangle with given angles α and β and given perimeter p.

Problem 111. Determine a constant k such that the polynomial

$$P(x, y, z) = x^5 + y^5 + z^5$$
$$+ k(x^3 + y^3 + z^3)(x^2 + y^2 + z^2)$$

has the factor $x + y + z$. Show that, for this value of k, $P(x, y, z)$ has the factor $(x + y + z)^2$.

Problem 112. Show that, for all positive real numbers p, q, r, s,

$$(p^2 + p + 1)(q^2 + q + 1)(r^2 + r + 1) \times$$
$$(s^2 + s + 1) \geq 81\, pqrs.$$

Problem 113. Prove that, for any positive integer n and any real number x,

$$\left[\frac{[nx]}{n} \right] = [x].$$

Problem 114. Observe the following sets of equations:

A.

$$1 = 1$$

$$2 \cdot 1 - \frac{1}{2} = 1 + \frac{1}{2}$$

$$3 \cdot 1 - 3 \cdot \frac{1}{2} + \frac{1}{3} = 1 + \frac{1}{2} + \frac{1}{3}$$

$$4 \cdot 1 - 6 \cdot \frac{1}{2} + 4 \cdot \frac{1}{3} - \frac{1}{4} = 1 + \frac{1}{2} + \frac{1}{3} + \frac{1}{4}$$

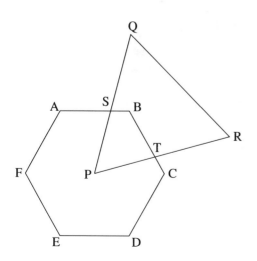

FIGURE 17

B.

$$1 = 1$$

$$\frac{1}{2} = 2 \cdot 1 - \left(1 + \frac{1}{2}\right)$$

$$\frac{1}{3} = 3 \cdot 1 - 3\left(1 + \frac{1}{2}\right) + \left(1 + \frac{1}{2} + \frac{1}{3}\right)$$

$$\frac{1}{4} = 4 \cdot 1 - 6\left(1 + \frac{1}{2}\right) + 4\left(1 + \frac{1}{2} + \frac{1}{3}\right)$$

$$- \left(1 + \frac{1}{2} + \frac{1}{3} + \frac{1}{4}\right).$$

State and prove a generalization for each set. Generalize the relationship between the two sets of equations.

Problem 115. $2n + 3$ points ($n \geq 1$) are given in the plane, no three on a line and no four on a circle. Prove that there exists a circle through three of them such that, of the remaining $2n$ points, n are in the interior and n are in the exterior of the circle.

Problem 116. From a fixed point P not in a given plane, three mutually perpendicular line segments are drawn terminating in the plane. Let a, b, c denote the lengths of the three segments. Show that $\frac{1}{a^2} + \frac{1}{b^2} + \frac{1}{c^2}$ has a constant value for all allowable configurations.

Problem 117. If a, b, c denote the lengths of the sides of a triangle show that

$$3(bc + ca + ab) \leq (a + b + c)^2 < 4(bc + ca + ab).$$

Problem 118. Andy leaves at noon and drives at constant speed back and forth from town A to town B. Bob also leaves at noon, driving at 40 km per hour back and forth from town B to town A on the same highway as Andy. Andy arrives at town B twenty minutes after first passing Bob, whereas Bob arrives at town A forty-five minutes after first passing Andy. At what time do Andy and Bob pass each other for the nth time?

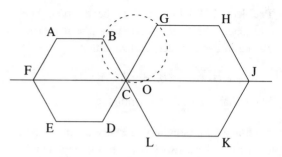

FIGURE 18

Problem 119. Two unequal regular hexagons $ABCDEF$ and $CGHJKL$ (shown in Figure 18) touch each other at C and are so situated that F, C and J are collinear. Show that:

i) the circumcircle of BCG bisects FJ (at 0 say);

ii) $\triangle BOG$ is equilateral.

Problem 120. Let n be a positive integer. Prove that the binomial coefficients

$$\binom{n}{1}, \ \binom{n}{2}, \ \binom{n}{3}, \ldots, \ \binom{n}{n-1}$$

are all even if and only if n is a power of 2.

Problem 121. Prove that, for any positive integer n,

$$1492^n - 1770^n - 1863^n + 2141^n$$

is divisible by 1946.

Problem 122. If $P(x)$ denotes a polynomial of degree n such that $P(k) = \frac{1}{k}$ for $k = 1$, $2, \ldots, n + 1$, determine $P(n + 2)$.

Problem 123. Prove that $\log x$ cannot be expressed in the form $\frac{f(x)}{g(x)}$, where $f(x)$ and $g(x)$ are polynomials in x.

Problem 124. A train leaves a station precisely on the minute, and after having travelled 8 miles, the driver consults his watch and sees the hour-hand is directly over the minute-hand. The average speed over the 8 miles is 33 miles per hour. At what time did the train leave the station?

FIGURE 19. $n = 13$

Problem 125. Describe a construction of a quadrilateral $ABCD$ given:
(a) the lengths of all four sides;
(b) that segments AB and CD are parallel;
(c) that segments BC and DA do not intersect.

Problem 126. You have a large number of congruent equilateral triangular tiles on a table and you want to fit n of them together to make a convex equiangular hexagon (i.e., one whose interior angles are all $120°$). Obviously, n cannot be any positive integer. The smallest feasible n is 6, the next smallest is 10 and the next 13 (Figure 19). Determine conditions for a possible n.

Problem 127. Let a, b, c denote three distinct integers and $P(x)$ a polynomial with integral coefficients. Show that it is impossible that $P(a) = b$, $P(b) = c$ and $P(c) = a$. (USAMO 1974)

Problem 128. Suppose the polynomial $x^n + a_1 x^{n-1} + a_2 x^{n-2} + \cdots + a_n$ can be factored into

$$(x + r_1)(x + r_2) \cdots (x + r_n),$$

where $r_1, r_2, \ldots, r_n$ are real numbers. Prove that $(n - 1)a_1^2 \geq 2na_2$.

Problem 129. For each positive integer n, determine the smallest positive number $k(n)$ such that

$$k(n) + \sin \frac{A}{n}, \quad k(n) + \sin \frac{B}{n}, \quad k(n) + \sin \frac{C}{n}$$

are the sides of a triangle whenever A, B, C are the angles of a triangle.

Problem 130. Prove that, for $n = 1, 2, 3, \ldots$,
(a) $(n + 1)^n \geq 2^n n!$;
(b) $(n + 1)^n (2n + 1)^n \geq 6^n (n!)^2$.

Problem 131. Let z_1, z_2, z_3 be complex numbers satisfying:
(1) $z_1 z_2 z_3 = 1$,
(2) $z_1 + z_2 + z_3 = \dfrac{1}{z_1} + \dfrac{1}{z_2} + \dfrac{1}{z_3}$.
Show that at least one of them is 1.

Problem 132. Let m_a, m_b, m_c and w_a, w_b, w_c denote, respectively, the lengths of the medians and angle bisectors of a triangle. Prove that

$$\sqrt{m_a} + \sqrt{m_b} + \sqrt{m_c} \geq \sqrt{w_a} + \sqrt{w_b} + \sqrt{w_c}.$$

Problem 133. Let n and r be integers with $0 \leq r \leq n$. Find a simple expression for

$$S = \binom{n}{0} - \binom{n}{1} + \binom{n}{2} - \cdots + (-1)^r \binom{n}{r}.$$

Problem 134. If x, y, z are positive numbers, show that

$$\frac{x^2}{y^2} + \frac{y^2}{z^2} + \frac{z^2}{x^2} \geq \frac{y}{x} + \frac{z}{y} + \frac{x}{z}.$$

Problem 135. Prove that all chords of parabola $y^2 = 4ax$ which subtend a right angle at the vertex of the parabola are concurrent. See Figure 20.

Problem 136. $ABCD$ is an arbitrary convex quadrilateral for which

$$\frac{\overline{AM}}{\overline{AB}} = \frac{\overline{NC}}{\overline{DC}},$$

as shown in Figure 21. Prove that the area of

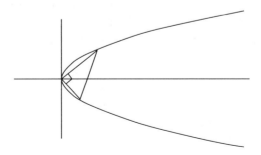

FIGURE 20

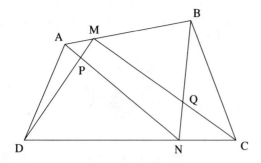

FIGURE 21

quadrilateral $PMQN$ equals the sum of the areas of triangles APD and BQC.

Problem 137. Show how to construct a sphere which is equidistant from five given non-cospherical points, no four in a plane. Is the solution unique?

Problem 138. Prove that $5^{2n+1} + 11^{2n+1} + 17^{2n+1}$ is divisible by 33 for every nonnegative integer n.

Problem 139. A polynomial $P(x)$ of the nth degree satisfies $P(k) = 2^k$ for $k = 0, 1, 2, \ldots, n$. Determine $P(n+1)$.

Problem 140. Suppose that $0 \le x_i \le 1$ for $i = 1, 2, \ldots, n$. Prove that
$$2^{n-1}(1 + x_1 x_2 \cdots x_n)$$
$$\ge (1 + x_1)(1 + x_2) \cdots (1 + x_n),$$
with equality if and only if $n - 1$ of the x_i's are equal to 1.

Problem 141. Sherwin Betlotz, the tricky gambler, will bet even money that you can't pick three cards from a 52-card deck without getting at least one of the twelve face cards. Would you bet with him?

Problem 142. Given a, b, c, d, find all x, y, z, w for which
(1) $y^2 z^2 w^2 x = a^7$,
(2) $z^2 w^2 x^2 y = b^7$,
(3) $w^2 x^2 y^2 z = c^7$,
(4) $x^2 y^2 z^2 w = d^7$.

Problem 143. Prove that if all plane cross-sections of a bounded solid figure are circles, then the solid is a sphere.

Problem 144. In how many ways can we stack n different coins so that two particular coins are not adjacent to each other?

Problem 145. Two fixed, unequal, nonintersecting and non-nested circles are touched by a variable circle at P and Q. Prove that there are two fixed points, through one of which PQ must pass.

Problem 146. If $S = x_1 + x_2 + \cdots + x_n$, where $x_i > 0 (i = 1, \ldots, n)$, prove that
$$\frac{S}{S - x_1} + \frac{S}{S - x_2} + \cdots + \frac{S}{S - x_n} \ge \frac{n^2}{n - 1},$$
with equality if and only if $x_1 = x_2 = \cdots = x_n$.

Problem 147. Factor
$$a^5(c - b) + b^5(a - c) + c^5(b - a).$$

Problem 148. In a mathematical competition, a contestant can score 5, 4, 3, 2, 1, or 0 points for each problem. Find the number of ways he can score a total of 30 points for 7 problems.

Problem 149. Observe that
$$6^2 - 5^2 = 11,$$
$$56^2 - 45^2 = 1111,$$
$$556^2 - 445^2 = 111111,$$
$$5556^2 - 4445^2 = 11111111.$$
State a generalization suggested by these examples and prove it.

Problem 150. Solve for x, y, z (in terms of a, r, s, t):
$$yz = a(y + z) + r$$
$$zx = a(z + x) + s$$
$$xy = a(x + y) + t.$$

Problem 151. Let ABC be an equilateral triangle, and let P be a point within the triangle.

Perpendiculars PD, PE, PF are drawn to the three sides of the triangle. Show that, no matter where P is chosen,

$$\frac{\overline{PD} + \overline{PE} + \overline{PF}}{\overline{AB} + \overline{BC} + \overline{CA}} = \frac{1}{2\sqrt{3}}.$$

Problem 152. Solve

$$\sqrt[3]{13x + 37} - \sqrt[3]{13x - 37} = \sqrt[3]{2}.$$

Problem 153. If A denotes the number of integers whose logarithms (to base 10) have the characteristic a, and B denotes the number of integers the logarithms of whose reciprocals have characteristic $-b$, determine $(\log A - a) - (\log B - b)$. (The characteristic of $\log x$ is the integer $[\log x]$.)

Problem 154. Show that three solutions, (x_1, y_1), (x_2, y_2), (x_3, y_3), of the four solutions of the simultaneous equations

$$(x - h)^2 + (y - k)^2 = 4(h^2 + k^2)$$
$$xy = hk$$

are vertices of an equilateral triangle. Give a geometrical interpretation.

Problem 155. Prove that, for each positive integer m, the smallest integer which exceeds $(\sqrt{3} + 1)^{2m}$ is divisible by 2^{m+1}.

Problem 156. Suppose that r is a nonnegative rational taken as an approximation to $\sqrt{2}$. Show that $\dfrac{r + 2}{r + 1}$ is always a better rational approximation.

Problem 157. Find all rational numbers k such that $0 \leq k \leq \frac{1}{2}$ and $\cos k\pi$ is rational.

Problem 158. Solve the simultaneous equations:

$$x + y + z = 0,$$
$$x^2 + y^2 + z^2 = 6ab,$$
$$x^3 + y^3 + z^3 = 3(a^3 + b^3).$$

Problem 159. Prove that the sum of the areas of any three faces of a tetrahedron is greater than the area of the fourth face.

Problem 160. Let a, b, c be the lengths of the sides of a right-angled triangle, the hypotenuse having length c. Prove that $a + b \leq \sqrt{2}\,c$. When does equality hold?

Problem 161. Determine all θ such that $0 \leq \theta \leq \frac{\pi}{2}$ and $\sin^5 \theta + \cos^5 \theta = 1$.

Problem 162. If the pth and qth terms of an arithmetic progression are q and p respectively, find the $(p + q)$th term.

Problem 163. Let ABC be a triangle with sides of lengths a, b, and c. Let the bisector of the angle C cut AB in D. Prove that the length of CD is

$$\frac{2ab \cos \frac{C}{2}}{a + b}.$$

Problem 164. For which positive integral bases b is 1367631, here written in base b, a perfect cube?

Problem 165. If x is a positive real number notequal to unity and n is a positive integer, prove that

$$\frac{1 - x^{2n+1}}{1 - x} \geq (2n + 1)x^n.$$

Problem 166. The pth, qth and rth terms of an arithmetic progression are q, r and p respectively. Find the difference between the $(p + q)$th and the $(q + r)$th terms.

Problem 167. Given a circle Γ and two points A and B in general position in the plane, construct a circle through A and B which intersects Γ in two points which are ends of a diameter of Γ.

Problem 168. Find the polynomial whose roots are the cubes of the roots of the polynomial $t^3 + at^2 + bt + c$ (where a, b, c are constants).

Problem 169. If a, b, c, d are positive real numbers, prove that

$$\frac{a^2 + b^2 + c^2}{a + b + c} + \frac{b^2 + c^2 + d^2}{b + c + d} + \frac{c^2 + d^2 + a^2}{c + d + a}$$
$$+ \frac{d^2 + a^2 + b^2}{d + a + b} \geq a + b + c + d$$

with equality only if $a = b = c = d$.

Problem 170. (a) Find all positive integers with initial digit 6 such that the integer formed by deleting this 6 is $\frac{1}{25}$ of the original integer.

(b) Show that there is no integer such that deletion of the first digit produces a result which is $\frac{1}{35}$ of the original integer.

Problem 171. Prove that if a convex polygon has three of its angles equal to $60°$ then it must be an equilateral triangle.

Problem 172. Prove that, for real numbers x, y, z,

$$x^4(1+y^4) + y^4(1+z^4) + z^4(1+x^4) \geq 6x^2y^2z^2.$$

When is there equality?

Problem 173. How many integers from 1 to 10^{30} inclusive are not perfect squares, perfect cubes, or perfect fifth powers?

Problem 174. What is the greatest common divisor of the set of numbers

$$\{16^n + 10n - 1 \mid n = 1, 2, 3, \ldots\}?$$

Problem 175. If $a_i \geq 1$ for $i = 1, 2, \ldots$, prove that, for each positive integer n,

$$n + a_1 a_2 \ldots a_n \geq 1 + a_1 + a_2 + \cdots + a_n$$

with equality if and only if no more than one of the a_i's is different from 1.

Problem 176. A certain polynomial $p(x)$ when divided by $x - a$, $x - b$, $x - c$ leaves remainders a, b, c respectively. What is the remainder when $p(x)$ is divided by $(x - a)(x - b)(x - c)$? (a, b, c distinct).

Problem 177. Determine the function $F(x)$ which satisfies the functional equation

$$x^2 F(x) + F(1 - x) = 2x - x^4$$

for all real x.

Problem 178. Prove that there are no positive integers a, b, c such that

$$a^2 + b^2 + c^2 = a^2 b^2.$$

Problem 179. A sequence $\{a_n\}$ of real numbers is defined by

$$a_1 = 1, \quad a_{n+1} = 1 + a_1 a_2 \ldots a_n \quad (n \geq 1).$$

Prove that

$$\sum_{n=1}^{\infty} \frac{1}{a_n} = 2.$$

Problem 180. Show how to construct the straight line joining two given points A, B with only a straight edge whose length is less than $\overline{AB}$.

Problem 181. Show that, if two circles, not in the same plane, either intersect in two points or are tangent, then they are cospherical (i.e., there is a sphere which contains the two circles).

Problem 182. Let x, y, z be the cube roots of three distinct prime integers. Show that x, y, z are never three terms (not necessarily consecutive) of an arithmetic progression.

Problem 183. OBC is a triangle in space, and A is a point not in the plane of the triangle. AO is perpendicular to the plane BOC and D is the foot of the perpendicular from A to BC. (See Figure 22.) Prove that $OD \perp BC$.

Problem 184. Let A, B, C, D be four points in space. Determine the locus of the centers of all parallelograms having one vertex on each of the four segments AB, BC, CD, DA.

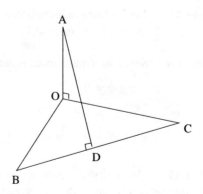

FIGURE 22

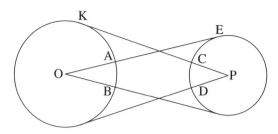

FIGURE 23

Problem 185. From the centers of two "exterior" circles draw the tangents to the other circle, as in Figure 23. Prove that
$$\overline{AB} = \overline{CD}.$$

Problem 186. P, Q, R, S denote points respectively on the sides AB, BC, CD, DA of a skew quadrilateral $ABCD$ such that P, Q, R, S are coplanar. Suppose also, that P', Q', R', S' are points on the sides $A'B'$, $B'C'$, $C'D'$, $D'A'$ respectively of a second skew quadrilateral $A'B'C'D'$. Assume that

$$\overline{AB} = \overline{A'B'}, \quad \overline{BC} = \overline{B'C'}, \quad \overline{CD} = \overline{C'D'},$$

$$\overline{DA} = \overline{D'A'} \quad \overline{AP} = \overline{A'P'}, \quad \overline{BQ} = \overline{B'Q'},$$

$$\overline{CR} = \overline{C'R'}, \quad \overline{DS} = \overline{D'S'}$$

Prove that P', Q', R', S' are also coplanar.

Problem 187. Find a positive number k such that, for some triangle ABC (with sides of length a, b, c opposite angles A, B, C respectively):
    (a) $a + b = kc$,
    (b) $\cot \frac{A}{2} + \cot \frac{B}{2} = k \cot \frac{C}{2}$.

Problem 188. A, B, C, D are four points in space such that

$$\angle ABC = \angle BCD = \angle CDA = \angle DAB = 90°.$$

Prove that A, B, C, D are coplanar.

Problem 189. If, in triangle ABC, $\angle B = 18°$ and $\angle C = 36°$, show that $a - b$ is equal to the circumradius.

Problem 190. Let $ABCD$ be a concyclic convex quadrilateral for which $AB \perp CD$. Denote by a, b, c, d the lengths of the edges AB, BC, CD, DA respectively. Prove that

$$(ab + cd)^2 + (ad + bc)^2 = (b^2 - d^2)^2.$$

Problem 191. Suppose that P, Q, R, S are points on the sides AB, BC, CD, DA respectively, of a tetrahedron $ABCD$, such that the lines PS and QR intersect. Show that the lines PQ, RS and AC are concurrent.

Problem 192. Let ABC and ABD be equilateral triangles which lie in two planes making an angle θ with each other. Find $\angle CAD$ (in terms of θ).

Problem 193. ABC is a triangle for which $\overline{BC} = 4$, $\overline{CA} = 5$, $\overline{AB} = 6$. Determine the ratio $\angle BCA/\angle CAB$.

Problem 194. Show how to construct the radius of a given solid sphere, given a pair of compasses, a straight-edge and a plane piece of paper.

Problem 195. Let n be positive integer not less than 3. Find a direct combinatorial interpretation of the identity

$$\binom{\binom{n}{2}}{2} = 3\binom{n + 1}{4}.$$

Problem 196. Prove that a convex polyhedron P cannot satisfy either (a) or (b):
(a) P has exactly seven edges;
(b) P has all its faces hexagonal.

Problem 197. Determine x in the equilateral triangle shown in Figure 24.

Problem 198. A, B, C, D are four points in space such that line AC is perpendicular to line BD. Suppose that A', B', C', D' are any four points such that

$$\overline{AB} = \overline{A'B'}, \quad \overline{BC} = \overline{B'C'},$$

$$\overline{CD} = \overline{C'D'}, \quad \overline{DA} = \overline{D'A'}.$$

Prove that line $A'C'$ is perpendicular to line $B'D'$.

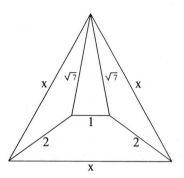

FIGURE 24

Problem 199. Let $P(x)$, $Q(x)$, and $R(x)$ be polynomials such that $P(x^5)+x\,Q(x^5)+x^2R(x^5)$ is divisible by $x^4 + x^3 + x^2 + x + 1$. Prove that $P(x)$ is divisible by $x - 1$. (USAMO 1976)

Problem 200. If $ABCDEFGH$ is a cube, as shown in Figure 25, determine the minimum perimeter of a triangle PQR whose vertices P, Q, R lie on the edges AB, CG, EH respectively.

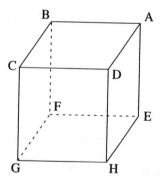

FIGURE 25

Problem 201. A man drives six kilometers to work every morning. Leaving at the same time each day, he must average exactly 36 kph in order to arrive on time. One morning, however, he gets behind a street-washer for the first two kilometers and this reduces his average speed for that distance to 12 kph. Given that his car can travel up to 150 kph, can he get to work on time?

Problem 202. A desk calendar consists of a regular dodecahedron with a different month on each of its twelve pentagonal faces. How many essentially different ways are there of arranging the months on the faces?

Problem 203. (a) Show that 1 is the only positive integer equal to the sum of the squares of its digits (in base 10).

(b) Find all the positive integers, besides 1, which are equal to the sum of the cubes of their digits (in base 10).

Problem 204. Find all the essentially different ways of placing four points in a plane so that the six segments determined have just two different lengths.

Problem 205. Three men play a game with the understanding that the loser is to double the money of the other two. After three games, each has lost just once, and each has $24.00. How much did each have at the start of the games?

Problem 206. (a) In a triangle ABC, $\overline{AB} = 2\overline{BC}$. Prove that BC must be the shortest side. If the perimeter of the triangle is 24, prove that $4 < \overline{BC} < 6$.

(b) If one side of a triangle is three times another and the perimeter is 24, find bounds for the length of the shortest side.

Problem 207. Show that, if k is a nonnegative integer:

(a) $1^{2k} + 2^{2k} + 3^{2k} \geq 2 \cdot 7^k$;
(b) $1^{2k+1} + 2^{2k+1} + 3^{2k+1} \geq 6^{k+1}$.

When does equality occur?

Problem 208. Solve

$$(x + 1)(x + 2)(x + 3) = (x - 3)(x + 4)(x + 5).$$

Problem 209. What is the smallest integer, which, when divided in turn by $2, 3, 4, \ldots, 10$ leaves remainders of $1, 2, 3, \ldots, 9$ respectively?

Problem 210. Two cars leave simultaneously from points A and B on the same road in opposite directions. Their speeds are constant, and in the ratio 5 to 4, the car leaving A being faster. The cars travel to and fro between A and B. They meet for the second time at the 145th milestone and for the third time at the 201st. What milestones are at A and B?

Problem 211. What are the last three digits of the number 7^{9999}?

Problem 212. ABC is a triangle such that $\overline{AB} = \overline{AC}$ and $\angle BAC = 20°$. The point X on AB is such that $\angle XCB = 50°$; the point Y on AC is such that $\angle YBC = 60°$. Determine $\angle AXY$.

Problem 213. Prove that the volume of the tetrahedron determined by the endpoints of two line segments lying on two skew lines is unaltered by sliding the segments (while leaving their lengths unaltered) along their lines.

Problem 214. (a) A file of men marching in a straight line one behind another is one kilometer long. An inspecting officer starts at the rear, moves forward at a constant speed until he reaches the front, then turns around and travels at the same speed until he reaches the last man in the rear. By this time, the column, marching at a constant speed, has moved one kilometer forward so that the last man is now in the position the front man was when the whole movement started. How far did the inspecting officer travel?

(b) Answer the question (a) if, instead of a column, we have a phalanx one kilometer square which the inspecting officer goes right around.

Problem 215. Let $ABCD$ be a tetrahedron whose faces have equal areas. Suppose O is an interior point of $ABCD$ and L, M, N, P are the feet of the perpendiculars from O to the four faces. Prove that

$$\overline{OA} + \overline{OB} + \overline{OC} + \overline{OD}$$
$$\geq 3(\overline{OL} + \overline{OM} + \overline{ON} + \overline{OP}).$$

Problem 216. A boat has sprung a leak. Water is coming in at a uniform rate and some has already accumulated when the leak is detected. At this point, 12 men of equal skill can pump the boat dry in 3 hours, while 5 men require 10 hours. How many men are needed to pump it dry in 2 hours?

Problem 217. Each move of a knight on a checkerboard takes it two squares parallel to one side of the board, and one square in a perpendicular direction. A *knight's tour* is a succession of knight's moves such that each square of the checkerboard is visited exactly once. The tour is *closed* when the last square occupied is a knight's move away from the first square. Show that if m, n are both odd, then a closed knight's tour is not possible on an $m \times n$ checkerboard.

Problem 218. According to the Dominion Observatory Time Signal, the hour and minute hand of my watch coincide every 65 minutes exactly. Is my watch fast or slow? By how much? How long will it take for my watch to gain or lose an hour?

Problem 219. Sketch the graph of the inequality

$$|x^2 + y| \leq |y^2 + x|.$$

Problem 220. Prove that the inequality

$$3a^4 - 4a^3b + b^4 \geq 0$$

holds for all real numbers a and b.

Problem 221. Find all triangles with integer side-lengths for which one angle is twice another.

Problem 222. (a) Show that if n is a triangular number, then so is $9n + 1$. (Triangular numbers are: $1, 3, 6, 10, \ldots, \dfrac{k(k+1)}{2}, \ldots$; see Tool Chest, B. 9)

(b) Find other numbers a, b such that $an + b$ is triangular whenever n is.

Problem 223. Prove that, for any positive integer n,

$$\left[\sqrt{n} + \sqrt{n+1}\right] = \left[\sqrt{4n+2}\right],$$

where $[\,\cdot\,]$ denotes the greatest integer function.

Problem 224. Prove or disprove the following statement. Given a line l and two points A and B not on l, the point P on l for which $\angle APB$ is largest must lie between the feet of the perpendiculars from A and B to l.

Problem 225. Determine all triangles ABC for which

$$\cos A \cos B + \sin A \sin B \sin C = 1.$$

Problem 226. Let

$$a_1 = 2^2 + 3^2 + 6^2, \quad a_2 = 3^2 + 4^2 + 12^2,$$
$$a_3 = 4^2 + 5^2 + 20^2,$$

and so on. Generalize these in such a way that the number a_n is always a perfect square.

Problem 227. Suppose that x, y, and z are non-negative real numbers. Prove that

$$8(x^3 + y^3 + z^3)^2 \geq 9(x^2 + yz)(y^2 + zx)(z^2 + xy).$$

Problem 228. Every person who has ever lived has, up to this moment, made a certain number of handshakes. Prove that the number of people who have made an odd number of handshakes is even. (Do not consider handshakes a person makes with himself.)

Problem 229. Given a, b, c, solve the following system of equations for x, y, z:

$$x^2 - yz = a^2 \tag{1}$$
$$y^2 - zx = b^2 \tag{2}$$
$$z^2 - xy = c^2. \tag{3}$$

Problem 230. Show that, for each positive integer n,

$$1^2 - 2^2 + 3^2 - 4^2 + \cdots + (-1)^n (n-1)^2$$
$$+ (-1)^{n+1} n^2$$
$$= (-1)^{n+1} (1 + 2 + 3 + \cdots + n).$$

Problem 231. If a b, c are the lengths of the sides of a triangle, prove that

$$abc \geq (a + b - c)(b + c - a)(c + a - b).$$

Problem 232. Prove that a longest chord of a centrally-symmetric region must pass through the center.

Problem 233. A disc is divided into k sectors and a single coin is placed in each sector. In any move, two coins (not necessarily in the same sector) are shifted, one clockwise and the other counter-clockwise, into neighboring sectors. Determine whether a sequence of moves is possible which will make all the coins end up in the same sector.

Problem 234. Suppose $\sin x + \sin y = a$ and $\cos x + \cos y = b$. Determine $\tan \frac{x}{2}$ and $\tan \frac{y}{2}$.

Problem 235. Two fixed points A and B and a moving point M are taken on the circumference of a circle. On the extension of the line segment AM a point N is taken, outside the circle, so that $\overline{MN} = \overline{MB}$. Find the locus of N.

Problem 236. Find the real values of x which satisfy the equation

$$(x + 1)(x^2 + 1)(x^3 + 1) = 30x^3.$$

A student wrote the following prescription in the fly leaf of an algebra text:
      If there should be another flood
      Hither for refuge fly
      Were the whole world to be submerged
      This book would still be dry.

Problem 237. Exactly enough gas to enable a race car to get around a circular track once is split up into a number of portions which are distributed at random to points around the track. Show that there is a point on the track at which the race car, with an empty gas tank, can be placed, so that it will be able to complete the circuit in one direction or the other.

Problem 238. Show that, for all real values of x (radians), $\cos(\sin x) > \sin(\cos x)$.

Problem 239. Prove that the equation

$$x^2 + y^2 + 2xy - mx - my - m - 1 = 0,$$

m is a positive integer, has exactly m solutions (x, y) for which x and y are both positive integers.

Problem 240. $PQRS$ is an arbitrary convex quadrilateral inscribed in a convex quadrilateral $ABCD$, as shown below.

$P'Q'R'S'$ is another quadrilateral, inscribed in $ABCD$, such that P', Q', R', S' are the "mirror" images of P, Q, R, S with respect to the midpoints of AB, BC, CD, DA respectively. Determine the entire class of convex quadrilaterals $ABCD$ such that the areas of $PQRS$ and $P'Q'R'S'$ are (necessarily) equal.

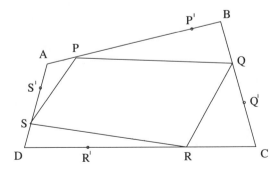

FIGURE 26

Problem 241. Given numbers a, b, c, d, no two equal, solve the system

$$x + ay + a^2 z + a^3 w = a^4,$$
$$x + by + b^2 z + b^3 w = b^4,$$
$$x + cy + c^2 z + c^3 w = c^4,$$
$$x + dy + d^2 z + d^3 w = d^4,$$

for x, y, z, w.

Problem 242. Let a, b, c, be integer sides of a right-angled triangle, where $a < b < c$. Show that $ab(b^2 - a^2)$ is divisible by 84.

Problem 243. If A, B, C denote the angles of a triangle, determine the maximum value of

$$\sin^2 A + \sin B \sin C \cos A.$$

Problem 244. If three points are chosen at random, uniformly with respect to arc length, on the circumferences of a given circle, determine the probability that the triangle determined by the three points is acute.

Problem 245. Is it possible to color the points (x, y) in the Cartesian plane for which x and y are integers with three colors in such a way that
(a) each color occurs infinitely often in infinitely many lines parallel to the x-axis, and
(b) no three points, one of each color, are collinear?

Problem 246. A man walks North at a rate of 4 kilometers per hour and notices that the wind appears to blow from the West. He doubles his speed, and now the wind appears to blow from the Northwest. What is the velocity of the wind?

Problem 247. Four solid spheres lie on the top of a table. Each sphere is tangent to the other three. If three of the spheres have the same radius R, what is the radius of the fourth sphere?

Problem 248. The equation $a^2 + b^2 + c^2 + d^2 = abcd$ has the solution $(a, b, c, d) = (2, 2, 2, 2)$. Find infinitely many other solutions in positive integers.

Problem 249. One side of a triangle is 10 feet longer than another and the angle between them is $60°$. Two circles are drawn with these sides as diameter. One of the points of intersection of the two circles is the vertex common to the two sides. How far from the third side of the triangle, produced, is the other point of intersection?

Problem 250. Given the equal sides of an isosceles triangle, what is the length of the third side which will provide the maximum area of the triangle?

Problem 251. Let $ABCD$ be a square, F be the midpoint of DC, and E be any point on AB such that $\overline{AE} > \overline{EB}$. Determine H on BC such that $DE \parallel FH$. Prove that EH is tangent to the inscribed circle of the square.

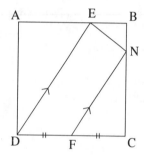

FIGURE 27

Problem 252. Given that a and b are two positive real numbers for which $a^a = b$ and $b^b = a$, show that $a = b = 1$.

Problem 253. What is the smallest perfect square that ends with the four digits 9009?

Problem 254. Two right-angled triangles ABC and FDC are such that their hypotenuses AB and FD intersect in E as shown in Figure 28.

Find x (the distance of the point E from the side FC) in terms of $\alpha = \angle BAC$, $\beta = \angle DFC$ and the lengths of the two hypotenuses.

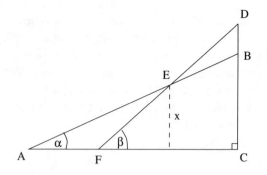

FIGURE 28

Problem 255. Observe that

$$\frac{1 - \frac{1}{2^3}}{1 + \frac{1}{2^3}} = \frac{2}{3}\left(1 + \frac{1}{2\cdot 3}\right)$$

and

$$\frac{1 - \frac{1}{2^3}}{1 + \frac{1}{2^3}} \cdot \frac{1 - \frac{1}{3^3}}{1 + \frac{1}{3^3}} = \frac{2}{3}\left(1 + \frac{1}{3\cdot 4}\right).$$

State and prove a general law suggested by these examples.

Problem 256. Let n be a positive integer. Show that $(x - 1)^2$ is a factor of $x^n - n(x - 1) - 1$.

Problem 257. Notice that, in the accompanying table, each of the 36 numbers in the square array is equal to the sum of the number at the head of its column and the number at the left of its row.

	1	9	3	2	4	8
2	**3**	11	<u>5</u>	4	6	10
7	8	<u>16</u>	**10**	9	11	15
3	4	**12**	6	<u>5</u>	7	11
5	6	14	8	7	**9**	<u>13</u>
8	<u>9</u>	17	11	10	12	**16**
7	8	16	10	**9**	<u>11</u>	15

For example, $17 = 9 + 8$ and $13 = 8 + 5$.

The six "bold" numbers are selected so that there is one in each row and in each column. The underlined numbers are selected in a similar way. Observe that the sum of the "bold" numbers is $3+10+12+9+16+9 = 59$ and that the sum of the underlined numbers is $5+16+5+13+9+11 = 59$. Show that the sum of any six of the 36 numbers, chosen so that there is exactly one in each of the six columns and exactly one in each of the six rows, is 59.

Problem 258. Equal circles are arranged in a regular pattern throughout the plane so that each circle touches six others. What percentage of the plane is covered by the circles?

Problem 259. Show that if a, b, c are integers which satisfy $a + b\sqrt{2} + c\sqrt{3} = 0$, then $a = b = c = 0$.

Problem 260. Prove that, for any distinct rational values of a, b, c, the number

$$\frac{1}{(b-c)^2} + \frac{1}{(c-a)^2} + \frac{1}{(a-b)^2}$$

is the square of some rational number.

Problem 261. Let ABC be an equilateral triangle. Let E be an arbitrary point on AC produced. Let D be chosen, as in Figure 29, so that CDE is an equilateral triangle. If M is the midpoint of segment AD, and N is the midpoint of segment BE, show that $\triangle CMN$ is equilateral.

Problem 262. Suppose that

$$a_1 > a_2 > a_3 > a_4 > a_5 > a_6$$

and that

$$p = a_1 + a_2 + a_3 + a_4 + a_5 + a_6$$

$$q = a_1a_3 + a_3a_5 + a_5a_1 + a_2a_4 + a_4a_6 + a_6a_2$$

$$r = a_1a_3a_5 + a_2a_4a_6.$$

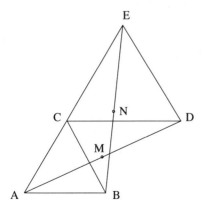

FIGURE 29

Are all the roots of the cubic $2x^3 - px^2 + qx - r$ real?

Problem 263. Show that $4n^3 + 6n^2 + 4n + 1$ is composite for $n = 1, 2, 3, \ldots$.

Problem 264. In the multiplicative "magic" square shown below, the products of the elements in each of the three rows, and in each of the three columns, and in each of the two diagonals, are all the same, i.e., abc, def, ghi, adg, beh, cfi, aei, ceg are all equal, to k say. If all entries of the square are integers, show that k must be a perfect cube.

$$\begin{array}{ccc} a & b & c \\ d & e & f \\ g & h & i \end{array}$$

Problem 265. Four suspects of a crime made the following statements to the police:

Andy: Carl did it.

Bob: I did not do it.

Carl: Dave did it.

Dave: Carl lied when he said I did it.

(a) Given that exactly one of the four statements is true, determine who did it.

(b) Given that exactly one of the four statements is false, determine who did it.

Problem 266. Can you load two dice (not necessarily in the same way) so that all outcomes $2, 3, \ldots, 12$ are equally likely?

Problem 267. (a) What is the area of the region in the Cartesian plane whose points (x, y) satisfy

$$|x| + |y| + |x + y| \le 2?$$

(b) What is the volume of the region in space whose points (x, y, z) satisfy

$$|x| + |y| + |z| + |x + y + z| \le 2?$$

Problem 268. In how many essentially different ways can three couples be seated around a circular dinner table so that no husband sits next to his wife? ("Essentially different" means that one arrangement is not a rotation of another. It is not assumed in this problem that men and women must sit in alternate seats.)

Problem 269. AB and AC are two roads with rough ground in between. (See Figure 30.) The distances AB and AC are both equal to p, while the distance BC is equal to q. A man at point B wishes to walk to C. On the road he walks with speed v, and on the rough ground his walking speed is w. Show that, if he wishes to take minimum time, he may do so by picking one of two particular routes. In fact, argue that he should go:

(a) by road through A if $2pw \le qv$;

(b) along the straight path BC if $2pw \ge qv$.

Problem 270. Show that there is no polynomial $p(x)$ such that, for each natural number n,

$$p(n) = \log 1 + \log 2 + \cdots + \log n.$$

Problem 271. For positive integers n define

$$f(n) = 1^n + 2^{n-1} + 3^{n-2} + 4^{n-3}$$
$$+ \cdots + (n-2)^3 + (n-1)^2 + n.$$

What is the minimum value of $\dfrac{f(n+1)}{f(n)}$?

Problem 272. Let a, b, c, d be natural numbers not less than 2. Write down, using parentheses, the various interpretations of

$$a^{b^{c^d}}.$$

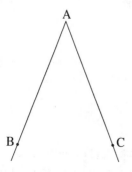

FIGURE 30

For example, we might have $a^{((b^c)^d)} = a^{(b^{cd})}$ or $(a^b)^{(c^d)} = a^{b(c^d)}$. In general, these interpretations will not be equal to each other.

For what pairs of interpretations does an inequality always hold? For pairs not necessarily satisfying an inequality in general, give numerical examples to illustrate particular instances of either inequality.

Problem 273. 35 persons per thousand have high blood pressure. 80% of those with high blood pressure drink, and 60% of those without high blood pressure drink. What percentage of drinkers have high blood pressure?

Problem 274. There are $n!$ permutations $(s_1, s_2, \ldots, s_n)$ of $(1, 2, 3, \ldots, n)$. How many of them satisfy $s_k \ge k - 2$ for $k = 1, 2, \ldots, n$?

Problem 275. Prove that, for any quadrilateral with sides a, b, c, d, it is true that

$$a^2 + b^2 + c^2 > \frac{1}{3} d^2.$$

Problem 276. Given 8 distinct positive integers, a_1, $a_2, \ldots, a_8$ from the set $\{1, 2, \ldots, 15, 16\}$, prove that there is a number k for which

$$a_i - a_j = k$$

has at least three distinct solutions (a_i, a_j).

Problem 277. Is there a fixed integer k for which the image of the mapping

$$(x, y) \longrightarrow x^2 + kxy + y^2, \qquad x, y \text{ integers}$$

includes (i) all integers, (ii) all positive integers? If so, find one. If not, give a proof.

Problem 278. Let a, b, c be integers, not all 0. Show that if $ax^2 + bx + c$ has a rational root then at least one of a, b, c is even.

Problem 279. Barbeau says, "I am heavier than Klamkin, and Klamkin is heavier than Moser." Klamkin says, "Moser is heavier than I am, and Moser is also heavier than Barbeau." Moser says, "Klamkin is heavier than I am, and Barbeau weighs the same as I do." Assuming that a lighter man makes true statements more often than a heavier man, arrange Barbeau, Klamkin, and Moser in increasing order of weight.

Problem 280. Let $f(x, y)$ be a function of two real variables which is not identically zero. If $f(x, y) = kf(y, x)$ for all values of x and y, what are the possible values of k?

Problem 281. Find the point which minimizes the sum of its distances from the vertices of a given convex quadrilateral.

Problem 282. At the winter solstice (usually December 22), the earth's axis is tilted about $23°27'$ away from the normal to the plane of its orbit, with the north pole pointed away from the sun. Find, approximately, the length of time elapsing between sunrise and sunset on the date of the winter solstice at a place whose latitude is $43°45'$ N. (How long is the day in your home town?)

Problem 283. A trapezoid is divided into four triangles by its diagonals. Let A and B denote the areas of the triangles adjacent to the parallel sides. (See Figure 31.) Find, in terms of A and B, the area of the trapezoid.

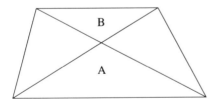

FIGURE 31

Problem 284. Let n be any natural number. Find the sum of the digits appearing in the integers

$$1, 2, 3, \ldots, 10^n - 2, 10^n - 1.$$

Problem 285. Find an expression in terms of a and b for the area of the hatched region in the right triangle in Figure 32.

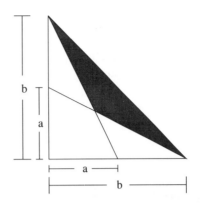

FIGURE 32

Problem 286. Given a point inside a regular pentagon, find 9 other points inside the pentagon, the sum of whose perpendicular distances from the sides of the pentagon (produced, if necessary) will be the same as for the given point.

Problem 287. Factor $(x + y)^7 - (x^7 + y^7)$.

Problem 288. $ABCDE$ is a regular pentagon. BE intersects AC and AD in H and K respectively. The line through H parallel to AD meets AB in F. The line through K parallel to AC meets AE in G. Prove that $AFHKG$ is a regular pentagon.

Problem 289. Although the addition given below might appear valid, show that, in fact, there is no substitution of distinct digits for the various letters which will give a numerically correct statement:

$$
\begin{array}{ccccc}
 & T & H & R & E & E \\
+ & & F & I & V & E \\
\hline
 & E & I & G & H & T \\
\end{array}
$$

Problem 290. Let a, b, c be any three positive integers, and let

x be the greatest common divisor of b and c,
y be the greatest common divisor of a and c,
z be the greatest common divisor of a and b.

Show that the greatest common divisor of a, b and c is equal to the greatest common divisor of x, y and z.

Problem 291. A culture of bacteria doubles in size every 12 hours. A dish which will contain 1,000,000 bacteria is full after 13 days. How long will it take to fill a dish whose capacity is 2,000,000 bacteria?

Problem 292. Let $f(x)$ be a nondecreasing function of a real variable, so that the slope of the line through any two points on the curve $y = f(x)$ is not negative. Let c be any real number. Solve the equation

$$x = c - f(x + f(c)).$$

Problem 293. Let E be the midpoint of the side BC of triangle ABC, and let F be chosen in segment AC so that $\overline{AC} = 3\overline{FC}$. Determine the ratio of the areas of the triangle FEC and the quadrilateral $ABEF$.

Problem 294. Find $(\log_3 169) \times (\log_{13} 243)$ without use of tables.

Problem 295. A right-angled triangle ABC with side lengths a, b, c $(c^2 = a^2 + b^2)$ determines a hexagon (see Figure 33) whose vertices are the "outside" corners of the squares on the sides AB, BC, CA. Find the area of this hexagon in terms of a, b, c.

Problem 296. Show that, for any positive integer n, the triangle with side lengths

$$6 \cdot 10^{n+2}, \quad 1125 \cdot 10^{2n+1} - 8,$$
$$1125 \cdot 10^{2n+1} + 8$$

is right-angled.

Problem 297. A tennis club invites 32 players of equal ability to compete in an elimination tournament (the players compete in pairs, with anyone losing a match prohibited from further play). What is the chance of a particular pair playing each other during the tournament?

Problem 298. The following construction was proposed for a straight-edge-and-compasses trisection of an arbitrary acute angle POQ. (See Figure 34.)

"From any point B on OQ, drop a perpendicular to meet OP at A. Construct an equilateral triangle ABC with C and O on opposite sides of the line AB. Then $\angle POC = \frac{1}{3}\angle POQ$."

Find the acute angles POQ for which the method works, and show that it is not valid for any other angle.

(Based on an idea of John and Stuart Rosenthal, while pupils of Forest Hill Senior Public School, Toronto.)

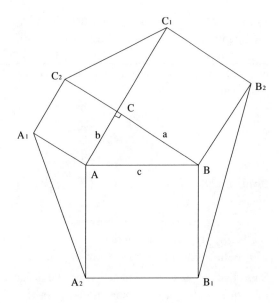

FIGURE 33

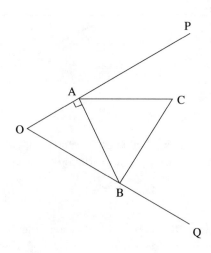

FIGURE 34

Problem 299. Given the number $111\ldots 11$, here expressed in base 2, find its square, also expressed in base 2.

Problem 300. Show that for $k = 1, 2, 3, \ldots$

$$\sin\frac{\pi}{2k}\,\sin\frac{3\pi}{2k}\,\sin\frac{5\pi}{2k}\,\ldots\,\sin\left(2\left[\frac{k+1}{2}\right]-1\right)\frac{\pi}{2k}$$
$$= \frac{1}{\sqrt{2}^{k-1}}.$$

Problem 301. (a) Verify that

$$1 = \frac{1}{2} + \frac{1}{5} + \frac{1}{8} + \frac{1}{11} + \frac{1}{20} + \frac{1}{41} + \frac{1}{110} + \frac{1}{1640}.$$

(b) Show that any representation of 1 as the sum of distinct reciprocals of numbers drawn from the arithmetic progression $\{2, 5, 8, 11, 14, 17, 20, \ldots\}$, such as is given in (a), must have at least eight terms.

Problem 302. Three of four corners of a square are cut off so that three isosceles right triangles are removed, as shown in Figure 35. Draw two straight lines in the pentagon which remains, dividing it into three parts which will fit together to form a square.

Problem 303. A pollster interviewed a certain number, N, of persons as to whether they used radio, television and/or newspapers as a source of news. He reported the following findings:

50 people used television as a source of news, either alone or in conjunction with other sources;
61 did not use radio as a source of news;
13 did not use newspapers as a source of news;
74 had at least two sources of news.

Find the maximum and minimum values of N consistent with this information.

Give examples of situations in which the maximum and in which the minimum values of N could occur.

FIGURE 35

Problem 304. Let V and F be the vertex and focus, respectively, of a parabola. Suppose that P is a point on the parabola, distinct from V, and that Q is a point within the parabola such that:
(i) PQ is normal to the parabola, and
(ii) $\overline{PQ} = \overline{VF}$.

Show that the segment PQ does not intersect the axis of the parabola.

Problem 305. x, y, and z are real numbers such that

$$x + y + z = 5 \quad \text{and} \quad xy + yz + zx = 3.$$

Determine the largest value that any one of the three numbers can be.

Problem 306. Prove that, for all positive integer values of n,

$$7^{2n} - 2352n - 1 \quad \text{is divisible by} \quad 2304.$$

Problem 307. Determine the volume of a tetrahedron $ABCD$ if

$$\overline{AB} = \overline{AC} = \overline{AD} = 5$$

and

$$\overline{BC} = 3, \ \overline{CD} = 4, \ \overline{DB} = 5.$$

Problem 308. Express

$$2(x^4 + y^4 + z^4 + w^4) - (x^2 + y^2 + z^2 + w^2)^2 + 8xyzw$$

as a product of nonconstant real polynomials.

Problem 309. Q is a point outside of a circle with center O. A second circle with center Q and radius OQ is drawn. Rays from Q intersecting the two circles in R and S are drawn, as shown in Figure 36. Show that the locus of P, the midpoint of RS, is not a straight line segment.

Problem 310. Observe that

$$1 = 1^2$$
$$2 + 3 + 4 = 3^2$$
$$3 + 4 + 5 + 6 + 7 = 5^2$$
$$4 + 5 + 6 + 7 + 8 + 9 + 10 = 7^2.$$

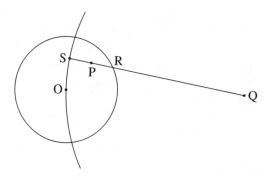

FIGURE 36

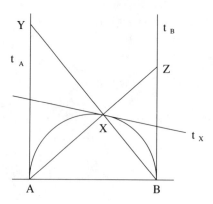

FIGURE 37

State and prove a generalization suggested by these examples.

Problem 311.   Let

$$f(x) = a_0 + a_1 x + a_2 x^2 + \cdots + a_n x^n$$

be a polynomial whose coefficients satisfy the conditions $0 \le a_i \le a_0 \quad (i = 1, 2, \ldots, n)$. Let

$$(f(x))^2$$

$$= b_0 + b_1 x + \cdots + b_{n+1} x^{n+1} + \cdots + b_{2n} x^{2n}.$$

Prove that

$$b_{n+1} \le \frac{1}{2}(f(1))^2.$$

Problem 312.   Prove that:

(i) $\displaystyle\sum_{k=1}^{\infty} \frac{1}{(2k-1)(2k+1)}$

$$= \frac{1}{1 \cdot 3} + \frac{1}{3 \cdot 5} + \frac{1}{5 \cdot 7} + \cdots = \frac{1}{2};$$

(ii) $\displaystyle\sum_{k=1}^{\infty} \frac{1}{k(k+1)(k+2)}$

$$= \frac{1}{1 \cdot 2 \cdot 3} + \frac{1}{2 \cdot 3 \cdot 4} + \frac{1}{3 \cdot 4 \cdot 5} + \cdots = \frac{1}{4}.$$

Problem 313.   Given a set of $(n + 1)$ positive integers, none of which exceeds $2n$, show that at least one member of the set must divide another member of the set.

Problem 314.   Given a circle with diameter AB, and a point X, other than A or B, on the circle, let t_A, t_B and t_X be the tangents to the circle at A, B and X respectively. Suppose AX produced

meets t_B at Z, and BX produced meets t_A at Y. Show that, either the three lines YZ, t_X and AB all pass through the same point, or else the three lines are parallel.

Problem 315.   Show that it is never possible to partition a set of six consecutive integers into two subsets in such a way that the least common multiple of the numbers in one subset is equal to the least common multiple of the numbers in the other.

Problem 316.   For any positive integer n, let $f(n)$ denote the nth positive nonsquare integer, i.e.,

$$f(1) = 2, \ f(2) = 3, \ f(3) = 5, \ f(4) = 6,$$
$$f(5) = 7, \ f(6) = 8, \ f(7) = 10, \ \ldots .$$

Prove that

$$f(n) = n + \{\sqrt{n}\},$$

where $\{x\}$ denotes the integer closest to x. (For example $\{\sqrt{1}\} = 1$, $\{\sqrt{2}\} = 1$, $\{\sqrt{3}\} = 2$, $\{\sqrt{4}\} = 2$.)

Problem 317.   I invite you to play the following card game: Shuffle an ordinary deck of cards, and turn them face up in pairs. If both cards of a pair are black, you get them. If both are red, I get them. If one is red and one is black, the pair belongs to neither one of us.

You pay one dollar for the privilege of playing the game. When the deck is exhausted, the game is over, and you pay nothing if you have no more cards than I have. On the other hand, for every card that you have more than I, I will pay you 3 dollars. Would you care to play with me?

Problem 318. Let ABO and $A'B'O$ be two right isosceles triangles with the common vertices at the right angles (see Figure 38). Prove that $\overline{AA'} = \overline{BB'}$ and that $AA' \perp BB'$.

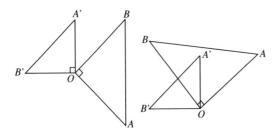

FIGURE 38

Problem 319. Given an arbitrary triangle ABC, let P and Q be the centers of squares on AB and AC, respectively (as in Figure 39). Show that, if M is the midpoint of BC, then triangle MPQ is right angled and isosceles.

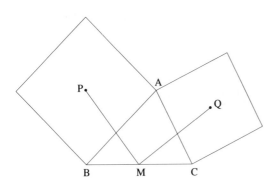

FIGURE 39

Problem 320. (a) Let P, Q, R be the centers of squares on the sides BC, CA, AB, respectively, of the triangle ABC. (See Figure 40.) Prove that $\overline{AP} = \overline{QR}$ and that $AP \perp QR$.

(b) Given an arbitrary convex quadrilateral $ABCD$ and the centers P, Q, R, S of the external squares on the sides AB, BC, CD, DA,

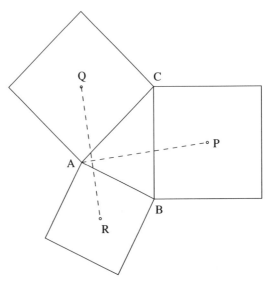

FIGURE 40

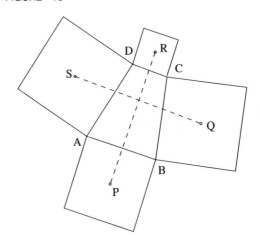

FIGURE 41

respectively, show that $\overline{PR} = \overline{QS}$ and $PR \perp QS$. (See Figure 41.)

Problem 321. ABC is a triangle whose angles satisfy $\angle A \geq \angle B \geq \angle C$. Circles are drawn such that each circle cuts each side of the triangle internally in two distinct points (see Figure 42).

(a) Show that the lower limit to the radii of such circles is the radius of the inscribed circle of the triangle ABC.

(b) Show that the upper limit to the radii of such circles is not necessarily equal to R, the ra-

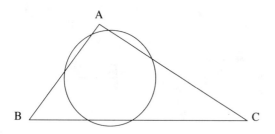

FIGURE 42

dius of the circumscribed circle of triangle ABC. Find this upper limit in terms of R, A and B.

Problem 322. (a) Let $f(n)$ denote the number of solutions (x, y) of $x + 2y = n$ for which x and y are both nonnegative integers. Show that

$$f(0) = f(1) = 1,$$

$$f(n) = f(n - 2) + 1, \quad n = 2, 3, 4, \ldots.$$

Find a simple explicit formula for $f(n)$.

(b) Let $g(n)$ denote the number of solutions (x, y, z) of $x + 2y + 3z = n$ for which x, y, and z are nonnegative integers. Show that

$$g(0) = g(1) = 1, \quad g(2) = 2,$$

$$g(n) = g(n - 3) + \left[\frac{n}{2}\right] + 1, \quad n = 3, 4, 5, \ldots$$

where

$$\left[\frac{n}{2}\right] = \begin{cases} \frac{n}{2} & \text{for } n \text{ even,} \\ \frac{n-1}{2} & \text{for } n \text{ odd.} \end{cases}$$

Problem 323. Solve for real x, y, z:

$$3x^2 + y^2 + z^2 = 2x(y + z).$$

Problem 324. A pencil, eraser, and notebook together cost one dollar. A notebook costs more than two pencils, three pencils cost more than four erasers, and three erasers cost more than a note-book. How much does each item cost (assuming that each item costs an integral number of cents)?

Problem 325. Let A be a square array of numbers, and let s be any number greater than or equal to every row sum and every column sum. Show that it is possible to replace each entry in the array by a number no smaller than the entry so that the new array B has every row sum and every column sum equal to s.

For example (we use integer entries for convenience)

$$A = \begin{pmatrix} 3 & -5 & 2 \\ -6 & 4 & 1 \\ 1 & 0 & 8 \end{pmatrix} \quad B = \begin{pmatrix} 14 & -5 & 2 \\ -4 & 14 & 1 \\ 1 & 2 & 8 \end{pmatrix}$$

$$s = 11$$

row sums : $0, -1, 9$ row sums : $11, 11, 11$

col. sums : $-2, -1, 11$ col. sums : $11, 11, 11$.

Problem 326. Observe that

$$2^2 + 3^2 + 4^2 + 14^2 = 15^2,$$

$$4^2 + 5^2 + 6^2 + 38^2 = 39^2,$$

$$6^2 + 7^2 + 8^2 + 74^2 = 75^2,$$

$$8^2 + 9^2 + 10^2 + 122^2 = 123^2.$$

State and prove a general result suggested by these examples.

Problem 327. Let three concentric circles be given such that the radius of the largest is less than the sum of the radii of the two smaller. Construct an equilateral triangle whose vertices lie one on each circle.

Problem 328. In how many zeros does 10000! end?

> There was a young man from old Trinity
> Who found the square root of infinity
> While counting the digits
> He was seized by the figits
> So he chucked Math and took up Divinity

Problem 329. A man is on a railway bridge joining A to B, $\frac{3}{8}$ of the way across from A. He hears a train approaching A; it is travelling 80 kph. If he runs towards A, he will meet the train at A. If he runs towards B, the train will overtake him at B. How fast can he run?

Problem 330. Let the sequence $u_1, u_2, u_3, \ldots$ be defined by

$$u_1 = 1, \quad u_{n+1} = u_n + 8n \quad (n = 1, 2, 3, \ldots).$$

Prove that $u_n = (2n - 1)^2$.

Problem 331. Show that each of the following polynomials is nonnegative for all real values of the variables, but that neither can be written as a sum of squares of real polynomials:

(a) $x^2y^2 + y^2z^2 + z^2x^2 + w^4 - 4xyzw$;

(b) $x^4y^2 + y^4z^2 + z^4x^2 - 3x^2y^2z^2$.

Problem 332. (a) For which real values of p and q are the roots of the polynomial $x^3 - px^2 + 11x - q$ three successive integers? Give the roots in these cases.

(b) For which real values of p and q does $x^3 - px^2 + 11x - q$ have exactly one root? What is the root?

Problem 333. Prove that, for all natural numbers n, $2^{2n} + 24n - 10$ is divisible by 18.

Problem 334. Let $a_1, a_2, \ldots, a_n$ be any distinct integers chosen from the set $\{1, 2, \ldots, 2n - 2, 2n - 1\}$. Prove that for some indices i and j (not necessarily distinct) $a_i + a_j = 2n$.

Problem 335. Given any $n + 2$ integers, show that for some pair of them either their sum or their difference is divisible by $2n$.

Problem 336. Let ABC be a triangle and D any point distinct from A, B, C on its circumcircle. Show that the feet of the perpendiculars dropped from D to the three sides (produced if necessary) of the triangle are collinear.

Problem 337. Suppose u and v are two real numbers such that u, v and uv are the three roots

of a cubic polynomial with rational coefficients. Show that at least one root is rational.

Problem 338. Let n be an integer. Show that the greatest common divisor of $n^2 + 1$ and $(n + 1)^2 + 1$ is either 1 or 5.

Problem 339. Without using tables, evaluate:

(a) $\cos 36° - \cos 72°$;

(b) $\cos 36° \cdot \cos 72°$.

Problem 340. Show without using a calculator that $7^{1/2} + 7^{1/3} + 7^{1/4} < 7$ and $4^{1/2} + 4^{1/3} + 4^{1/4} > 4$.

Problem 341. Two players play the following game. The first player selects any integer from 1 to 11 inclusive. The second player adds any positive integer from 1 to 11 inclusive to the number selected by the first player. They continue in this manner alternately. The player who reaches 56 wins the game. Which player has the advantage?

Problem 342. A circle of radius r is inscribed in a sector of a circle of radius R. The length of the chord of the sector is equal to $2a$. Find a relation between r, R and a in which each variable occurs once.

Problem 343. Each vertex of a parallelogram is connected with the midpoints of its two opposite sides by straight lines. (See Figure 43.) What portion of the area of the parallelogram is the area of the figure bounded by these lines?

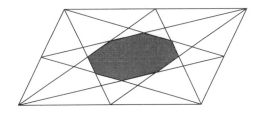

FIGURE 43

Problem 344. Let u be an arbitrary but fixed number between 0 and 1, i.e., $0 < u < 1$. Form the sequence $u_1, u_2, u_3, \ldots$ as follows:

$$u_1 = 1 + u$$

$$u_2 = \frac{1}{u_1} + u$$

$$u_3 = \frac{1}{u_2} + u$$

and so on, i.e., $u_n = \dfrac{1}{u_{n-1}} + u$ for $n = 2, 3, 4, \ldots$. Does it ever happen that $u_n \leq 1$?

Problem 345. (a) Show that any two positive consecutive integers are relatively prime. (Note: Two integers are relatively prime if their only common divisor is 1. More generally, (a, b) denotes the greatest common divisor of a and b; thus, that a, b are relatively prime is denoted by $(a, b) = 1$.)

(b) Find a positive integer relatively prime to $2, 3, 4, \ldots, n$.

(c) Let r be a positive integer. Find a positive integer $t \geq r$ for which

$$(r + i,\ t + r + i) = 1 \qquad \text{for} \quad i = 1, 2, \ldots, r.$$

Problem 346. Your calculator is not working properly—it cannot perform multiplications. But it can add (and subtract) and it can compute the reciprocal $\frac{1}{x}$ of any number x. Can you nevertheless use this defective calculator to multiply numbers?

Problem 347. Is it possible for a proper nonempty subset of the plane to have at least three noncurrent axes of symmetry? (An axis of symmetry is a line about which the subset reflects onto itself.)

Problem 348. Observe that

$$3 + 5 = 8 = 2^3$$

$$5 + 7 = 12 = 2^2 \cdot 3$$

$$7 + 11 = 18 = 2 \cdot 3^2$$

$$11 + 13 = 24 = 2^3 \cdot 3.$$

Show that if p and q are any two consecutive odd primes, then $p + q$ is a product of at least 3 (not necessarily distinct) primes.

Problem 349. (a) Show that $\sqrt{2} + \sqrt{3}$ is not rational.

(b) Given the positive integers m and n, under what conditions is $\sqrt{m} + \sqrt{n}$ rational?

Problem 350. Let $P_1, P_2, \ldots, P_m$ be m points on a line and $Q_1, Q_2, \ldots, Q_n$ be n points on a distinct and parallel line. All segments P_iQ_j are drawn. What is the maximum number of points of intersection?

Problem 351. Factor $(x+y+z)^5 - x^5 - y^5 - z^5$.

Problem 352. Without using "long" multiplication, a computer or a pocket calculator, verify that
(a) $13! = 112296^2 - 79896^2$,
(b) $240^4 + 340^4 + 430^4 + 599^4 = 651^4$.

Problem 353. For $n = 1, 2, 3, \ldots$ find a "closed" expression for the sum

$$\frac{1}{2} + \frac{3}{2^2} + \frac{5}{2^3} + \cdots + \frac{2n-1}{2^n}.$$

Problem 354. Let k, m, and n be positive integers with the property: for some number $x \neq 1$, the numbers $\log_k x$, $\log_m x$, $\log_n x$ are consecutive terms of an arithmetic progression. Show that

$$n^2 = (kn)^{\log_k m}.$$

Problem 355. Solve the following system of 100 equations in 100 unknowns:

$$
\begin{array}{ccccccc}
x_1 & + & x_2 & + & x_3 & = & 0 \\
x_2 & + & x_3 & + & x_4 & = & 0 \\
& & & \vdots & & & \\
x_{98} & + & x_{99} & + & x_{100} & = & 0 \\
x_{99} & + & x_{100} & + & x_1 & = & 0 \\
x_{100} & + & x_1 & + & x_2 & = & 0 \,.
\end{array}
$$

Problem 356. Show that for any real numbers x, y, and any positive integer n,
(a) $0 \leq [nx] - n[x] \leq n - 1$,
(b) $[x] + [y] + (n-1)[x + y] \leq [nx] + [ny]$.
($[z]$ denotes the greatest integer not exceeding z.)

Problem 357. Bisect a straight line segment with a "try-square". (With a try-square one can draw a line through pairs of points, and erect the perpendicular to a line from any point on the line.)

Problem 358. Let p be the perimeter and m the sum of the lengths of the three medians of any triangle. Prove that $\frac{3}{4}p < m < p$.

Problem 359. (a) Which is larger, $29\sqrt{14} + 4\sqrt{15}$ or 124?

(b) Which is larger, $759\sqrt{7}+2\sqrt{254}$ or 2040? (No calculators please.)

Problem 360. Observe that

$$1 = 1^2,$$
$$2 = -1^2 - 2^2 - 3^2 + 4^2,$$
$$3 = -1^2 + 2^2,$$
$$4 = -1^2 - 2^2 + 3^2,$$
$$5 = 1^2 + 2^2,$$
$$6 = 1^2 - 2^2 + 3^2.$$

This suggests the conjecture: any positive integer n can be expressed in the form

$$n = \varepsilon_1 1^2 + \varepsilon_2 2^2 + \varepsilon_3 3^2 + \cdots + \varepsilon_m m^2,$$

with m a positive integer and $\varepsilon_i = 1$ or -1, $i = 1, 2, \ldots, m$. Prove this conjecture.

Problem 361. Let M be the center of a circle and A, B two points on the circumference not diametrically opposed. The tangents at A and B intersect at C. Let CM intersect the circle in D, and suppose that the tangent through D intersects

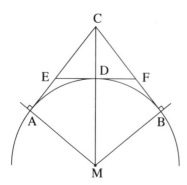

FIGURE 44

AC and BC at E and F respectively. See Figure 44.

(a) Show that the area of quadrilateral $ADBM$ is the geometric mean of the areas of triangle ABM and quadrilateral $ACBM$.

(b) Show that the area of pentagon $AEDFBM$ is the harmonic mean of the areas of quadrilaterals $ADBM$ and $ACBM$.

Problem 362. Two positive numbers with distinct first digits are multiplied together. Is it possible for the first digit of the product to fall strictly between the first digits of the two numbers?

Problem 363. A man whose clock had stopped running wound it up, but did not have access to the correct time to reset it. Leaving the clock at home, he walked to the home of a friend whose clock was correct, stayed for some time and then walked home (in the same time as he took earlier). Upon arriving at home, he set his clock to the correct time even though he did not know how long he had walked! Explain.

Problem 364. A gambler played the following game with a friend. He bet half the money in his pocket on the toss of a coin; he won on HEADS and lost on TAILS. The coin was tossed and the money handed over. The game was repeated, each time for half the money held by the gambler. At the end, the number of times the gambler lost was equal to the number of times he won. Did he gain, lose, or break even?

Problem 365. Encountering a man on the porch of his house, a census taker asked, "What are the ages of the persons living here?" The man replied, "All our ages are square integers. My age is the sum of the ages of my wife, son and daughter. My father's age is the sum of my age and the ages of my wife and daughter. Although he has passed the prime of his life, his age is a prime number." What ages did the census taker record and what obvious remark did he make about the wife's age?

Problem 366. (a) Find the pentagon for which five given points are the midpoints of the edges.

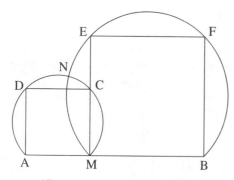

FIGURE 45

(b) Given the midpoints of the sides of an n-gon, is it always possible to determine the n-gon?

Problem 367. Let M be any point on segment AB. Construct squares $AMCD$ and $MBFE$ as in Figure 45, and let N be the second point of intersection of the circles circumscribing the squares. Prove that the lines BC and AE both pass through N.

Problem 368. (a) Show that for any positive integer n, the integers $21n + 4$ and $14n + 3$ are relatively prime.

(b) For what integers a and b is $7an + 4$ relatively prime to $7bn + 3$ for every positive integer n?

Problem 369. Show that among those people at the recent concert in the Ottawa National Arts Centre, there were two people who have the same number of acquaintances present at the concert.

Problem 370. Let n be a positive integer. Let

a_1 denote the number of solutions (x, y), in nonnegative integers, of $x + 2y = n$,

a_2 the number of solutions (x, y), in nonnegative integers, of $2x + 3y = n - 1$,

a_3 the number of solutions (x, y), in nonnegative integers, of $3x + 4y = n - 2$,

(and so on until)

a_n the number of solutions, in nonnegative integers, of $nx + (n + 1)y = 1$.

Show that $a_1 + a_2 + \cdots + a_n = n$. For example, in the case $n = 7$:

$x + 2y = 7$ has 4 solutions, $(1, 3)$, $(3, 2)$, $(5, 1)$, $(7, 0)$,

$2x + 3y = 6$ has 2 solutions, $(3, 0)$, $(0, 2)$,

$3x + 4y = 5$ has 0 solutions,

$4x + 5y = 4$ has 1 solution, $(1, 0)$,

$5x + 6y = 3$ has 0 solutions,

$6x + 7y = 2$ has 0 solutions,

$7x + 8y = 1$ has 0 solutions,

and

$$4 + 2 + 0 + 1 + 0 + 0 + 0 = 7.$$

Problem 371. Let $a(n)$ denote the number of ways of expressing the positive integer n as an *ordered* sum of 1's and 2's, e.g., $a(5) = 8$ because

$$5 = 1 + 1 + 1 + 1 + 1 = 2 + 1 + 1 + 1$$
$$= 1 + 2 + 1 + 1 = 1 + 1 + 2 + 1$$
$$= 1 + 1 + 1 + 2 = 2 + 2 + 1 = 2 + 1 + 2$$
$$= 1 + 2 + 2.$$

Let $b(n)$ denote the number of ways of expressing n as an *ordered* sum of integers greater than 1, e.g., $b(7) = 8$ because

$$7 = 3 + 2 + 2 = 2 + 3 + 2 = 2 + 2 + 3$$
$$= 3 + 4 = 4 + 3 = 2 + 5 = 5 + 2 = 7.$$

Prove that

$$a(n) = b(n + 2) \qquad \text{for} \quad n = 1, 2, \ldots.$$

Problem 372. Five gamblers A, B, C, D, E play together a game which terminates with one of them losing and then the loser pays to each of the other four as much as each has. Thus, if they start a game possessing α, β, γ, δ, ϵ dollars respectively, and say for example that B loses, then B gives A, C, D, E respectively α, γ, δ, ϵ dollars, after which A, B, C, D, E have 2α, $\beta - \alpha - \gamma - \delta - \epsilon$, 2γ, 2δ, 2ϵ dollars respectively. They play five games: A loses the first game, B loses the second, C loses the third, D the fourth and E the fifth. After the final payment, made by E, they find that they are equally wealthy, i.e., each has the same integral number of dollars as the others. What is the smallest amount that each could have started with?

Problem 373. Consider a square array of numbers consisting of m rows and m columns. Let a_{ij} be the number entered in the ith row and jth column. For each i, let r_i denote the sum of the numbers in the ith row, and c_i the sum of the numbers in the ith column. Show that there are distinct indices i and j for which $(r_i - c_i)(r_j - c_j) \leq 0$.

Problem 374. The function f has the property that

$$|f(a) - f(b)| \leq |a - b|^2$$

for any real numbers a and b. Show that f is a constant function.

Problem 375. A rocket car accelerates from 0 kph to 240 kph in a test run of one kilometer. If the acceleration is not allowed to increase (but it may decrease) during the run, what is the longest time the run can take?

Problem 376.

> Big fleas have little fleas
> Upon their backs that bite 'em,
> And little ones have lesser ones,
> And so ad infinitum.

If the flea on the bottom weighs $\sqrt{2}$ grams, and every other flea weighs $\sqrt{2-x}$, where x represents the weight of the flea on whose back it rests (while biting, of course), how much does the flea on top weigh?

Problem 377. Let r be one of the roots of the quadratic equation $x(1-x) = 1$; the other root is $1-r$. Show that for $n = 1, 2, 3, \ldots$

$$r^n + (1-r)^n = \begin{cases} 2(-1)^n & \text{if } 3 \mid n, \\ (-1)^{n-1} & \text{if } 3 \nmid n. \end{cases}$$

Problem 378. Mr. Laimbrain makes up in perseverance what he lacks in wit. Early this morning he set out to sod an area of his lawn of the shape indicated in Figure 46. He has 31 sods each 2 feet by 1 foot. He made up his mind to do it without cutting a sod. This afternoon he was still feverishly arranging and rearranging the pieces of turf. Can you help him out?

Problem 379. The Fibonacci sequence f_1, f_2, $f_3, \ldots$ is defined by

$$f_1 = f_2 = 1, \qquad f_n = f_{n-1} + f_{n-2}, \quad n \geq 3.$$

Thus, the sequence begins

$$1, \ 1, \ 2, \ 3, \ 5, \ 8, \ 13, \ 21, \ 34, \ 55, \ldots.$$

Let

$$Q = \begin{pmatrix} 1 & 1 \\ 1 & 0 \end{pmatrix}.$$

Prove that

$$Q^n = \begin{pmatrix} f_{n+1} & f_n \\ f_n & f_{n-1} \end{pmatrix}, \quad n = 2, 3, 4, \ldots$$

Establish the identity:

$$f_{3n} = f_{n+1}^3 + f_n^3 - f_{n-1}^3, \quad n = 1, 2, 3, \ldots.$$

Problem 380. Prove that the function

$$f(x, y) = \frac{(x+y)(x+y+1)}{2} + x$$

is a <u>one-to-one</u> map from the set

$$\{(x, y) \mid x, y \text{ integers} \geq 0, \quad x^2 + y^2 > 0\}$$

(the lattice points other than $(0, 0)$ in the first quadrant) onto the set

$$\{m \mid m \text{ integral and } > 0\}$$

of positive integers.

Problem 381. Two equal regular tetrahedra intersect in such a way that each face of either passes

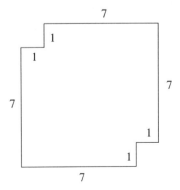

FIGURE 46

through the midpoints of three concurrent edges of the other. The union U of the two tetrahedra is a three-dimensional "star". Describe the intersection V of the two tetrahedra. Determine the ratio of the volumes of U and V.

Problem 382.   Prove that for $n = 1, 2, 3, \ldots$

$$n^4 + 2n^3 + 2n^2 + 2n + 1$$

is not the square of an integer.

Problem 383.   A checker club having 8 players decided to split them into two evenly matched teams. The players had 1, 2, 3, 5, 8, 10, 11 and 12 years of playing experience respectively. They decided to split them so that the sum of the years of experience on either side would be the same. Curiously, they found that when this was done, the sum of the squares of the years of experience on either side was the same, and similarly for the cubes. How were the teams set up?

Problem 384.   A manufacturer had to ship 150 washing machines to a neighboring town. Upon inquiring he found that two types of trucks were available. One type was large and would carry 18 machines, the other type was smaller and would carry 13 machines. The cost of transporting a large truckload was \$35, that of a small one \$25. What is the most economical way of shipping the 150 machines?

Problem 385.   Two ships S and T are steaming on straight courses with constant speeds. At 10:00 hours, they are 5 kilometers apart, at 11:00, they are 4 km apart, and at 13:00, they are 10 km apart. At 7:00, S was due west of T.

(a) How far apart were they at 7:00?
(b) When are they 26 km apart?
(c) How near do they pass to one another? At what time are they nearest?
(d) When is S due north of T?
(e) When is S southwest of T?
(f) Suppose S and T have the same speed and that T is heading due south. What is the speed and direction of S?

Problem 386.   (a) If a regular hexagon and an equilateral triangle have the same perimeter, determine the ratio of their areas.

(b) Given a circle, determine the ratio of the area of the circumscribed regular hexagon to the area of the inscribed regular hexagon.

Problem 387.   The point P divides side BC of triangle ABC into ratio $\dfrac{\overline{BP}}{\overline{PC}} = \dfrac{1}{2}$, and $\angle CBA = 45°$ while $\angle APC = 60°$. (See Figure 47.) Find $\angle ACB$ without the use of trigonometry.

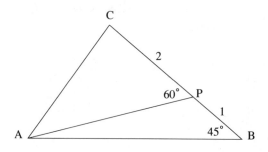

FIGURE 47

Problem 388.   Let ℓ and m be parallel lines and P a point between them. Find the triangle APB of smallest area, with A on ℓ, B on m, and $\angle APB = 90°$.

Problem 389.   Sketch the graph of the curve

$$|3x^2 + y^2 - 12| = |x^2 - y^2 + 4|.$$

Problem 390.   Let p be a prime greater than 3. Show that the sum of the quadratic residues (of p) which lie between 1 and $p-1$ inclusive is divisible by p. (See Tool Chest, B. 11 for definitions.)

Problem 391.   The exterior and interior bisectors of the angle at vertex A of triangle ABC meet side BC produced in points D and E respectively. (See Figure 48.) If $\overline{AD} = \overline{AE}$, find $\angle BCA - \angle CBA$.

Problem 392.   (a) Observe that

$$9^3 + 15^3 + 12^3 = 18^3,$$
$$28^3 + 53^3 + 75^3 = 84^3,$$
$$65^3 + 127^3 + 248^3 = 260^3.$$

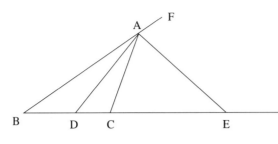

FIGURE 48

Find a generalization.

(b) Observe that

$$3^3 + 4^3 + 5^3 = 6^3,$$

$$12^3 + 19^3 + 53^3 = 54^3,$$

$$27^3 + 46^3 + 197^3 = 198^3.$$

Find a generalization.

(c) Observe that

$$3^3 + 10^3 + 18^3 = 19^3,$$

$$12^3 + 31^3 + 102^3 = 103^3,$$

$$27^3 + 64^3 + 306^3 = 307^3.$$

Find a generalization.

Problem 393. Let P be a non–self-intersecting polygon with n sides (Figure 49a). Let m other points $Q_1, Q_2, \ldots, Q_m$ interior to P be given (Figure 49b). A triangulation is obtained by joining some pairs of the $n + m$ points creating a dissection of the polygon into triangles none of which contains a P_i or Q_j in its interior, nor does any side of a triangle contain a Q_j. There are many

triangulations. Figure 49c and Figure 49d show two of the many triangulations of the polygon of Figure 49b. Show that the number of triangles is the same for all triangulations and find a formula in terms of n and m.

Problem 394. Show that if A, B, C are the angles of any triangle, then

$$3(\sin^2 A + \sin^2 B + \sin^2 C)$$

$$- 2(\cos^3 A + \cos^3 B + \cos^3 C) \leq 6.$$

Problem 395. Given equilateral triangle ABC, choose D on side AB and E on side AC so that $\overline{AD} = \overline{AE}$. Erect equilateral triangles PCD, QAE, and RAB as in Figure 50. Show that

(a) triangle PQR is equilateral;

(b) the midpoints of PE, AQ and RD are vertices of an equilateral triangle.

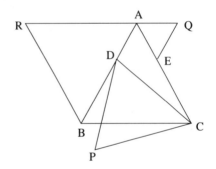

FIGURE 50

Problem 396. Let C be the circle of radius 3 centered at $(3,0)$. For any h, $0 < h < 6$, the circle of radius h centered at O meets the positive

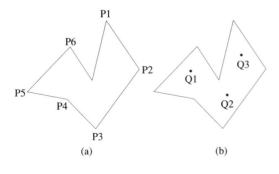

(a)   (b)

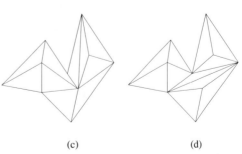

(c)   (d)

FIGURE 49

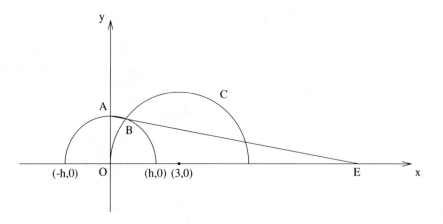

FIGURE 51

y-axis, in A say, and it meets the upper half of C in B say. (See Figure 51.) Let E be the point at which the line AB produced intersects the x-axis. What happens to E as h is taken smaller and smaller? Does E recede from O? If yes, how far? Or does E come closer to O? How close?

Problem 397. For any positive rational number u, let us agree to call the numbers $u+1$ and $\dfrac{u}{u+1}$ the *children* of u. Show that every positive rational number is the descendant of 1 in a unique way.

Problem 398. Let Q be a quadrilateral whose side lengths are positive integers for which the sum of any three is a multiple of the fourth. Show that some pair of sides of Q have the same length.

Problem 399. Consider a rectangular array of dots with an even number of rows and an even number of columns. Color the dots, each one red or blue, subject to the condition that in each row half the dots are red and half are blue, and in each column half the dots are red and half are blue. Now, if two points are adjacent (in a row or in a column) and like-colored, join them by an edge of their color. Show that the number of blue segments is equal to the number of red segments.

Problem 400. At a party there are more than 3 people. Every four of the people have the property that one of the four is acquainted with the other three. Show that with the possible exception of three of the people, everyone at the party is acquainted with all of the others at the party.

Problem 401. It is intuitive that the smallest regular n-gon which can be inscribed in a given regular n-gon will have its vertices at the midpoints of the sides of the given n-gon. Give a proof!

Problem 402. The real numbers x, y, z are such that

$$x^2 + (1 - x - y)^2 + (1 - y)^2$$
$$= y^2 + (1 - y - z)^2 + (1 - z)^2$$
$$= z^2 + (1 - z - x)^2 + (1 - x)^2.$$

Determine the minimum value of $x^2 + (1 - x - y)^2 + (1 - y)^2$.

Problem 403. Determine all the roots of the quartic equation $x^4 - 4x = 1$.

Problem 404. $P(x)$ and $Q(x)$ are two polynomials that satisfy the identity $P(Q(x)) \equiv Q(P(x))$ for all real numbers x. If the equation $P(x) = Q(x)$ has no real solution, show that the equation $P(P(x)) = Q(Q(x))$ also has no real solution. (1980 Canadian Mathematical Olympiad)

Problem 405. Determine the maximum value of
$$P = \frac{(b^2 + c^2 - a^2)(c^2 + a^2 - b^2)(a^2 + b^2 - c^2)}{(abc)^2},$$

where a, b, c are real and

$$\frac{b^2 + c^2 - a^2}{bc} + \frac{c^2 + a^2 - b^2}{ca} + \frac{a^2 + b^2 - c^2}{ab} = 2.$$
(1)

Problem 406. Can one find triplets of real numbers (a, b, c) such that none of the numbers is a cube of an integer and such that

$$S_n = a^{\frac{n}{3}} + b^{\frac{n}{3}} + c^{\frac{n}{3}}$$

is integral for all positive integral n?

Problem 407. If $S = a_1 + a_2 + \cdots + a_n$, where $a_1, a_2, \ldots, a_n$ are sides of a polygon, prove that

$$\frac{n+2}{S - a_k} \geq \sum_{i=1}^{n} \frac{1}{S - a_i} \quad \text{for } k = 1, 2, \ldots, n.$$

Problem 408. It is intuitive that if a rectangle is inscribed in an ellipse, the sides must be parallel to the axes of the ellipse. Give a proof.

Problem 409. Determine the maximum area of a rectangle inscribed in the ellipse $\dfrac{x^2}{a^2} + \dfrac{y^2}{b^2} = 1$.

Problem 410. If w and z are complex numbers, prove that

$$2\,|w|\,|z|\,|w - z| \geq \{|w| + |z|\}\,|\,w|z| - z|w|\,|.$$

Problem 411. If a and b are real and unequal, prove that the equation

$$(a - b)x^n + (a^2 - b^2)x^{n-1} + \cdots + (a^n - b^n)x$$
$$+ a^{n+1} - b^{n+1} = 0$$
(1)

has at most one real root.

Problem 412. If $a_0 \geq a_1 \geq a_2 \geq \cdots \geq a_n > 0$, prove that any root r of the polynomial

$$P(z) \equiv a_0 z^n + a_1 z^{n-1} + \cdots + a_n$$

satisfies $|r| \leq 1$, i.e., all the roots lie inside or on the unit circle centered at the origin in the complex plane.

Problem 413. There are many sums for which nice formulae are known, e.g., arithmetic and geometric sums, as well as

$$1^2 + 2^2 + \cdots + n^2 = \frac{n(n + 1)(2n + 1)}{6},$$
$$n = 1, 2, 3, \ldots,$$
$$1^3 + 2^3 + \cdots + n^3 = \frac{n^2(n + 1)^2}{4},$$
$$n = 1, 2, 3, \ldots.$$

Establish formulae for:

$$S_n = \left[1^{1/2}\right] + \left[2^{1/2}\right] + \cdots + \left[(n^2 - 1)^{1/2}\right],$$
$$n = 2, 3, \ldots;$$
(1)
$$T_n = \left[1^{1/3}\right] + \left[2^{1/3}\right] + \cdots + \left[(n^3 - 1)^{1/3}\right],$$
$$n = 2, 3, \ldots.$$
(2)

($[x]$ denotes the largest integer $\leq x$.)

Problem 414. The seven integers in the circular arrangement of seven disks in Figure 52 have the property that every one of the integers $1, 2, 3, \ldots, 14$ is either in a disk or else is the sum of the integers in two adjacent disks. Can you re-

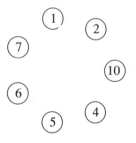

FIGURE 52

A. N. Whitehead once cautioned a student about a theory of logic. "You must take it with a grain of er ... um ... ah ...." For almost a minute, Whitehead groped for the word, until the student suggested, "Salt, Professor?" "Ah yes," Whitehead beamed, "I knew it was some chemical."

place the numbers in the disks with integers none of which is 5 and retain the same property?

Problem 415. Place the digits 0, 1, 2, 3, 4, 5, 6, 7, 8, 9 in the ten blank spaces in such an order that the indicated division will have a remainder of 1981

$$3168 \;\big|\; \overline{71\_543\_2\_985\_2\_356\_8\_7\_836\_7\_39997}.$$

(L. Moser)

Problem 416. Determine all triangles ABC for which

$$\tan(A - B) + \tan(B - C) + \tan(C - A) = 0.$$

Problem 417. Show that mth roots (m an integer > 1) of three distinct prime numbers cannot be terms (not necessarily consecutive) of a geometric progression.

Problem 418. For any simple closed curve there may exist more than one chord of maximum length. For example, in a circle all the diameters are chords of maximum length. In contrast, a proper ellipse has only one chord of maximum length (the major axis). Show that no two chords of maximum length of a given simple closed curve can be parallel.

Problem 419. Show that two bounded figures can have at most two centers of homotheticity.

Problem 420. A projectile in flight is observed simultaneously from three radar stations which are situated at vertices of an equilateral triangle of side a. The distances of the projectile from the three stations, taken in order around the triangle, are found to be R_1, R_2, and R_3. Determine the height of the projectile above the plane of the triangle.

Problem 421. In a given sphere, APB, CPD and EPF denote three mutually perpendicular and concurrent chords (at point P). If $\overline{AP} = 2a$, $\overline{PB} = 2b$, $\overline{CP} = 2c$, $\overline{PD} = 2d$, $\overline{EP} = 2e$ and $\overline{PF} = 2f$, determine the radius R of the sphere.

Problem 422. We start with 7 sheets of paper and a number of them are each cut into 7 smaller pieces. Then some of the smaller pieces are each cut into 7 still smaller pieces and so on repeatedly. Finally, the process is stopped and it turns out that the total number of pieces of paper is some number between 1988 and 1998. Can one determine the exact final number of pieces of paper?

Problem 423. What is the least number of plane cuts required to cut a block $a \times b \times c$ into abc unit cubes, if piling is permitted? (L. Moser)

Problem 424. Given a non-coplanar hexagon whose opposite sides are parallel. Prove that the midpoints of its six edges are coplanar.

Problem 425. Given two tangent congruent circles. From the point of tangency, two particles move on the two circles, both counter-clockwise with the same speed (not necessarily constant) at all times. (See Figure 53.) Prove that relative to one of the particles, the other one will appear to move on a circle whose radius is equal to the diameter $2R$ of the given circles.

Problem 426. Find the locus of midpoints of all chords of a parabola which pass through a given interior point of the parabola.

Problem 427. Let $G(n)$ denote the integer closest to $\dfrac{(n+3)^2}{12}$. For example,

$$G(1) = 1, \quad G(2) = 2, \quad G(3) = 3,$$
$$G(4) = 4, \quad G(5) = 5, \quad G(6) = 7.$$

Prove that

$$G(n) = G(n - 6) + n, \qquad n = 7, 8, 9, \ldots.$$

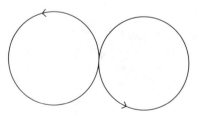

FIGURE 53

Problem 428. What is the largest value of n, in terms of m, for which the following statement is true? If from among the first m natural numbers any n are selected, among the remaining $m - n$ at least one will be a divisor of another. (Student-Faculty Colloquium, Carleton College)

Problem 429. Conjecture: If $f(t)$, $g(t)$, $h(t)$ are real-valued functions of a real variable, then there are numbers x, y, z such that $0 \leq x, y, z \leq 1$ and

$$|xyz - f(x) - g(y) - h(z)| \geq \frac{1}{3}.$$

Prove this conjecture. Show that if the number $\frac{1}{3}$ is replaced by a constant $c > \frac{1}{3}$, then the conjecture is false; i.e., the number $\frac{1}{3}$ in the conjecture is best possible.

Problem 430. It is known in Euclidean geometry that the sum of the angles of a triangle is constant. Prove, however, that the sum of the dihedral angles of a tetrahedron is not constant. (1979 Canadian Mathematical Olympiad)

Problem 431. The digital expression $x_n x_{n-1} \ldots x_1 x_0$ is the representation of the number A to base a as well as that of B to base b, while the digital expression $x_{n-1} x_{n-2} \ldots x_1 x_0$ is the representation of C to base a and also of D to base b. Here, a, b, n are integers greater than one.

Show that $\frac{C}{A} < \frac{D}{B}$ if and only if $a > b$.

Problem 432. Show that 5 or more great circles on a sphere, no 3 of which are concurrent, determine at least one spherical polygon having 5 or more sides. (L. Moser)

Problem 433. Determine all triplets (x, y, z) of integers for which

$$x^3 + y^3 + z^3 = (x + y + z)^3.$$

Problem 434. Find all x satisfying

$$16(\sin^5 x + \cos^5 x) = 11(\sin x + \cos x),$$

$$0 \leq x \leq 2\pi.$$

Problem 435. If $a_{n+1} = \dfrac{1 + a_n a_{n-1}}{a_{n-2}}$ and $a_1 = a_2 = a_3 = 1$, show that a_n is an integer for $n = 4, 5, \ldots$.

Problem 436. If a, b, c, d are positive integers for which $ab = cd$, show that $a^2 + b^2 + c^2 + d^2$ is composite. (West German Olympiad)

Problem 437. A pack of 13 distinct cards is shuffled in some particular manner and then repeatedly in exactly the same manner. What is the maximum number of shuffles required for the cards to return to their original positions?

Problem 438. If

$$\frac{a}{bc - a^2} + \frac{b}{ca - b^2} + \frac{c}{ab - c^2} = 0,$$

prove that also

$$\frac{a}{(bc - a^2)^2} + \frac{b}{(ca - b^2)^2} + \frac{c}{(ab - c^2)^2} = 0.$$

Problem 439. If a, a' and b, b' and c, c' are the lengths of the three pairs of opposite edges of an arbitrary tetrahedron, prove that
  (i) there exists a triangle whose sides have lengths $a + a'$, $b + b'$ and $c + c'$;
  (ii) the triangle in (i) is acute.

Problem 440. Determine the maximum value of

$$\sqrt[3]{4 - 3x + \sqrt{16 - 24x + 9x^2 - x^3}}$$
$$+ \sqrt[3]{4 - 3x - \sqrt{16 - 24x + 9x^2 - x^3}}$$

in the interval $-1 \leq x \leq 1$.

Problem 441. Let u_n be the nth term of the sequence

$$1, 2, 4, 5, 7, 9, 10, 12, 14, 16, 17, \ldots,$$

where one odd number is followed by two evens, then three odds, and so on. Prove that

$$u_n = 2n - \left[\frac{1}{2}(1 + \sqrt{8n - 7}) \right],$$

where square brackets denote the greatest integer function.

Problem 442. If e and f are the lengths of the diagonals of a quadrilateral of area F, show that $e^2 + f^2 \geq 4F$, and determine when there is equality.

Problem 443. Inside a cube of side 15 units there are 11000 given points. Prove that there is a sphere of unit radius within which there are at least 6 of the given points.

Problem 444. There is on the market a three-dimensional noughts and crosses game which is a distinct improvement over tic-tac-toe. The $4 \times 4 \times 4$ board (see Figure 54) is composed of 64 cells into which each player alternately places a counter of his own color. The first player to place four of his counters in a straight line wins. How many different ways of doing this are there?

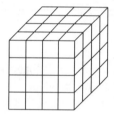

FIGURE 54

Problem 445. Prove that if the top 26 cards of an ordinary shuffled deck contain more red cards than there are black cards in the bottom 26, then there are in the deck at least three consecutive cards of the same color. (L. Moser)

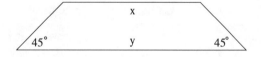

FIGURE 55

Problem 446. *The Carpenter's Problem.* A carpenter had 4 pieces of wood cut in the shape of an isosceles trapezoid as indicated in Figure 55. The x and y in each case are a whole number of inches. Although the x and y are different in each piece, the area is the same for all four pieces and is indeed a whole number of square inches, less than 40. What are the dimensions of the four pieces.

Problem 447. If m and n are positive integers, show that

$$\frac{1}{\sqrt[n]{m}} + \frac{1}{\sqrt[m]{n}} > 1$$

Problem 448. In a non-recent edition of *Ripley's Believe It Or Not,* it was stated that the number

$$N = 526315789473684210$$

is a *persistent* number, that is, if multiplied by any positive integer the resulting number always contains the ten digits $0, 1, 2, \ldots, 9$ in some order with possible repetitions.

(a) Prove or disprove the above statement.

(b) Are there any persistent numbers smaller than the above number?

Problem 449. If a, b, c, d are real, prove that

$$\left\{ \begin{array}{l} a^2 + b^2 = 2, \\ c^2 + d^2 = 2, \\ \quad\quad ac = bd, \end{array} \right\}$$

if and only if

$$\left\{ \begin{array}{l} a^2 + c^2 = 2, \\ b^2 + d^2 = 2, \\ \quad\quad ab = cd \end{array} \right\}.$$

Problem 450. How many permutations $a_1, a_2, \ldots, a_n$ of $1, 2, \ldots, n$ are there for which $a_1 \neq 1$ and/or $a_2 \neq 2$?

Problem 451. A man and his grandson were born on the same day of the year. One year on that day the man noted that his age was an integral multiple of his grandson's age and furthermore that this phenomenon would be repeated for the following five birthdays as well. How old was the man when he made these observations?

Problem 452. Prove that $\log_{10} 2$ is irrational.

Problem 453. Seventy-five coplanar points are given, no three collinear. Prove that, of all the triangles which can be drawn with these points as vertices, not more than seventy per cent are acute-angled.

Problem 454. Let T_1 and T_2 be two acute-angled triangles with respective side lengths a_1, b_1, c_1 and a_2, b_2, c_2, areas Δ_1 and Δ_2, circumradii R_1 and R_2 and inradii r_1 and r_2. Show that, if $a_1 \geq a_2$, $b_1 \geq b_2$, $c_1 \geq c_2$, then $\Delta_1 \geq \Delta_2$ and $R_1 \geq R_2$, but it is not necessarily true that $r_1 \geq r_2$.

Problem 455. Given two points, one on each of two given skew lines (lines in space not lying in a common plane), prove that there exists a unique sphere tangent to each of the given lines at each of the given points. (*Crux Mathematicorum*)

Problem 456. $\{2, 5, 11, 23, 47\}$ and $\{2, 3, 7, 13\}$ are examples of chains $\{p_1, p_2, \ldots, p_m\}$ of primes p_i such that, for $2 \leq i \leq m$, either $p_i = 2p_{i-1} - 1$ or $p_i = 2p_{i-1} + 1$. Show that any such chain must have at most a finite number of elements.

Problem 457. M is a point on the side AB of the triangle ABC. r_1, r_2, r are the radii of the inscribed circles of triangles AMC, BMC, ABC, respectively, and t_1, t_2, t are the radii of the escribed circles of the same triangles tangent to the sides AM, BM, AB, respectively. Prove that

$$\frac{r_1}{t_1} \cdot \frac{r_2}{t_2} = \frac{r}{t}.$$

Problem 458. Four nonnegative integers are written in a ring. A new ring of integers is formed, whose entries are the absolute values of the differences of adjacent integers in the first ring. Show that, if the process is repeated enough, four zeros are obtained. For example,

$$\begin{pmatrix} 9 & 3 \\ 1 & 0 \end{pmatrix} \to \begin{pmatrix} 6 & 3 \\ 8 & 1 \end{pmatrix} \to \begin{pmatrix} 3 & 2 \\ 2 & 7 \end{pmatrix} \to \begin{pmatrix} 1 & 5 \\ 1 & 5 \end{pmatrix}$$

$$\to \begin{pmatrix} 4 & 0 \\ 0 & 4 \end{pmatrix} \to \begin{pmatrix} 4 & 4 \\ 4 & 4 \end{pmatrix} \to \begin{pmatrix} 0 & 0 \\ 0 & 0 \end{pmatrix}$$

Problem 459. It is easy to see that there exists an infinite family of ellipses which can be inscribed in a given square. Prove, however, that only one ellipse can be inscribed in a given regular pentagon.

Problem 460. Determine all real x, y, z such that

$$xa^2 + yb^2 + zc^2 \leq 0$$

whenever a, b, c are sides of a triangle.

Problem 461. If three equal cevians of a triangle divide the sides in the same ratio and same sense, must the triangle be equilateral?

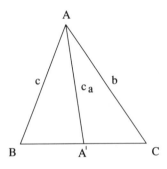

FIGURE 56

Problem 462. Determine the maximum value of

$$(\sin A_1)(\sin A_2) \cdots (\sin A_n)$$

if

$$(\tan A_1)(\tan A_2) \cdots (\tan A_n) = 1.$$

Problem 463. Two triangles have sides (a_1, b_1, c_1), (a_2, b_2, c_2) and respective areas Δ_1, Δ_2. Establish the Newberg–Pedoe inequality

$$a_1^2 \left(b_2^2 + c_2^2 - a_2^2\right) + b_1^2 \left(c_2^2 + a_2^2 - b_2^2\right)$$
$$+ c_1^2 \left(a_2^2 + b_2^2 - c_2^2\right) \geq 16\Delta_1\Delta_2,$$

and determine when there is equality.

Problem 464. It is easy to see that if the vertices of an inscribed n-gon in a given regular n-gon divide each side of the given n-gon in the same ratio and sense, then the inscribed n-gon is also regular. Prove the converse result, i.e., if $B_1, B_2, \ldots, B_n$ are the vertices of an inscribed regular n-gon in the regular n-gon of vertices $A_1, A_2, \ldots, A_n$, where the vertices are consecutive, have the same sense and B_i lies between A_i and A_{i+1}, then $A_i B_i =$ constant.

Problem 465. Let m and n be given positive numbers with $m \geq n$. Call a number x "good" (with respect to m and n) if:

$$m^2 + n^2 - a^2 - b^2 \geq (mn - ab)x$$

$$\text{for all} \quad 0 \leq a \leq m, \ 0 \leq b \leq n. \qquad (1)$$

Determine (in terms of m and n) the largest good number.

Problem 466. Prove that, for any quadrilateral (simple or not, planar or not) of sides a, b, c, d

$$a^4 + b^4 + c^4 \geq \frac{d^4}{27}.$$

Problem 467. Determine the maximum of $x^2 y$, subject to constraints

$$x + y + \sqrt{2x^2 + 2xy + 3y^2} = k$$

$$\text{(constant)}, \quad x, y \geq 0.$$

Problem 468. Prove

$$\frac{4^m}{2\sqrt{m}} < \binom{2m}{m} < \frac{4^m}{\sqrt{3m+1}} \qquad (1)$$

for natural numbers $m > 1$.

Problem 469. Determine all pairs of rational numbers (x, y) such that

$$x^3 + y^3 = x^2 + y^2.$$

Problem 470. What is the probability of an odd number of sixes turning up in a random toss of n fair dice?

Problem 471.

> Ma and Pa and brother and me
> The sum of our ages is eighty-three
> Six times Pa's is seven times Ma's
> And Ma's is three times me.

Find the ages; no fractions please.

Problem 472. Let $(n_1, n_2, \ldots, n_r)$ and $[n_1, n_2, \ldots, n_r]$ denote the greatest common divisor and least common multiple, respectively, of the set of positive integers $n_1, n_2, \ldots, n_r$. For example, $(6, 12, 15) = 3$ and $[2, 3, 4] = 12$. Prove

$$\frac{(a,b,c)^2}{(b,c)(c,a)(a,b)} = \frac{[a,b,c]^2}{[b,c][c,a][a,b]}.$$

(1972 USAMO)

Problem 473. In setting the type for the multiplication

$$(abc)(bca)(cab) = 234235286,$$

with $a > b > c$, the setter pied all the digits except the units digit 6. Restore them to their proper order. (*Math. Mag.* 1959)

Problem 474. Sketch the curve given by the equation $(2x + 3y)^2(y - x) = x + y$.

Problem 475. Determine A_{1990} if

$$A_{n+1} = \frac{A_n}{1 + nA_n},$$

$$n = 0, 1, 2, \ldots \text{ and } A_0 = A.$$

Problem 476. The numbers $1, 2, 3, \ldots, 100$ are multiplied by the numbers $1, 2, 3, \ldots, 100$ in some order. Show that the 100 products obtained cannot represent all positive remainders on division by 101.

Problem 477. In this question, a *real function* means a function f such that $f(x)$ exists and is real for all real numbers x. Show that there is only one value of the constant b for which there exists a real function f with the property that, for all real x and y,

$$f(x - y) = f(x) - f(y) + bxy.$$

Problem 478. A cute girl only eats pie cut in the shape of acute triangles. Show how to cut a pie of any polygonal shape so as to serve cute girls, with no pieces left over. (*J. Recreational Math.* 1972)

Problem 479. In a regular (equilateral) triangle, the circumcenter O, the incenter I, and centroid G all coincide. Conversely, if any two of O, I, G coincide, the triangle is equilateral. Also, for a regular tetrahedron, O, I and G coincide. Prove or disprove the converse result that if O, I and G all coincide for the tetrahedron, the tetrahedron must be regular.

Problem 480. Prove that there is an infinite set of positive integers of the form $2^n - 3$ with the property: no two members of the set have a common prime factor.

Problem 481. Prove that the sum of the squares of the reciprocals of the lengths of the segments joining one vertex of a regular polygon of n sides to the other $n - 1$ vertices is $\dfrac{n^2 - 1}{12R^2}$, where R is the radius of the circumscribing circle.

Problem 482. If n^2 coins, of which exactly n are silver, are arranged at random in n rows, each containing n coins, prove that the chance that at least one row occurs in which there is no silver coin is

$$1 - \frac{(n-1)!(n^2-n)!n^{n-1}}{(n^2-1)!}.$$

(*Math. Gazette* 1904)

Problem 483. Given an integer of n nonzero digits in base b, show that it is always possible to replace a certain r $(0 \leq r \leq n - 1)$ of these digits by zeros in such a way that the resulting number is divisible by n. (Leo Moser)

Problem 484. Find the rhombus of minimum area which can be inscribed (one vertex to a side) within a given parallelogram. (*Math. Gazette* 1904)

Problem 485. A chord of a right triangle, which is parallel to the hypotenuse and passes through

the incenter I, is divided by I into two segments of length m and n. Determine the area of the triangle (in terms of m and n).

Problem 486. If

$$a + b + c + d + e + f = 0$$

and

$$a^3 + b^3 + c^3 + d^3 + e^3 + f^3 = 0,$$

prove that

$$(a + c)(a + d)(a + e)(a + f)$$
$$= (b + c)(b + d)(b + e)(b + f).$$

(*Math. Gazette*)

Problem 487. Prove that the sum of the digits of every multiple of 2739726 up to the 72nd is 36. (E. M. Langley, *Math. Gazette* 1896)

Problem 488. Determine the largest real number k, such that

$$|z_2 z_3 + z_3 z_1 + z_1 z_2| \geq k|z_1 + z_2 + z_3|$$

for all complex numbers z_1, z_2, z_3 with unit absolute value.

Problem 489. If the odds against a horse in a race be a to b, then we may call $\dfrac{b}{a + b}$ the apparent chance of that horse winning. Prove that, if the sum of the apparent chances of all the horses in a race be less than unity, one can arrange bets so as to make sure of winning the same sum of money whatever be the outcome of the race. (R. W. Genese, *Math. Gazette* 1896)

Problem 490. If the roots of the equation

$$a_0 x^n - n a_1 x^{n-1} + \frac{n(n-1)}{1.2} a_2 x^{n-2}$$
$$- \cdots + (-1)^n a_n = 0$$

are all positive, show that $a_r a_{n-r} > a_0 a_n$ for all values of r between 1 and $n - 1$ inclusive, unless the roots are all equal. (A. Lodge, *Math. Gazette* 1896)

Problem 491. Suppose $u \le 1 \le w$. Determine all values of v for which $u + vw \le v + wu \le w + uv$.

Problem 492. Find the shortest distance between the plane $Ax + By + Cz = 1$ and the ellipsoid

$$\frac{x^2}{a^2} + \frac{y^2}{b^2} + \frac{z^2}{c^2} = 1.$$

You can assume A, B, C are all positive and that the plane does not intersect the ellipsoid. (No calculus please.)

Problem 493. One of the problems on the first William Lowell Putnam Mathematical Competition, was to find the length of the shortest chord that is normal to the parabola $y^2 = 2ax$ at one end. (Assume $a > 0$.) A calculus solution is straight forward. Give a completely "no calculus" solution.

Problem 494. Alice and Bob play a fair game repeatedly for one nickel each game. If originally Alice has a nickels and Bob has b nickels, what is Alice's chance of winning all of Bob's money, assuming the play goes on until one person has lost all her or his money?

Problem 495. If P, Q, R are any three points inside or on a unit square, show that the smallest of the three distances determined by them is at most $2\sqrt{2 - \sqrt{3}}$, i.e., show

$$\min(\overline{PQ}, \overline{QR}, \overline{RP}) \le 2\sqrt{2 - \sqrt{3}}.$$

Also determine when there is equality.

Problem 496. Any 5 points inside or on a 2×1 rectangle determine 10 segments (joining the pairs of points). Show that the smallest of these 10 segments has a length at most $2\sqrt{2 - \sqrt{3}}$. (Leo Moser)

Problem 497. Show that the Diophantine equation

$$\frac{1}{x_1} + \frac{1}{x_2} + \cdots + \frac{1}{x_n} + \frac{1}{x_1 x_2 \ldots x_n} = 1$$

has at least one solution for every n. (Leo Moser)

Problem 498. If the sum of the face angles at each vertex of a tetrahedron is $180°$, prove that the tetrahedron is isosceles, i.e., the opposite edges are equal in pairs.

Problem 499. Prove that every positive integer which is not a member of the infinite set below is equal to the sum of two or more distinct members of the set:

$$\{3, -2, 2^2 3, -2^3, \ldots, 2^{2k} 3, -2^{2k+1}, \ldots\}$$
$$= \{3, -2, 12, -8, 48, -32, 192, \ldots\}.$$

Problem 500. One is given $n + 1$ rays emanating from a common vertex in n-dimensional Euclidean space R^n. If all the acute angles between pairs of the rays are congruent, determine the common angle.

E. Kummer, the German algebraist, was rather poor at arithmetic. Whenever he had occasion to do simple arithmetic in class he would get his students to help him. Once he had occasion to find 7×9. "7×9", he began, "7×9 is er—ah—ah—7×9 is . . . ." 61, a student suggested. Kummer wrote 61 on the board. "Sir", said another student, "It should be 69." "Come, come gentlemen," said Kummer, "it can't be both—it must be one or the other."

SOLUTIONS

Problem 1. The three lengths are of the form $a - d$, a, $a + d$ with $(a - d)^2 + a^2 = (a + d)^2$. This reduces to $a(a - 4d) = 0$, or $a = 4d$. Thus the sides are $3d$, $4d$, $5d$.

Problem 2. Let $A(a, 2a)$ be an arbitrary point on the line $y = 2x$, and let $B(b, b)$ be an arbitrary point on the line $y = x$. The midpoint of segment AB is the point (x, y) with $x = \dfrac{a + b}{2}$ and $y = \dfrac{2a + b}{2}$, and the length of the segment AB is 4. Hence $(a - b)^2 + (2a - b)^2 = 4^2$ or $5a^2 - 6ab + 2b^2 = 4^2$ with $a = 2(y - x)$, $b = 2(2x - y)$, so

$$20(y - x)^2 - 24(y - x)(2x - y) + 8(2x - y)^2 = 4^2$$

or

$$25x^2 - 36xy + 13y^2 = 4.$$

This is an ellipse whose axes must be (by the obvious symmetry of the locus) the bisectors of the angles formed by the two given lines.

Problem 3. It is easy to deduce that the rectangle is 9 by 16. The square therefore has area 144, so its side is of length 12 and its perimeter is 48.

Problem 4. A suggested law is

$$(2n + 1)^2 + (2n(n + 1))^2 = (2n(n + 1) + 1)^2,$$
$$n = 1, 2, 3, \ldots,$$

and this identity is easy to establish by squaring out or by using a difference-of-squares factorization.

Problem 5. Call the required sum S_n, so that for $n \geq 2$

$$S_n = 6 + 66 + 666 + \cdots + \underbrace{666 \cdots 6}_{n\ 6's},$$

$$= (0 + 6) + (60 + 6) + (660 + 6)$$
$$+ \cdots + (666 \cdots 60 + 6),$$

$$= 10(6 + 66 + \cdots + \underbrace{666 \cdots 6}_{n-1\ 6's}) + 6n,$$

$$= 10 S_{n-1} + 6n,$$

$$= 10(S_n - \underbrace{666 \cdots 6}_{n\ 6's}) + 6n.$$

Solving for S_n we obtain

$$9 S_n = \underbrace{666 \cdots 6}_{n\ 6's} 0 - 6n$$

$$= \frac{2}{3}(999 \cdots 90 - 9n),$$

$$S_n = \frac{2}{3}(\underbrace{111 \cdots 1}_{n\ 1's} 0 - n).$$

Since

$$\underbrace{111 \cdots 1}_{n\ 1's} 0 = 10 + 10^2 + \cdots + 10^n = \frac{10(10^n - 1)}{9},$$

we may write

$$S_n = \frac{2}{3}\left(\frac{10(10^n - 1)}{9} - n\right)$$

$$= \frac{2}{27}\left(10^{n+1} - 10 - 9n\right)$$

$$\text{for} \quad n = 1, 2, 3, \ldots.$$

Problem 6. *First solution.* We may assume that $x > y > z \geq 1$. Letting N denote the number of examinations, we have $N > 1$ (implied by the word "series" in the statement of the problem) and

$$(x + y + z)N = 20 + 10 + 9 = 39.$$

Since $x + y + z \geq 3 + 2 + 1 = 6$, we know that $N \leq 6$, and since N divides 39, we deduce $N = 3$, so that $x + y + z = 13$. The solutions (x, y, z) of the equation $x + y + z = 13$ with $x, y, z \geq 1$, and x, y, z distinct integers, are:

$$(x, y, z) = (10, 2, 1),\ (9, 3, 1),\ (8, 4, 1),\ (8, 3, 2),$$
$$(7, 5, 1),\ (7, 4, 2),\ (6, 5, 2),\ (6, 4, 3).$$

Except for $(8, 4, 1)$, all these possibilities are eliminated by the fact that Alice's marks sum to 20. Now we know that Betty's algebra mark is 8 (the largest of 8, 4, 1) and the problem is to fill in the table (see below) so that each row is a permutation of 8, 4, 1 and the column sums are 20, 10, 9.

	Alice	Betty	Carol
algebra		8	
geometry			
other subject			
	20	10	9

The only solution is easily seen to be

	Alice	Betty	Carol
algebra	4	8	1
geometry	8	1	4
other subject	8	1	4

Thus, Carol placed second in geometry.

Second solution. As before we deduce that there are three examinations and $x + y + z = 13$. Not all of Alice's marks are the same, since 20 is not divisible by 3. Suppose, if possible, Alice's highest mark is y. Then she must have either $\{y, z, z\}$ or $\{y, y, z\}$. But

$$y + z + z < y + y + z < x + y + z = 13 < 20.$$

Hence Alice has at least one x.

We will show that Alice cannot have precisely one x. If she had, since $x + y + z = 13$ and Alice's total is 20, her marks must be $\{x, y, y\}$.

Then $x + 2y = 20$ and $y = z + 7$, so $x + 2z = 6$. Betty therefore cannot have $\{x, z, z\}$; nor can she have $\{x, y, z\}$ or better, which excludes all possibilities. Hence Alice has two x's. Because $x + y + z = 13$ and Betty has one x and a total mark 10, Betty must have $\{x, z, z\}$ and the marks awarded must be

	Algebra	$\star$	$\star$
Alice	?	x	x
Betty	x	z	z
Carol	?	y	y

Hence Carol has y in all examinations except algebra. Thus there is no ambiguity, and Carol placed second in geometry.

Remarks. In fact, it can be deduced that Alice's algebra mark is y and that $x = 8$, $y = 4$, and $z = 1$, so that the situation is indeed feasible.

If the conditions are weakened to require merely that x, y, z are nonnegative, then it is possible that $n = 13$. In this case $x = 2$, $y = 1$, $z = 0$ and there are many possibilities, so that further information would be required to solve the problem.

Problem 7. We see that

$$f_0(x) = \frac{1}{1-x},$$

$$f_1(x) = f_0(f_0(x)) = \frac{1}{1 - \frac{1}{1-x}} = \frac{1-x}{-x},$$

$$f_2(x) = f_0(f_1(x)) = \frac{1}{1 - \frac{1-x}{-x}} = x,$$

$$f_3(x) = f_0(f_2(x)) = \frac{1}{1-x} = f_0(x).$$

Hence, in general

$$f_{n+3}(x) = f_n(x), \qquad n = 0, 1, 2, 3\ldots,$$

and, in particular

$$f_{1976}(1976) = f_2(1976) = 1976.$$

Problem 8. On division by 3, each of the numbers leaves a remainder of 0, 1, or 2. If three of these five remainders are equal, then the sum of

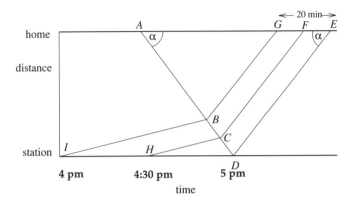

FIGURE 57

the three corresponding numbers is divisible by 3. If no three of the remainders are equal then the remainders include 0, 1, and 2, and again the result follows.

Comments. The following generalization appeared in the West German Olympiad, 1981 (see *Crux Mathematicorum* 12 (1986) 43).

Let n be a positive power of 2. Prove that from any set of $2n - 1$ positive integers, one can choose a subset of n integers such that their sum is divisible by n.

More generally, any set S of $2n - 1$ integers contains a subset of n integers whose sum is divisible by n. This result is due to Erdős, Ginsburg and Ziv (Theorem in additive number theory, *Bull. Res. Council Israel* 10 (1961)). Further proofs of this are given by R. Graham (*Mathematical Intelligencer* 1 (1979) 250) and also by T. Redmond and C Ryavec (*ibid* 2 (1980) 106).

Problem 9. *First solution.* In the trip in which Smith got home 20 minutes earlier than usual, the chauffeur saved 10 minutes on each leg of this trip, and consequently picked up Smith at 4:50 P.M. (This part is a well-known problem in which we are assuming that Smith and the chauffeur walk and ride, respectively, at constant rates.) Also, it follows now that the chauffeur's speed is five times that of Smith. Suppose Smith meets the chauffeur the second time t minutes after 4:30 P.M. The chauffeur is saved a journey of $\frac{t}{5}$ min-

utes from the meeting place to the station (which he normally reaches at 5:00 P.M. Hence $t + \frac{t}{5} = 30$, so $t = 25$ and Smith arrives at home 10 minutes earlier than usual.

Second solution (by "world lines"; see Figure 57). We plot the distance vs time lines for Smith and his chauffeur (see diagram next page). The chauffeur's world line is ADE, where A and E are not known but $\angle DAE = \angle DEA$. Smith's world line after arriving at 4 P.M. is IB, BG where $BG \parallel DE$. Then $\overline{GE} = 20$ minutes. Smith's world line after arriving at 4:30 P.M. is HC, CF where $HC \parallel IB$ and $CF \parallel DE$. It then follows that $\overline{BC} = \overline{CD}$ and $\overline{GF} = \overline{FE}$. Thus $\overline{FE} = 10$ minutes.

Problem 10. Let P be the position of the center of gravity of the system. As long as P is above the surface of the water, P falls as the surface rises. Hence at some stage P must be at the surface. At this instant P is at its lowest because the addition of more water must raise the water level and the level of P.

Problem 11. Advise him to play in the first game. We show that Father wins the tournament with greater probability when he plays in the first game than he does by sitting out the first game.

Let F, M, and S denote father, mother, and son respectively and let $X > Y$ denote that X wins the game playing against Y.

If F and M play the first game, then F wins the tournament in the following three sequences:

(1) $F > M, \quad F > S$;

(2) $F > M, \quad S > F, \quad M > S, \quad F > M$;

(3) $M > F, \quad S > M, \quad F > S, \quad F > M$.

If F and S play the first game, the winning sequences for F are the same as those above with M and S interchanged.

If M and S play the first game, then F wins the tournament in the following two sequences:

(4) $S > M, \quad F > S, \quad F > M$;

(5) $M > S, \quad F > M, \quad F > S$.

Let $\overline{AB}$ denote the probability that $A > B$. Note that $\overline{AB} + \overline{BA} = 1$.

If F and M play the first game, the probability P_{FM} that F wins the tournament is

$$P_{FM} = \overline{FM} \cdot \overline{FS} + \overline{FM} \cdot \overline{SF} \cdot \overline{MS} \cdot \overline{FM}$$
$$+ \overline{MF} \cdot \overline{SM} \cdot \overline{FS} \cdot \overline{FM}.$$

If F and S play the first game, the probability P_{FS} that F wins the tournament is

$$P_{FS} = \overline{FS} \cdot \overline{FM} + \overline{FS} \cdot \overline{MF} \cdot \overline{SM} \cdot \overline{FS}$$
$$+ \overline{SF} \cdot \overline{MS} \cdot \overline{FM} \cdot \overline{FS}.$$

If M and S play the first game, the probability P_{MS} that F wins the tournament is

$$P_{MS} = \overline{SM} \cdot \overline{FS} \cdot \overline{FM} + \overline{MS} \cdot \overline{FM} \cdot \overline{FS}$$
$$= (\overline{SM} + \overline{MS}) \cdot (\overline{FS} \cdot \overline{FM}) = \overline{FS} \cdot \overline{FM}.$$

It is clear that $P_{FM} > P_{MS}$ and $P_{FS} > P_{MS}$.

Remark. It is intuitively clear that F will be best off playing (in the first game) against the weaker of M and S.

Problem 12. *First solution.* Our solution is based on the following simple lemma:

Let $WXYZ$ be a quadrilateral with $\overline{WZ} = \overline{XY}$ and $\angle WXY = \frac{\pi}{2}$; then $\angle XYZ \le \frac{\pi}{2}$ and equality holds if and only if $WXYZ$ is a rectangle.

We apply this lemma as follows. Rotate $\triangle DGH$ about point H and bring G and E into coincidence. (Note that if D falls on D' then $\angle AHD' = \frac{\pi}{2}$.) Applying the lemma to the resulting quadrilateral $AHD'E$, we obtain $\alpha \le \frac{\pi}{2}$. Continuing in this manner, we obtain $\beta \le \frac{\pi}{2}$, $\gamma \le \frac{\pi}{2}$ and $\delta \le \frac{\pi}{2}$; since $\alpha + \beta + \gamma + \delta = 2\pi$, we must conclude that $\alpha = \beta = \gamma = \delta = \frac{\pi}{2}$. Then, from the second part of the lemma, we obtain $\overline{AE} = \overline{BF} = \overline{CG} = \overline{DH}$. Hence $ABCD$ is a square.

Second solution. Rotate the figure counterclockwise (through $\frac{\pi}{2}$) about the center of the square so that H falls on G, G falls on F, A falls on A' and D falls on D'. Since $\angle FGC + \angle GFD' = \angle FGC + \angle HGD = \frac{\pi}{2}$, FD' is perpendicular to GC. Since $\overline{FD'} = \overline{GD} = \overline{FC}$, D' is on either GC or on the side of line GC opposite to F, and hence A' is on the same side of GD as F. Hence $\angle EHA = \angle HGA'$ does not exceed $\angle HGD$. Thus, applying the argument all around

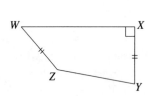

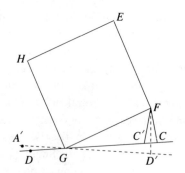

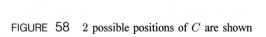

FIGURE 58   2 possible positions of C are shown

the square,

$$\angle EHA \leq \angle HGD \leq \angle GFC$$

$$\leq \angle FEB \leq \angle EHA$$

with the result that all these angles are equal. Hence

$$\angle HGD + \angle DHG = \angle EHA + \angle DHG = \frac{\pi}{2},$$

so $\angle D = \frac{\pi}{2}$. Likewise, $\angle A = \angle B = \angle C = \angle D = \frac{\pi}{2}$. The result now follows easily.

Remark. For the similar problem with equilateral triangles see *Math. Mag.* 43 (1970) 280–282; for the general n-gonal case see *ibid* 44 (1971) 296.

Problem 13. If the integers $1, 2, 3, \ldots, 126$ are split into 6 sets, then by the Pigeonhole Principle, one of the sets will contain two (or more) of the seven chosen integers. Thus, the problem is to arrange the splitting so that in each of the 6 sets the largest integer is at most twice the smallest. Clearly, the following splitting does the trick:

$$\{1, 2\}, \quad \{3, 4, 5, 6\}, \quad \{7, 8, \ldots, 13, 14\},$$

$$\{15, 16, \ldots, 29, 30\}, \quad \{31, 32, \ldots, 61, 62\},$$

$$\{63, 64, \ldots, 125, 126\}.$$

(Find a set of 7 positive integers not greater than 127 such that no two of them satisfy the inequalities $1 < \frac{y}{x} \leq 2$.)

Problem 14. Divide the square into four squares of side $\frac{1}{2}$. By the Pigeonhole Principle, one of these four squares contains at least two of the points, whose distance apart must be no greater than the diagonal of the square of side $\frac{1}{2}$, namely $\frac{\sqrt{2}}{2}$.

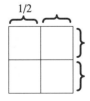

FIGURE 59

Problem 15. Suppose there are N parties, and their promises are the sets $S_1, S_2, \ldots, S_N$. We know that no two of these sets are equal (because no two parties make exactly the same promises). Furthermore $S_i \cap S_j \neq \emptyset$ for $i \neq j$ (because every pair of parties makes a promise in common). Thus, no S_i is the complement of an S_j, and hence there are at most $\frac{1}{2}2^n = 2^{n-1}$ subsets (of the set of all promises) in the list $S_1, S_2, \ldots, S_N$, i.e., $N \leq 2^{n-1}$.

Let $p_1, p_2, \ldots, p_n$ be the n promises, and let A_i, $i = 1, 2, \ldots, 2^{n-1}$, be the subsets of $\{p_2, p_3, \ldots, p_n\}$. The 2^{n-1} sets $\{p_1\} \cup A_i$ show that there can be as many as 2^{n-1} parties.

Problem 16. We shall refer to such a $(2m+1) \times (2n+1)$ checkerboard with one red square and two black squares removed as a *deleted checkerboard*. First, we note that the case $m = n = 1$ is easily handled by exhaustion. Owing to the symmetry there are only six cases that need to be considered, and these are shown below.

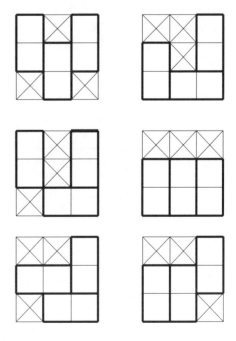

FIGURE 60

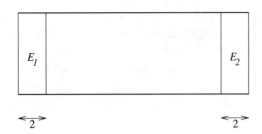

FIGURE 61

We now proceed by induction. We are given a $(2m+1) \times (2n+1)$ deleted checkerboard C and we may assume that any smaller $(2k+1) \times (2l+1)$ deleted checkerboard which is contained in C may be covered with dominoes. Since at least one of the two dimensions of C is of length at least five, C has two oppositely placed, non-overlapping ends E_1 and E_2 of width two (see Figure 61).

Clearly, we can choose an end containing at most one of the deleted squares of C. Let this end be E_1 and consider the following two cases.

Case 1. E_1 *contains no deleted square of C.* Then $C - E_1$ contains all three of the deleted squares. By the induction assumption, $C - E_1$ can be covered with dominoes. This covering, together with the obvious one for E_1, yields the desired covering of C.

Case 2. E_1 *contains exactly one deleted square of C.* In this case, with the deleted square in E_1 we identify the *associated square* of the same color in $C - E_1$ as shown in Figure 62.

Now delete the associated square in $C - E_1$. By the induction assumption, there is a domino covering of $C - E_1$ with this deletion. Now C, with its original deletions, may be covered by making use of the covering just found, together with the scheme shown in Figure 63.

This procedure would fail only in the case where the only choice for the associated square in $C - E_1$ was also deleted. This is impossible in the case of a red square. In the case of a black square, we infer that the one deleted red square is in E_2 and proceed as before.

Problem 17. We show the more general result that for any positive integers m and n, $D(9m + n) = D(n)$. If

$$9m + n = a_s 10^s + a_{s-1} 10^{s-1} + \cdots + a_1 10 + a_0$$

then

$$9m + n - \{a_s(10^s - 1) + a_{s-1}(10^{s-1} - 1)$$
$$+ \cdots + a_1(10 - 1)\} = a_0 + a_1 + \cdots + a_s.$$

Since the number within the braces is divisible by 9, there is a number $m_1 < m$ for which

$$a_0 + a_1 + \cdots + a_s = 9m_1 + n.$$

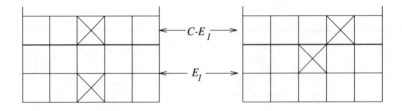

FIGURE 62

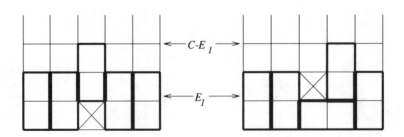

FIGURE 63

Repeated application of this yields integers $m > m_1 > m_2 > \cdots > m_k = 0$ such that

$$D(9m + n) = D(9m_1 + n) = D(9m_2 + n)$$
$$= \cdots = D(9m_k + n) = D(n).$$

When $m = (137)n$ we have $9m+n = 9(137)n + n = (1234)n$, so $D((1234)n) = D(n)$.

Problem 18. *First solution.* To see how this problem is solved, consider a square with center A, and points B, C on two adjoining sides (possibly extended), as in Figure 64a. A rotation of 90° counterclockwise about center A takes B onto B', on the side through C. Thus, to solve the problem, find the image B' of B in the 90° counterclockwise rotation with center A. Then line $B'C$ contains one side of the square, and the rest is easy (see Figure 64b), i.e., B' is located so that $B'A$ and BA are equal in length and perpendicular. The line $B'C$ contains one side of the square, which is easily constructed.

Another square is obtained by locating B'' as the image of B in the *clockwise* quarter turn with center A. Then line $B''C$ contains one side of a square with the desired properties.

There are special cases: if A is the midpoint of segment BC, there is precisely one square (with diagonal BC); if $\angle BAC$ is right and $\overline{AC} = \overline{AB}$ there are infinitely many squares. If $\angle BAC$ is right and $\overline{AC} \neq \overline{AB}$ then the problem is not feasible.

Second solution. The facts that the angle subtended at the circumference by a diameter of a circle is right, and that equal arcs subtend equal angles at the circumference, permit the construction first of a vertex of the square, then of its diagonal. With diameter BC, construct a circle. Let D be the midpoint of *either* arc BC (there are two possibilities). Let P be the intersection of AD with the circle which is distinct from D. Then PA is part of the diagonal of a square, with BP and CP parts of the two sides produced. See Figure 65.

Since $\angle BPC = \frac{\pi}{2}$ and $\overset{\frown}{BD} = \overset{\frown}{CD}$, $\angle BPD = \angle CPD = \frac{\pi}{4}$, and the proof of the construction is clear. Again special cases can be delineated. If A is the midpoint of BC, then P, D are both midpoints of the two arcs BC and $BPCD$ is the unique square; if $\angle BAC = \frac{\pi}{2}$ and $\overline{AC} \neq \overline{AB}$, then A lies on the circle and the

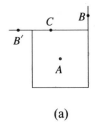

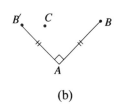

(a)   (b)

FIGURE 64

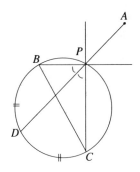

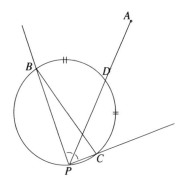

FIGURE 65

square degenerates to a single point (since A is both vertex and center). However, if $\angle BAC = \frac{\pi}{2}$ and $\overline{AC} = \overline{AB}$, taking D opposite to A yields the degenerate case, while if D is taken to be A, P can be chosen arbitrarily on the arc BC opposite to A and there are infinitely many possible squares.

Problem 19. The given inequality is equivalent (on squaring) to

$$n^{\sqrt{n+1}} > (n+1)^{\sqrt{n}}$$

or (raising to the power $\sqrt{n}$)

$$n^{\sqrt{n+1}\sqrt{n}} > (n+1)^n$$

or (dividing by n^n)

$$n^{\sqrt{n+1}\sqrt{n}-n} > \left(1+\frac{1}{n}\right)^n$$

or

$$n^{1/(1+\sqrt{1+\frac{1}{n}})} > \left(1+\frac{1}{n}\right)^n.$$

Now

$$\left(1+\frac{1}{n}\right)^n = 1 + n\cdot\frac{1}{n} + \frac{n(n-1)}{2\cdot 1}\cdot\frac{1}{n^2}$$
$$+ \cdots + \frac{n(n-1)\cdots 1}{n(n-1)\cdots 1}\cdot\frac{1}{n^n}$$

so

$$\left(1+\frac{1}{n}\right)^n < 1 + 1 + \frac{1}{2!} + \cdots + \frac{1}{n!}$$
$$< 1 + 1 + \frac{1}{2} + \frac{1}{2^2} + \cdots < 3.$$

Consequently, all we need satisfy is

$$n > 3^{2+\frac{1}{2n}} > 3^{(1+\sqrt{1+\frac{1}{n}})}.$$

Clearly the latter is satisfied for $n \geq 10$. Finally, it can be shown that $n = 7$ is the minimum allowable value for n.

Remark. Using calculus (which is not as elementary), it is easy to show that $(\sqrt{x})^{\frac{1}{\sqrt{x}}}$, or equivalently that $\frac{\log^2 x}{x}$ is monotonically decreasing for $x > e^2 = 7.389\ldots$. A similar calculus solution appears in D. S. Mitrinović, *Elementary Inequalities,* P. Nordhoff, Groningen, 1964, p. 60.

Problem 20. *First solution.* Reflect Z and C across OXY as shown in Figure 66. Since $\angle DYZ' = \angle Z'ZD = \angle C'XD$, C', D, X, Y all lie on a circle. Since $\angle C'XY = \frac{\pi}{2}$, the diameter of the latter circle is $\overline{C'Y} = \overline{CY}$. Thus $\overline{CY} \geq \overline{XD}$ with equality if and only if D coincides with Y (since $\angle DYX$ is strictly less than $\frac{\pi}{2}$ when $D \neq Y$). We now show more by proving that $\frac{\overline{CY}}{\overline{DX}}$ is an increasing function of $\overline{XC}$. Let $\overline{OY} = 1$, $\overline{OX} = a$, and $\overline{XC} = t$. Then

$$\overline{CY}^2 = (1-a)^2 + t^2,$$
$$\overline{AX}^2 = 1 - a^2 = \overline{DX}\cdot\overline{ZX}$$

$$\overline{CY}\cdot\overline{CZ} = \left(\sqrt{1-a^2}+t\right)\left(\sqrt{1-a^2}-t\right)$$
$$= 1 - a^2 - t^2.$$

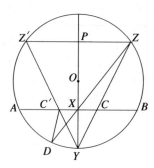

FIGURE 66

By similar triangles,

$$\overline{PZ} = t\left(1+\frac{\overline{CZ}}{\overline{CY}}\right), \qquad \overline{PX} = (1-a)\frac{\overline{CZ}}{\overline{CY}},$$

where P is the intersection of lines ZZ' and YX. Then

$$\overline{ZX}^2 = \overline{PZ}^2 + \overline{PX}^2$$

or

$$\frac{(1-a^2)^2}{\overline{DX}^2}$$
$$= \frac{1}{\overline{CY}^2}\left\{t^2(\overline{CY}+\overline{CZ})^2 + (1-a)^2\overline{CZ}^2\right\}.$$

Simplifying:

$$\left(\frac{\overline{CY}}{\overline{DX}}\right)^2 = \frac{1}{(1-a^2)^2}\{t^2\overline{CY}^2 + 2t^2\overline{CY}\cdot\overline{CZ} + (t^2+(1-a)^2)\overline{CZ}^2\}$$

$$= \frac{1}{(1-a^2)^2}\{t^2(1-a)^2 + t^4 + 2t^2(1-a^2-t^2) + \overline{CY}^2\cdot\overline{CZ}^2\}$$

$$= \frac{t^2}{(1+a)^2} + \frac{1}{(1-a^2)^2}\{(1-a^2-t^2)(1-a^2+t^2)+t^4\}$$

$$= \frac{t^2}{(1+a)^2} + \frac{1}{(1-a^2)^2}\{(1-a^2)^2 - t^4 + t^4\} = \frac{t^2}{(1+a)^2} + 1.$$

It is now easy to show that $\overline{CY}^2 - \overline{XD}^2$ and $\overline{CY} - \overline{XD}$ are also monotonic increasing functions of t.

Second solution. (Mark Kleiman) Draw DY and choose H on DY so that XH is perpendicular to DY. We have that $\angle XDH = \angle ZDY = \frac{\pi}{2} - \angle XYZ$ since the intercepted arcs form a semi-circle. Thus, right triangle XDH is similar to right triangle XCY and so $\frac{XH}{DX} = \frac{XY}{CY}$. Since $\overline{XH} \le \overline{XY}$, $\overline{DX} \le \overline{CY}$.

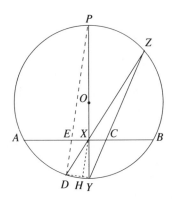

FIGURE 67

Problem 21. Consider the n sums

$$s_1 = a_1,\ s_2 = a_1 + a_2,\ s_3 = a_1 + a_2 + a_3,\ \ldots,$$

$$s_n = a_1 + a_2 + \cdots + a_n.$$

Let s_i leave a remainder of r_i on division by n, i.e.,

$$s_i = q_i n + r_i,\quad 0 \le r_i \le n-1,\quad i = 1, 2, \ldots, n.$$

If for some i, $r_i = 0$, then s_i has the desired properties. If $r_j \ne 0$ for $j = 1, 2, \ldots, n$ then we have n integers $r_1, \ldots, r_n$ all in the set $\{1, 2, \ldots, n-1\}$, which contains only $n-1$ integers. By the Pigeonhole Principle two of the r_i's must be equal, say $r_l = r_m$ with $m > l$; in this case

$$a_{l+1} + \cdots + a_m = s_m - s_l = (q_m - q_l)n.$$

Problem 22. Consider any three points A, B, C. Assume without loss of generality that AB is the longest side of triangle ABC. Then A and B cannot be joined.

Problem 23. *First solution.* Let $m = \dfrac{p}{q}$, where p and q are positive integers having no common prime factors. Then

$$m + \frac{1}{m} = \frac{p^2 + q^2}{pq}$$

so that, if this is an integer, then p and q both divide $p^2 + q^2$. But then p divides q^2 and q divides p^2. Since p and q have no common prime factors, this means that $p = q = 1$ and $m = 1$.

Second solution. It suffices to show that the only positive integer k for which the equation $x + \frac{1}{x} = k$ has a positive rational root is $k = 2$. This equation, namely

$$x^2 - kx + 1 = 0,$$

has roots

$$\frac{k + \sqrt{k^2 - 4}}{2},\qquad \frac{k - \sqrt{k^2 - 4}}{2}.$$

If these roots are rational, then $k^2 - 4$ must be a square. But

$$(k-1)^2 < k^2 - 4 < k^2 \qquad \text{when} \quad k \geq 3,$$

so $k^2 - 4$ cannot be a square when $k \geq 3$, and, when $k = 1$, $k^2 - 4$ is negative and so cannot be a square. Thus the only possible value for k is 2.

Problem 24. Since $\angle APC$ and $\angle ABC$ are both right, the circle on diameter AC passes through B and P. Since AP and PC are equal chords of this circle they subtend equal angles at the circumference, so $\angle ABP = \angle CBP$.

FIGURE 68

Remark. Note the relation to the second solution of problem 18.

Problem 25. Let a be the distance between L_1 and L_3; let b be the distance between L_2 and L_4. Clearly, the sum of the distances from a point to the 4 lines is at least $a + b$. Now let K be positive, and let L denote the locus of points whose sum of distances (from the four lines) equals K.

Case 1. $K < a + b$. Then L is empty.

Case 2. $K = a + b$. Then L is the parallelogram $ABCD$ and its interior.

Case 3. $K > a + b$. Then L is a centrally symmetric octagon, as in Figure 69.

This is an immediate consequence of the following: If P is any point on the side BC of the isosceles triangle ABC ($a = \overline{AB} = \overline{AC}$) then the sum of the distances from P to AB and AC is a constant. For if these distances are d_1 and d_2, then the area of $\triangle ABC$ equals

$$\frac{1}{2}ad_1 + \frac{1}{2}ad_2 = \frac{1}{2}a(d_1 + d_2),$$

as in Figure 70.

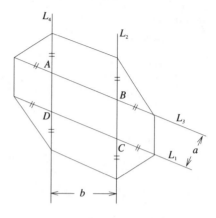

FIGURE 69

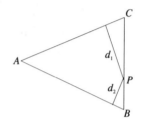

FIGURE 70

Remark. It may happen that all four lines are parallel. In this case, let c be the sum of the distance between the inner two and the distance between the outer two. If $K < c$, L is empty. If $K = c$, then L is the region between the inner two lines with the two lines. If $K > c$, then L consists of two parallel straight lines, one on each side of the inner two lines.

Problem 26. Choose two of the points, say A and B, so that the three remaining points (call them P_1, P_2, P_3) lie on the same side of line AB. (How would you do this?) The angles $\angle AP_1B$, $\angle AP_2B$, $\angle AP_3B$ are all different (in size), for if two of them, say $\angle AP_1B$ and $\angle AP_2B$ were equal, then A, B, P_1, P_2 would be on a circle. Assume without loss of generality that

$$\angle AP_1B < \angle AP_2B < \angle AP_3B.$$

The circle through A, B and P_2 contains P_3 in its interior and P_1 in its exterior. (Generalize to 3 dimensions.)

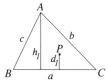

FIGURE 71

Comment. See R. Honsberger, *Mathematical Morsels,* MAA, Problem 23, p. 48.

Problem 27. Let sides BC, CA, and AB have lengths a, b, and c respectively (Figure 71). Then

$$\frac{1}{2}ah_1 = \frac{1}{2}bh_2 = \frac{1}{2}ch_3 = \frac{1}{2}d_1a + \frac{1}{2}d_2b + \frac{1}{2}d_3c,$$

since all four expressions are equal to the area Δ of triangle ABC. Hence

$$\frac{d_1}{h_1} + \frac{d_2}{h_2} + \frac{d_3}{h_3}$$

$$= \frac{ad_1}{ah_1} + \frac{bd_2}{bh_2} + \frac{cd_3}{ch_3}$$

$$= \frac{ad_1}{2\Delta} + \frac{bd_2}{2\Delta} + \frac{cd_3}{2\Delta}$$

$$= \frac{ad_1 + bd_2 + cd_3}{2\Delta} = \frac{2\Delta}{2\Delta} = 1.$$

Comment. For generalizations see M. Klamkin, *Crux Mathematicorum* 13 (1987) 274.

Problem 28. Observe that the meeting place should be between the first and last houses. For, given any point beyond either end house, the sum of the distances would exceed the sum of the distances to the closer end house. Now note that for all points chosen between the first and last houses, the sum of the two distances from the meeting place is constant, so we can remove these houses from further consideration and minimize the sum of the distances to the remaining houses. Repeating the above argument, we deduce that the meeting place should be between the second and second-last house, beween the third and third-last house, and so on. Thus, if n is even, the boys should meet anywhere between the two middle houses; if n is odd, at the middle house.

An "analytic" version goes as follows. On a coordinate line, let the houses have coordinates $a_1, a_2, \ldots, a_n$ with

$$0 \le a_1 < a_2 < \cdots < a_n,$$

and let the point P where they meet have coordinate x. Then the total distance walked is

$$D(x) = |x - a_1| + |x - a_2| + \cdots + |x - a_n|.$$

If n is even , say $n = 2m$, then

$$D = \sum_{i=1}^{m}(|x - a_i| + |a_{2m+1-i} - x|)$$

(and by the Triangle Inequality)

$$D \ge \sum_{i=1}^{m} |x - a_i + a_{2m+1-i} - x|$$

$$= \sum_{i=1}^{m} (a_{2m+1-i} - a_i)$$

$$= a_{m+1} + \cdots + a_{2m} - (a_1 + \cdots + a_m),$$

with equality if $a_m \le x \le a_{m+1}$.

If n is odd, say $m = 2m + 1$, then

$$D = \sum_{i=1}^{m} (|x - a_i| + |a_{2m+2-i} - x|)$$

$$+ |x - a_{m+1}|$$

$$\ge \sum_{i=1}^{m} |x - a_i + a_{2m+2-i} - x| + |x - a_{m+1}|$$

$$= \sum_{i=1}^{m} |a_{2m+2-i} - a_i| + |x - a_{m+1}|$$

$$= \sum_{i=1}^{m} (a_{2m+2-i} - a_i) + |x - a_{m+1}|$$

$$\ge a_{m+2} + \cdots + a_{2m+1} - (a_1 + \cdots + a_n),$$

with equality if $x = a_{m+1}$.

If one plots the graph of $D(x)$ against x, $0 \le x < a_n$, one gets one of the two unbounded polygonal convex figures in Figure 72.

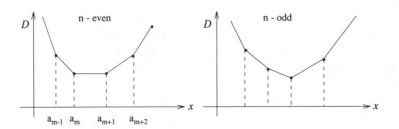

FIGURE 72

Problem 29. Looking at the desired result, note that Q_1 and Q_2 are related by a $180°$ rotation with center P. Hence Q_1 is on γ_1 and on the image of γ_2 in this rotation, i.e., Q_1 is on γ_1 and on the circle through P equal to and tangent to γ_2.

Problem 30. A counterclockwise rotation of $60°$ with center B carries P to C and A to R; hence $\overline{PA} = \overline{CR}$. Similarly, $\overline{CR} = \overline{BQ}$.

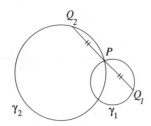

FIGURE 73

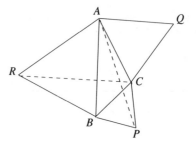

FIGURE 74

Comment. Construct a line such that the ratio of the two chords through P have a given value.

For a set of related constructions using half-turns see I. M. Yaglom, *Geometric Transformations I*, New Math. Library 8, Mathematical Association of America, pp. 21–40.

Comment. For other interesting properties of this configuration see R. A. Johnson *Advanced Euclidean Geometry*, Dover, NY, 1960, pp. 218–222.

Problem 31. Express each term of the sum as a fraction with denominator equal to the least common multiple of $2, 3, \ldots, n$. All numerators will be even except for the single term whose original

Said a monk as he swung by his tail
To his children both male and female
    "From your offsprings, my dears,
    In some millions of years
May emerge a professor at Yale."
        H. E. Salzer, *Scripta Mathematica* 21(1956)

denominator was the highest power of 2 not exceeding n. Thus the sum of the numerators is odd while the common denominator is even.

Remark. It can be shown that, when a, $d > 0$,

$$\frac{1}{a} + \frac{1}{a+d} + \cdots + \frac{1}{a + (n-1)d}$$

(the sum of the reciprocals of an arithmetic progression) is never an integer. A proof depends on Bertrand's Postulate: *There always exists a prime between m and $2m$.*

Problem 32. Let A and B denote the endpoints of the arc. Consider the plane π which passes through the center O and which is normal to the angle bisector of $\angle AOB$. See Figure 75. We shall show that the arc $\overset{\frown}{AB}$ must lie in the hemisphere produced by this plane which contains A and B.

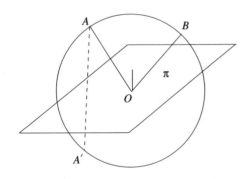

FIGURE 75

Consider the image A' of A across π. Note that $\overline{A'B} = 2$ since A', O, and B are collinear. Also, note that if X is any point in π, then $\overline{AX} = \overline{A'X}$. It is impossible for any arc $\overset{\frown}{AB}$ of length less than 2 to contain any point belonging to π because if $\overset{\frown}{AB}$ contains X, then

$$\overset{\frown}{AX} + \overset{\frown}{XB} \geq \overline{AX} + \overline{XB} = \overline{A'X} + \overline{XB},$$

and by the Triangle Inequality

$$\overline{A'X} + \overline{XB} \geq \overline{A'B}.$$

Similarly, we can show more generally that if two boundary points of a centrosymmetric body whose minimum central diameter has length 2 are joined by an arc of length less than 2, then the arc must lie in some half of the body bounded by a plane section through the center.

Problem 33. *First solution.* Express n in the form $6m + r$ with $0 \leq r \leq 5$, and compute $\left[\frac{n}{3}\right]$, $\left[\frac{n+2}{6}\right]$, $\left[\frac{n+4}{6}\right]$, $\left[\frac{n}{2}\right]$, and $\left[\frac{n+3}{6}\right]$. These are displayed in the table. It is easy to verify that in each row the sum of the first three entries (in the body of Table 1) equals the sum of the last two entries.

Second solution. For each positive integer r and integer $k = 0, 1, \ldots, r - 1$, $\left[\frac{n+k}{r}\right]$ is the number of positive multiples of r not exceeding $n + k$, and at the same time it is the number of positive integers not exceeding n which are congruent to $r - k$ modulo r. Thus $\left[\frac{n}{3}\right] + \left[\frac{n+2}{6}\right] + \left[\frac{n+4}{6}\right]$ is

n	$\left[\frac{n}{3}\right]$	$\left[\frac{n+2}{6}\right]$	$\left[\frac{n+4}{6}\right]$	$\left[\frac{n}{2}\right]$	$\left[\frac{n+3}{6}\right]$
$6m$	$2m$	m	m	$3m$	m
$6m + 1$	$2m$	m	m	$3m$	m
$6m + 2$	$2m$	m	$m + 1$	$3m + 1$	m
$6m + 3$	$2m + 1$	m	$m + 1$	$3m + 1$	$m + 1$
$6m + 4$	$2m + 1$	$m + 1$	$m + 1$	$3m + 2$	$m + 1$
$6m + 5$	$2m + 1$	$m + 1$	$m + 1$	$3m + 2$	$m + 1$

TABLE 1

the number of positive integers not exceeding n which are congruent to 0, 2, 3, 4 modulo 6, while $\left[\frac{n}{2}\right] + \left[\frac{n+3}{6}\right]$ is the number of positive integers not exceeding n which are either even or congruent to 3 modulo 6.

Problem 34. The smallest n digit integer is 10^{n-1} while the largest is $10^n - 1$. Hence we want the sum of the arithmetic progression

$$10^{n-1} + \left(10^{n-1} + 1\right) + \left(10^{n-1} + 2\right)$$
$$+ \cdots + \left(10^n - 2\right) + \left(10^n - 1\right).$$

The number of terms is $10^n - 1 - 10^{n-1} + 1 = 9 \cdot 10^{n-1}$, so the sum of the progression is

$$\frac{9 \cdot 10^{n-1}}{2}\left(10^{n-1} + 10^n - 1\right)$$
$$= 45 \cdot 10^{n-2}\left(10^{n-1} \cdot 11 - 1\right)$$
$$= 495 \cdot 10^{2n-3} - (100 - 55)10^{n-2}$$
$$= (494 + 1)10^{2n-3} - 10^n + 55 \cdot 10^{n-2}$$
$$= 494 \cdot 10^{2n-3} + 10^{2n-3} - 10^n + 55 \cdot 10^{n-2}$$
$$= 494 \cdot 10^{2n-3} + \left(10^{n-3} - 1\right)10^n + 55 \cdot 10^{n-2}$$
$$= 494 \cdot 10^{2n-3} + \underbrace{99\cdots9}_{n-3}\cdot 10^n + 55 \cdot 10^{n-2}$$
$$= 494 \underbrace{99\cdots9}_{n-3}\,55\underbrace{00\cdots0}_{n-2}.$$

Question. What happens if $n = 2$?

Problem 35. *First solution.* Let $x = \overline{BR} = \overline{RP} = \overline{QC}$; then $1 - x = \overline{RC} = \overline{PQ} = \overline{AQ}$. If $x \geq \frac{2}{3}$, then

$$\text{area } \triangle BRP = \frac{1}{2}x^2 \geq \frac{1}{2}\left(\frac{2}{3}\right)^2 = \frac{2}{9}.$$

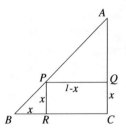

FIGURE 76

If $x \leq \frac{1}{3}$, then $1 - x \geq \frac{2}{3}$ and

$$\text{area } \triangle AQP = \frac{1}{2}(1 - x)^2 \geq \frac{1}{2}\left(\frac{2}{3}\right)^2 = \frac{2}{9}.$$

If $\frac{1}{3} < x < \frac{2}{3}$, then $-\frac{1}{6} < x - \frac{1}{2} < \frac{1}{6}$, $\left(x - \frac{1}{2}\right)^2 < \frac{1}{36}$, and

$$\text{area rectangle } PQCR = x(1 - x)$$
$$= \frac{1}{4} - \left(x - \frac{1}{2}\right)^2$$
$$> \frac{1}{4} - \frac{1}{36} = \frac{2}{9}.$$

Second solution. Plot the three curves $y = \frac{x^2}{2}$, $y = x(1 - x)$, and $y = \frac{1}{2}(1 - x)^2$ and observe that for any $0 \leq x \leq 1$ the largest of the three y's is at least $\frac{2}{9}$.

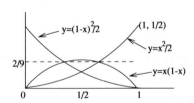

FIGURE 77

I'm glad I'm educated
I think its simply grand
To know so many facts and stuff
That I don't understand.

S. O. Barkerd

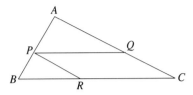

FIGURE 78 $PQ \parallel BC$ and $PR \parallel AC$

Remark. This problem can be extended to parallelograms inscribed in an arbitrary triangle.

Proof. By parallel projection of previous result.

Problem 36. *First solution.* The first result is apparent when a pair of 3, 4, 5 right triangles are placed "back-to-back" in two different ways, as in Figure 79.

The same idea applied to the right triangle with sides 5, 12, 13 shows that the triangle with sides 10, 13, 13 has the same area as triangle with sides 24, 13, 13. Indeed, a right triangle with sides a, b, c ($c^2 = a^2 + b^2$) leads to triangles of equal area having sides $2a$, c, c, and $2b$, c, c.

Second solution. Let the isoceles triangles have side lengths (u, u, v) and (x, x, y) respectively. By Heron's formula, the condition that their areas are equal is

$$v^2(2u + v)(2u - v) = y^2(2x + y)(2x - y)$$

or

$$4u^2v^2 + y^4 = 4x^2y^2 + v^4.$$

Some solutions can be found by imposing the additional condition $u = x$, whence $4u^2 = 4x^2 = v^2 + y^2$ (assuming $v^2 - y^2 \neq 0$). Thus $(v, y, 2x)$

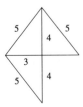

FIGURE 79

is a Pythagorean triple. A few solution pairs $\{(u, u, v), (x, x, y)\}$ are

$$\{(5, 5, 6),\ (5, 5, 8)\}$$

$$\{(13, 13, 10),\ (13, 13, 24)\}$$

$$\{(25, 25, 14),\ (25, 25, 48)\}.$$

Can solutions be found with $u \neq x$?

Problem 37.   First note that
$$a^2 + b^2 + c^2 + d^2 = x^2 + (1 - x)^2$$
$$+ y^2 + (1 - y)^2$$
$$+ u^2 + (1 - u)^2$$
$$+ v^2 + (1 - v)^2.$$

Now consider
$$x^2 + (1 - x)^2 = 2\left\{\left(x - \frac{1}{2}\right)^2 + \frac{1}{4}\right\}.$$

Since $0 \leq x \leq 1$, it easily follows that
$$\frac{1}{2} \leq x^2 + (1 - x)^2 \leq 1,$$

and similarly with x replaced by y, u, v. Adding the four inequalities yields the desired result.

It is interesting to observe that also $2\sqrt{2} \leq a + b + c + d \leq 4$.

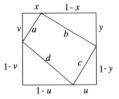

FIGURE 80

Comment. More generally, show that if the square is replaced by a $m \times n$ rectangle then

$$m^2 + n^2 \leq a^2 + b^2 + c^2 + d^2 \leq 2(m^2 + n^2).$$

Problem 38. If we let x denote the overlapping part, then the non-overlapping parts have areas $\pi R^2 - x$, $\pi r^2 - x$ and hence the difference is

$$(\pi R^2 - x) - (\pi r^2 - x) = \pi(R^2 - r^2).$$

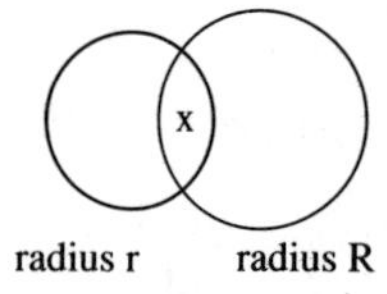

x
radius r
radius R

FIGURE 81

Problem 39. An ordered sum

$$a_1 + a_2 + \cdots + a_k = n, \qquad a_i \geq 1,$$

is represented by n 1's in a row separated by $k-1$ "strokes" /:

$$\underbrace{111\cdots1}_{a_1}/\underbrace{11\cdots1}_{a_2}/\underbrace{11\cdots1}_{a_3}/\ldots/\underbrace{11\cdots1}_{a_k}.$$

To obtain all such displays (for all $1 \leq k \leq n$), line up n 1's and for each of the n spaces between adjacent pairs, either put in a stroke or do not put in a stroke. This can be done in 2^{n-1} ways.

Problem 40. $T_1, T_2, \ldots, T_k$ play $\frac{1}{2}k(k-1)$ games amongst themselves and $k(n-k)$ games against $T_{k+1}, T_{k+2}, \ldots, T_n$. Now $s_1+s_2+\cdots+s_k$ is greatest when all the latter games are won by $T_1, T_2, \ldots, T_k$ so that

$$\begin{aligned} s_1 + s_2 + \ldots + s_k &\leq \frac{1}{2}k(k-1) + k(n-k) \\ &= nk - \frac{1}{2}k(k+1). \end{aligned}$$

Problem 41. A general law suggested is:

$$1^2 + 3^2 + 5^2 + \cdots + (2n-1)^2 = \frac{(2n-1)2n(2n+1)}{6}, \qquad n = 1, 2, 3, \ldots.$$

A proof by induction is easy, and we leave this to the reader.

The following proof assumes the formula for the sum of squares:

$$1^2 + 2^2 + 3^2 + \cdots + k^2 = \frac{k(k+1)(2k+1)}{6}, \qquad k = 1, 2, 3, \ldots.$$

From this we have

$$1^2+2^2+3^2+\cdots+(2n)^2 = \frac{2n(2n+1)(4n+1)}{6}$$

and

$$\begin{aligned} 2^2 + 4^2 + 6^2 + \cdots + (2n)^2 &= 2^2(1^2 + 2^2 + 3^2 + \cdots + n^2) \\ &= \frac{4n(n+1)(2n+1)}{6}, \end{aligned}$$

so (taking the difference)

$$\begin{aligned} 1^2 + 3^2 + 5^2 + \cdots + (2n-1)^2 &= \frac{2n(2n+1)(4n+1)}{6} - \frac{4n(n+1)(2n+1)}{6} \\ &= \frac{(2n-1)(2n)(2n+1)}{6}. \end{aligned}$$

Problem 42. Solving the equation $x = \dfrac{x^2+1}{198}$, or $x^2 - 198x + 1 = 0$, we obtain the roots

$$\alpha = 99 + \sqrt{99^2 - 1} = 99 + 70\sqrt{2},$$
$$\beta = 99 - \sqrt{99^2 - 1} = 99 - 70\sqrt{2},$$

with

$$\alpha + \beta = 198, \qquad \alpha\beta = 1.$$

Hence

$$\alpha = \frac{\alpha^2+1}{198} > \frac{1}{198}, \qquad \beta = \frac{\beta^2+1}{198} > \frac{1}{198},$$

$$\alpha = 198 - \beta < 198 - \frac{1}{198} = 197.99494949\ldots,$$

$$\beta = 198 - \alpha < 198 - \frac{1}{198} = 197.99494949\ldots,$$

and

$$\sqrt{2} = \frac{\alpha - 99}{70} < \frac{197.99\dot{4}\dot{9} - 99}{70} = 1.41\dot{4}2135\dot{6}.$$

Finally , $(1.41421356)^2 = 1.9999999932\ldots < 2$, so $\sqrt{2} > 1.41421356$.

Problem 43. Partition the triangle into 4 equilateral triangles of side 1 (see Figure 82). By the Pigeon Hole Principle, one of these four triangles must contain 2 (or more) of the 5 points, and these 2 points are within distance 1 of each other.

FIGURE 82

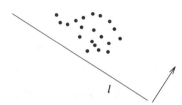

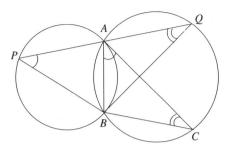

FIGURE 84

FIGURE 83

gent to circle PAB. Then $\angle APB = \angle CAB$ and $\angle AQB = \angle ACB$. Hence $\triangle BPQ$ is similar to $\triangle BAC$, so that $\dfrac{\overline{BP}}{\overline{BQ}} = \dfrac{\overline{BA}}{\overline{BC}}$, a constant.

Problem 44. Yes. Start with a line l not parallel to any of the (finite number of) lines determined by the given points and such that all the given points lie on one side of l (see Figure 83). Translate l in the perpendicular direction. It will pass over the points one at a time. Stop when you have passed over half the points.

Problem 45. *First solution.* $\angle BPA$ remains constant, as does $\angle AQB$, since these angles are subtended at the circumference of the two circles by the constant arc AB. Hence the sines of these angles are constant. By the law of sines applied to $\triangle BPQ$

$$\frac{\overline{BP}}{\overline{BQ}} = \frac{\sin Q}{\sin P} = \quad \text{a constant.}$$

Second solution. This solution is motivated by looking at the extreme case where P coincides with A and PQ is tangent to the circle PAB. Let C be on circle BAQ such that AC is tan-

Problem 46. (a) If n is even,

$$f(n) = 0 + 1 + 2 + 3 + \cdots + \left(\frac{n}{2} - 1\right)$$
$$+ 1 + 2 + 3 + \cdots + \frac{n}{2}$$
$$= \frac{1}{2}\left(\frac{n}{2} - 1\right)\frac{n}{2} + \frac{1}{2}\left(\frac{n}{2}\right)\left(\frac{n}{2} + 1\right) = \frac{n^2}{4}.$$

If n is odd,

$$f(n) = 0 + 1 + 2 + \cdots + \frac{n-1}{2}$$
$$+ 1 + 2 + \cdots + \frac{n-1}{2}$$
$$= 2\left(\frac{1}{2}\right)\left(\frac{n-1}{2}\right)\left(\frac{n-1}{2} + 1\right) = \frac{n^2 - 1}{4}.$$

Thus

$$f(n) = \begin{cases} \frac{n^2}{4} & \text{if } n \text{ is even,} \\ \frac{n^2 - 1}{4} & \text{if } n \text{ is odd.} \end{cases}$$

There was a mathematician named Ben
Who could only count modulo ten.
He said, "When I go
Past my last little toe
I shall have to start over again."

(b) Note that $s + t$ and $s - t$ differ by $2t$, which is even. Hence $s + t$ and $s - t$ are either both even or both odd. In the "even" case,

$$f(s + t) - f(s - t) = \frac{(s + t)^2}{4} - \frac{(s - t)^2}{4} = st.$$

In the "odd" case,

$$f(s + t) - f(s - t) = \frac{(s + t)^2 - 1}{4}$$
$$- \frac{(s - t)^2 - 1}{4} = st.$$

Problem 47. Through P, Q, and R draw lines parallel to QR, RP, and QP respectively, thus obtaining the triangle ABC with the desired properties.

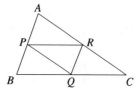

FIGURE 85

Problem 48. In general if n is an odd positive integer then $a + b$ is a factor of $a^n + b^n$. Hence 5 divides $1^{99} + 4^{99}$ and also divides $2^{99} + 3^{99}$.

Problem 49. *First solution.* Every integer has one of the forms $4n$, $4n + 1$, $4n + 2$, $4n + 3$, so their squares leave a remainder of $0, 1$ or 4 on division by 8. Hence $a^2 + b^2$ cannot equal $8c + 6$, which leaves a remainder of 6 on division by 8.

Second solution. Since

$$a^2 + b^2 = 2(4c + 3)$$

is even, a and b must have the same parity, i.e., both are even or both are odd. In the first case, $a = 2m$, $b = 2n$ and hence

$$4m^2 + 4n^2 = 2(4c + 3)$$

or

$$2(m^2 + n^2) = 4c + 3,$$

and this is absurd because the left side is even while the right side is odd. In the second case, $a = 2m + 1$, $b = 2n + 1$ and hence

$$(2m + 1)^2 + (2n + 1)^2 = 2(4c + 3)$$

or

$$m(m + 1) + n(n + 1) = 2c + 1,$$

and again the left side is even while the right side is odd.

Problem 50. Let A, B denote the two sets of four numbers each. 1 and 2 cannot be in the same set, for otherwise their sum could not be realized in the other set; 1 and 3 cannot be in the same set, for otherwise their sum could not be realized in the other set. Hence we may assume $1 \in A$, $2 \in B$, $3 \in B$. Then 4 must be in A in order that the sum $2 + 3$ be realized in A. Thus we have $\{1, 4\} \subset A$, $\{2, 3\} \subset B$.

A similar argument shows that 5, 8, belong to the same set, while 6, 7 belong to the other. Hence the only possible splittings are:

(i) $\{1, 4, 5, 8\}$, $\{2, 3, 6, 7\}$;
(ii) $\{1, 4, 6, 7\}$, $\{2, 3, 5, 8\}$.

The first of these splittings doesn't have like sums, while the second one does. Hence the second splitting is the only solution.

(Formulate and solve a similar problem for splitting the first sixteen integers.)

Remark. For the two sets in (ii), the sum of the numbers in one and the sum of their squares are equal to the corresponding sums in the other. This too can be generalized for the first 2^n integers.

Problem 51. We have $x > 0, y > 0, x - 2y > 0$, and $2\log(x - 2y) = \log x + \log y$, so

$$\log(x - 2y)^2 = \log xy,$$
$$(x - 2y)^2 = xy,$$
$$\left(\frac{x}{y} - 2\right)^2 = \frac{x}{y},$$

$$\left(\frac{x}{y}\right)^2 - 5\left(\frac{x}{y}\right) + 4 = 0,$$

$$\left(\frac{x}{y} - 4\right)\left(\frac{x}{y} - 1\right) = 0,$$

$$x = 4y \quad \text{or} \quad x = y.$$

The latter situation is impossible, for then $x - 2y < 0$. Hence $\frac{x}{y} = 4$.

Problem 52. *First solution.* First observe that $2 = f(2) = f(1 \cdot 2) = f(1)f(2) = f(1)2$, so $f(1) = 1$. Furthermore

$$f(2^2) = f(2 \cdot 2) = f(2)f(2) = 2 \cdot 2 = 2^2,$$

$$f(2^3) = f(2 \cdot 2^2) = f(2)f(2^2) = 2 \cdot 2^2 = 2^3$$

and so on, i.e., $f(2^k) = 2^k$ for $k = 0, 1, 2, \ldots$. Now consider successive powers of 2 and the integers between:

$$2^k < 2^k + 1 < 2^k + 2 < \cdots < 2^k + 2^k - 1$$
$$= 2^{k+1} - 1 < 2^{k+1}.$$

Their f values satisfy

$$2^k < f(2^k + 1) < f(2^k + 2)$$
$$< \cdots < f(2^{k+1} - 1) < 2^{k+1}.$$

Thus, between 2^k and 2^{k+1} we have $2^k - 1$ distinct integers $f(2^k + j), j = 1, 2, \ldots, 2^k - 1$. Since there are exactly $2^{k+1} - 2^k - 1 = 2^k - 1$ integers between 2^k and 2^{k+1}, it follows that $f(2^k + j) = 2^k + j$, $j = 1, 2, \ldots$.

Second solution. As before, we observe that $f(2^k) = 2^k$ for $k = 1, 2, \ldots$. From property (5) we see that $f(m + 1) > f(m)$, i.e.,

$f(m + 1) \geq f(m) + 1$; hence $f(m) \geq m$ and $f(m + k) \geq f(m) + k$. It remains to show that for no n is $f(n) > n$. Suppose that $f(n) > n$ for some n. Then $2^n > n$ and

$$2^n = f(2^n) = f(n + 2^n - n) \geq f(n) + 2^n - n$$
$$> n + 2^n - n = 2^n,$$

which is absurd.

Third solution (by induction). First we observe that $f(1) = 1$. Now assume that $f(k) = k$ for $k = 1, 2, \ldots, n$. We show that $f(n + 1) = n + 1$. If $n + 1 = 2j$, then $1 \leq j < n$ and

$$f(n + 1) = f(2j) = 2j = n + 1.$$

If $n + 1 = 2j + 1$, then $1 \leq j < n$ and

$$2j = f(2j) < f(2j + 1) < f(2j + 2)$$
$$= f(2(j + 1)) = 2f(j + 1) = 2(j + 1)$$
$$= 2j + 2.$$

Thus

$$2j < f(2j + 1) < 2j + 2,$$

so $f(2j + 1) = 2j + 1 = n + 1$.

Problem 53. Let BC, BD, BE be congruent segments in π which are respectively parallel to the three lines in π; let A be a point in space such that AB is parallel to l (see Figure 86). It follows that $\triangle ABC \cong \triangle ABD \cong \triangle ABE$ and hence $\overline{AC} = \overline{AD} = \overline{AE}$. Thus A and B are respectively on the planes which perpendicularly bisect CD and DE, and hence AB is perpendicular to π.

Alternatively,

$$\overrightarrow{BA} \bullet \overrightarrow{CB} = \overrightarrow{BA} \bullet \overrightarrow{DB} = \overrightarrow{BA} \bullet \overrightarrow{EB},$$

I once picked up a copy of Casey's *Sequel to Euclid* on a second-hand bookstall, and on the inside of the front cover was one of the blue-edged labels that doctors stick on medicine bottles with the inscription "Poison— to be taken three times a day."

F. Bowman, *Mathematical Gazette,* 19 (1935) p. 273.

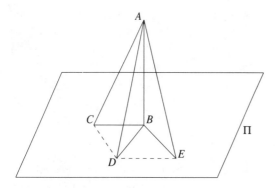

FIGURE 86

so that

$$0 = \overrightarrow{BA} \bullet (\overrightarrow{CB} - \overrightarrow{DB}) = \overrightarrow{BA} \bullet \overrightarrow{CD},$$
$$0 = \overrightarrow{BA} \bullet (\overrightarrow{CB} - \overrightarrow{EB}) = \overrightarrow{BA} \bullet \overrightarrow{CE}.$$

Hence $\overrightarrow{BA} \perp \overrightarrow{CD}$ and $\overrightarrow{BA} \perp \overrightarrow{CE}$, and it follows that $\overrightarrow{BA} \perp \pi$.

Problem 54. We have

$$a = \underbrace{11\ldots1}_{m}$$
$$= 1 + 10 + 10^2 + \cdots + 10^{m-1}$$
$$= \frac{10^m - 1}{9},$$
$$b = 1\underbrace{0\ldots0}_{m-1}5 = 5 + 10^m,$$
$$ab + 1 = \left(\frac{10^m - 1}{9}\right)(5 + 10^m) + 1$$
$$= \left(\frac{10^m + 2}{3}\right)^2$$
$$= \left(\frac{10^m - 10 + 12}{3}\right)^2$$
$$= \left(\frac{10(10^{m-1} - 1)}{3} + 4\right)^2$$
$$= (\underbrace{33\ldots3}_{m-1}4)^2,$$
$$\sqrt{ab + 1} = \underbrace{33\ldots3}_{m-1}4.$$

Problem 55. *First solution.* Set up a coordinate system so that the pole of height h is based at $(-a, 0)$ and the pole of height k at $(a, 0)$. Then a point (x, y) subtends equal elevations if

$$\frac{h}{\sqrt{(x+a)^2 + y^2}} = \frac{k}{\sqrt{(x-a)^2 + y^2}},$$

or

$$h^2(x^2 - 2ax + a^2 + y^2) = k^2(x^2 + 2ax + a^2 + y^2),$$

or

$$x^2(k^2 - h^2) + 2ax(k^2 + h^2) + y^2(k^2 - h^2)$$
$$+ a^2(k^2 - h^2) = 0.$$

If $k = h$, the equation becomes $x = 0$, so the locus is the y-axis.

If $k \neq h$, say $k > h$, then

$$x^2 + 2ax\left(\frac{k^2 + h^2}{k^2 - h^2}\right) + y^2 + a^2 = 0,$$

and the locus is a circle with a diameter joining the two "obvious" points on the x-axis with equal elevations.

Second solution. Let A, B be respectively the foot of the flagpoles (see Figure 87). A point P is on the locus if $\angle APA' = \angle BPB'$ or, equivalently $\frac{\overline{PA}}{k} = \frac{\overline{PB}}{h}$, i.e.,

$$\frac{\overline{PA}}{\overline{PB}} = \frac{k}{h} \qquad \text{(a constant)}.$$

Now (in the plane of APB) the bisectors of the internal angle and external angle at the vertex P of triangle APB meet the line AB in points C and D with

$$\frac{k}{h} = \frac{\overline{AP}}{\overline{BP}} = \frac{\overline{AC}}{\overline{BC}} = \frac{\overline{AD}}{\overline{BD}}$$

(see Figure 88), and such that $\angle DPC = 90°$.

The points C and D can be located as in Figure 89 (drawn in the plane of the flagpoles), and then the locus of P is seen to be the circle (in the "level" plane) on diameter CD. This circle is known as the Circle of Appolonius.

Problem 56. For $k = 3, 4, \ldots$,

$$k! = 2 \cdot 3 \cdot 4 \cdots k > 2^{k-1},$$

so

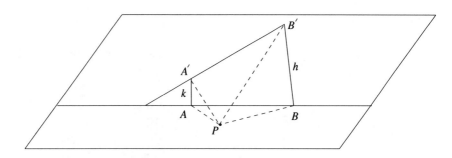

FIGURE 87

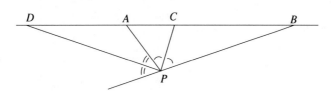

FIGURE 88

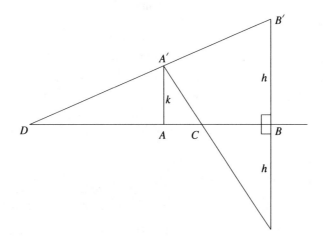

FIGURE 89 $\frac{\overline{AC}}{k} = \frac{\overline{BC}}{h}$ and $\frac{\overline{AD}}{k} = \frac{\overline{BD}}{h}$

$$\frac{1}{k!} < \frac{1}{2^{k-1}}, \quad k = 3, 4, \ldots.$$

Hence, for $n \geq 1$,

$$1 + \frac{1}{1!} + \frac{1}{2!} + \frac{1}{3!} + \cdots + \frac{1}{n!}$$

$$< 1 + 1 + \frac{1}{2} + \frac{1}{2^2} + \frac{1}{2^3} + \cdots + \frac{1}{2^{n-1}}$$

$$= 1 + \frac{1 - (1/2)^n}{1/2} = 3 - \frac{1}{2^{n-1}} < 3.$$

Problem 57. Because AX is parallel to BP, triangles AXP and AXB have equal areas. Similarly triangles DXP and DXC have equal areas. Hence

$$\text{area } APD = \text{area } AXD + \text{area } DXP + \text{area } AXP$$

$$= \text{area } AXD + \text{area } DXC + \text{area } AXB$$

$$= \text{area } ABCD.$$

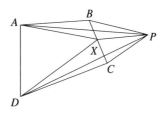

FIGURE 90

Problem 58. For $k = 1, 2, 3, \ldots$, we have

$$\sqrt{k+1} - \sqrt{k}$$

$$= \frac{(\sqrt{k+1} - \sqrt{k})(\sqrt{k+1} + \sqrt{k})}{\sqrt{k+1} + \sqrt{k}}$$

$$= \frac{1}{\sqrt{k+1} + \sqrt{k}},$$

so that

$$\frac{1}{2\sqrt{k+1}} < \sqrt{k+1} - \sqrt{k} < \frac{1}{2\sqrt{k}}.$$

Adding the left inequalities for $k = 1, 2, \ldots, n-1$ and the right inequalities for $k = 1, 2, \ldots, n$ yields the desired inequalities.

Exercise. In a similar fashion find bounds on

$$1 + \frac{1}{\sqrt[3]{2^2}} + \frac{1}{\sqrt[3]{3^2}} + \cdots + \frac{1}{\sqrt[3]{n^2}}.$$

Problem 59. The vertices of the quadrilateral partition the circumference of the circle into four arcs, whose lengths sum to 2π, so the smallest of the arcs has length at most $\frac{2\pi}{4} = \frac{\pi}{2}$. The corresponding segment (joining the ends of this smallest arc) has length $\leq \sqrt{2}$.

Problem 60. The exterior angles of any convex polygon sum to $360°$. The given polygon has 4 right angles, and hence 4 exterior right angles summing to $360°$. There is no "room" for more vertices!

Comment. More generally, if a convex polygon has n of its angles equal to $\dfrac{360°}{n}$ then the polygon is a regular n-gon.

Problem 61. Let the two red balls R_1 and R_2 weigh r_1 and r_2 respectively, and use analogous notation for the other colors. First, balance R_1 and W_1 against R_2 and B_1.

If $r_1 + w_1 = r_2 + b_1$, then either: $r_1 < r_2$ and $w_1 > b_1$, or $r_1 > r_2$ and $w_1 < b_1$. These cases can be distinguished by then balancing W_1 against W_2.

If $r_1 + w_1 > r_2 + b_1$, then certainly $r_1 > r_2$ and one of these possibilities occurs:

(i) $w_1 > w_2$ and $b_1 > b_2$;

(ii) $w_1 > w_2$ and $b_1 < b_2$;

(iii) $w_1 < w_2$ and $b_1 < b_2$.

For the second weighing, balance R_1 and B_1 against W_2 and B_2, and deduce:

if $r_1 + b_1 > w_2 + b_2$, then $b_1 > b_2$ and $w_1 > w_2$;

if $r_1 + b_1 = w_2 + b_2$, then $b_1 < b_2$ and $w_1 > w_2$;

if $r_1 + b_1 < w_2 + b_2$, then $b_1 < b_2$ and $w_1 < w_2$.

The final case $r_1 + w_1 < r_2 + b_1$ can be disposed of similarly.

Problem 62. Calling the distance D, the engine speed V and the wind speed W, $W < V$, the time for the round trip in still air is $\dfrac{2D}{V}$, while in a wind the trip takes the longer time

$$\frac{D}{V+W} + \frac{D}{V-W} = \frac{2DV}{V^2 - W^2}$$

$$= \frac{2D}{V} \cdot \frac{V^2}{V^2 - W^2} > \frac{2D}{V}.$$

This agrees with intuition, for the plane "fights" the wind for a longer time than it "rides" the wind.

Comment. For more general results see *Crux Mathematicorum* 12 (1986) 277–279.

Problem 63. *First solution.* Let P be the midpoint of AB, so that (because $\angle AOB$ is a right angle) $\overline{PO} = \overline{PA} = \overline{PB}$. Since $\angle POC$ is a right angle $\overline{PC} > \overline{PO}$, and hence $\overline{PC} > \overline{PA} = \overline{PB}$. Thus, $\angle ACB$ cannot be a right angle. Similarly $\angle BCA$ and $\angle BAC$ are also not right angles. (Must triangle ABC be acute angled?)

"I only took the regular course," said the Mock Turtle with a sigh—"Reeling and Writhing, of course, to begin with, and then the different branches of Arithmetic— Ambition, Distraction, Uglification and Derision."   Lewis Carroll

Second solution. Let lines OA, OB and OC be the x, y, z axes of a Cartesian coordinate system. Let A, B, and C have coordinates $(a,0,0)$, $(0,b,0)$, $(0,0,c)$ and without loss of generality assume $\angle CAB$ is a right angle. Then

$$\left(\sqrt{b^2+c^2}\right)^2 = \left(\sqrt{a^2+b^2}\right)^2 + \left(\sqrt{a^2+c^2}\right)^2,$$
$$b^2 + c^2 = a^2 + b^2 + a^2 + c^2,$$
$$2a^2 = 0,$$

so $a = 0$, a contradiction.

Problem 64. The system to be solved is:

$$x + yz = 2; \qquad y + zx = 2; \qquad z + xy = 2.$$

Subtracting the second (third) from the first (second) yields

$$(x - y)(1 - z) = 0, \qquad (y - z)(1 - x) = 0.$$

Each of the four cases

$$x - y = 0 = y - z, \qquad x - y = 0 = 1 - x,$$
$$1 - z = 0 = y - z, \qquad 1 - z = 0 = 1 - x,$$

implies $x = y = z = 1$ or $x = y = z = -2$.

Problem 65. Divide the square into 4 congruent squares of side $\frac{1}{2}$. By the Pigeonhole Principle, one of these will contain 3 (or more) of the points. They determine a triangle whose area is at most one-half of the area of the "small" square.

Problem 66. From

$$0 = a^2 + b^2 + c^2 - ab - bc - ca$$
$$= \frac{1}{2}\left((a-b)^2 + (b-c)^2 + (c-a)^2\right)$$

it follows that $a - b = b - c = c - a = 0$.

Comment. Similarly, if Z_1, Z_2, Z_3 are complex numbers such that

$$Z_1^2 + Z_2^2 + Z_3^2 = Z_2 Z_3 + Z_3 Z_1 + Z_1 Z_2$$

then the points (in the plane) representing these complex numbers are the vertices of an equilateral triangle.

Problem 67. Let Δ denote the area of triangle ABC. Then

$$\Delta = \frac{1}{2}ah_a = \frac{1}{2}bh_b = \frac{1}{2}ch_c.$$

Thus, $a + h_a \geq b + h_b$ is equivalent to

$$a - b \geq h_b - h_a = 2\Delta\left(\frac{1}{b} - \frac{1}{a}\right) = \frac{2\Delta(a-b)}{ab}$$

or $(a-b)(ab - 2\Delta) \geq 0$. Since $2\Delta = ab\sin C \leq ab$ it follows that $ab - 2\Delta \geq 0$; this with $a - b \geq 0$ yields the desired inequality.

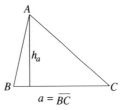

FIGURE 91

Problem 68. Let $n = x10^4 + y10^3 + z10^2 + u10 + v$, so that $m = x10^3 + y10^2 + u10 + v$ and $10m - n = (u - z)10^2 + (v - u)10 - v$ is a three-digit integer. Thus $\frac{10m-n}{m}$ is an integer which has a three-digit numerator and a four-digit denominator, and hence $10m - n = 0$, i.e., $u = v = z = 0$. Now we have $n = x10^4 + y10^3$ and $m = x10^3 + y10^2$, i.e., n is of the form $n = 10^3 N$, where $10 \leq N \leq 99$.

Problem 69. If

$$\frac{1}{x} + \frac{1}{y} + \frac{1}{z} = \frac{1}{x+y+z},$$

then (an easy manipulation shows that) $(x + y)(y + z)(z + x) = 0$.

If $x + y = 0$, then $x^{2n+1} = -y^{2n+1}$ so that $x^{2n+1} + y^{2n+1} = 0$ and both expressions

$$x^{2n+1} + y^{2n+1} + z^{2n+1}, \qquad (x + y + z)^{2n+1}$$

are equal to z^{2n+1}; similarly, in the cases $y + z = 0$ or $z + x = 0$ the result follows.

Problem 70. *First solution.* The points, if they exist, will be the centers of circles which pass through the two given points A, B and are tangent to the given line l. Clearly, if the points are on opposite sides of the line or both on l, no such circle exists.

For the possible cases, we have:

1. The line joining the two given points is parallel to the given line. Here the circle is unique and easily constructed.

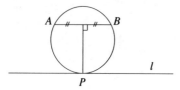

FIGURE 92

2. One point, A, is on the line. The circle is then unique. If BA is perpendicular to l, then BA is a diameter of the circle.

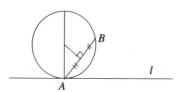

FIGURE 93

3. Neither A nor B is on l; there are two possible circles. Since $\overline{QP}^2 = \overline{QA} \cdot \overline{QB}$ is known, one can construct P. Then the centers of the two circles can be easily determined.

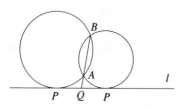

FIGURE 94

Second solution. The locus of points equidistant from the points A and B is a straight line b, the perpendicular bisector of BC; the locus of points equidistant from A and l is a parabola p with focus A and directrix l. The intersection of b and p yields the required points.

b and p will not intersect if A and B are on the opposite sides of l. (why?) In case 1, b is parallel to the axis of p and there is a unique intersection point; in case 3, b intersects p in two points.

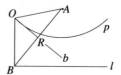

FIGURE 95

Case 2 is interesting. If A is on l and B off l, then p degenerates to a line through A perpendicular to l. and so has a single point of intersection with b. However, we could take A off l and B on l (see Figure 95). In this case, if R is the midpoint of AB, then $\angle ROA = \angle ROB$, so b is tangent to p and there is a unique point O of intersection: $OB \perp l$ and $\overline{OA} = \overline{OB}$.

Problem 71. *First solution.* Express n in base 2, i.e.,

$$n = \epsilon_0 + \epsilon_1 2 + \epsilon_2 2^2 + \epsilon_3 2^3 + \epsilon_4 2^4$$
$$+ \epsilon_5 2^5 + \epsilon_6 2^6 + \cdots,$$

where the ϵ's are each 0 or 1. Then it is easy to see that

> There was an old man who said, "Do
> Tell me how I add two and two.
>   I think more and more
>   That it makes about four,
> But I fear that is almost too few."

$$\left[\frac{n+1}{2}\right] = \epsilon_0 + \epsilon_1 + \epsilon_2 2 + \epsilon_3 2^2 + \epsilon_4 2^3 + \epsilon_5 2^4 + \epsilon_6 2^5 + \cdots$$

$$\left[\frac{n+2}{4}\right] = \quad\quad \epsilon_1 + \epsilon_2 \quad + \epsilon_3 2 \quad + \epsilon_4 2^2 + \epsilon_5 2^3 + \epsilon_6 2^4 + \cdots$$

$$\left[\frac{n+4}{8}\right] = \quad\quad\quad\quad \epsilon_2 \quad + \epsilon_3 \quad\quad + \epsilon_4 2 \quad + \epsilon_5 2^2 + \epsilon_6 2^3 + \cdots$$

$$\left[\frac{n+8}{16}\right] = \quad\quad\quad\quad\quad\quad \epsilon_3 \quad + \epsilon_4 \quad\quad + \epsilon_5 2 \quad + \epsilon_6 2^2 + \cdots$$

$$\vdots$$

Adding up the columns we obtain the desired result.

Second solution. $\left[\frac{n+1}{2}\right]$ is the number of even numbers between 1 and $n+1$ inclusive, which is the same as the number of odd numbers between 1 and n inclusive.

$\left[\frac{n+2}{4}\right]$ is the number of multiples of 4 between 1 and $n+2$ inclusive, which is the same as the number of multiples of 2 not divisible by 4 between 1 and n inclusive.

In general, for each k, $\left[\frac{n+2^{k-1}}{2^k}\right]$ is the number of multiples of 2^k between 1 and $n+2^{k-1}$ inclusive, which is the same as the number of multiples of 2^{k-1} not divisible by 2^k between 1 and n inclusive.

Since, for each number m between 1 and n inclusive, there is exactly one value of k for which m is divisible by 2^{k-1} but not by 2^k, the left-hand side of the equation counts each number from 1 to n exactly once.

Problem 72. *First solution.* Let M be the midpoint of the segment AB. Draw lines through A and B parallel to line MC. Draw a line through

C perpendicular to line MC. Suppose it meets the lines through A and B in points X, Y. Then $\overline{CX} = \overline{CY}$, so the circle with center C and radius CX has the desired properties.

Second solution. This problem is related to problem 29. Draw the circles with diameters AC and BC respectively. As in problem 29, construct segment PCQ so that

$$\overline{PC} = \overline{CQ}.$$

The circle with center C passing through P and Q is the required circle. AP and BQ, being perpendicular to the diameter PQ, are the parallel tangents.

Comment. A more difficult problem is: Given three noncollinear points A, B, C and an angle α, construct the circle with center C so that the tangents to it passing through A and B make the angle α. If r is the radius of the circle, one obtains an equation of the form

$$r^4[(b^2 - a^2)^2 + 4k^2 a^2 b^2] - 2r^2 k^2 a^2 b^2 (a^2 + b^2)$$
$$+ k^4 a^4 b^4 = 0, \quad a, b, k \text{ constants.}$$

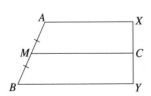

FIGURE 96

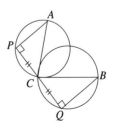

FIGURE 97

Problem 73.   Clearly

$$(x-1)f(x) = x^5 - 1.$$

Hence

$$f(x^5) = x^{20} + x^{15} + x^{10} + x^5 + 1$$
$$= (x^{20} - 1) + (x^{15} - 1) + (x^{10} - 1)$$
$$+ (x^5 - 1) + 5,$$

where each of the quantities in parentheses is seen to have the factor $x^5 - 1$ and hence also the factor $f(x)$. Thus

$$f(x^5) = (\text{ a multiple of } f(x)) + 5.$$

Remark.   A direct verification by long division is not difficult as the pattern of coefficients in the quotient is fairly evident.

Problem 74.   The polynomial $f(x) - 5$ has the 4 integral roots a, b, c, d, and hence

$$f(x) - 5 = (x-a)(x-b)(x-c)(x-d)g(x),$$

where $g(x) = x^m + b_1 x^{m-1} + \cdots + b_m$ and the $b_1, b_2, \ldots, b_m$ are integers. Thus, if x is an integer, the equation $f(x) = 8$, or

$$(x-a)(x-b)(x-c)(x-d)g(x) = 3,$$

implies that at least 3 of the 4 integers $x-a$, $x-b$, $x-c$, $x-d$ are equal to 1 or -1, so that two of them would be equal, a contradiction.

Problem 75.   For all i and j

$$a_{ij} \leq \max\{a_{i1}, a_{i2}, \ldots, a_{in}\} = M_i,$$

so

$$m_j = \min\{a_{1j}, a_{2j}, \ldots, a_{nj}\}$$
$$\leq \min\{M_1, M_2, \ldots, M_n\} = M$$

and hence

$$m = \max\{m_1, m_2, \ldots, m_n\} \leq M.$$

Problem 76.   Let us try to construct the longest possible geometric progression. Suppose it has n terms, that its common ratio (necessarily rational) is $\frac{p}{q}$ in its lowest terms and its smallest term is a. The common ratio should be as small as possible,

so p should be $q + 1$. In addition, the last term $\frac{ap^{n-1}}{q^{n-1}}$ is an integer, so q^{n-1} divides a. Thus a is divisible by a large power. Since $3^6 = 729$ we see that a sequence with at least 7 terms must start with 729 or a multiple of $2^6 = 64$. A little thought suggests that the best that can be done is to start with 128, take a common ratio $\frac{3}{2}$ and be content with 6 terms $\{128, 192, 288, 432, 648, 972\}$.

A more "formal" proof follows. The geometric progression

$$128, \quad 192, \quad 288, \quad 432, \quad 648, \quad 972$$

with common ratio $\frac{3}{2}$ shows that the longest sequence is at least of length 6. Consider now the geometric progression

$$100 \leq a < ar < ar^2 < \cdots < ar^{n-1} \leq 1000,$$

where the n terms are all integers and $r > 1$. Clearly r must be rational, say $r = \frac{p}{q}$ with $p > q \geq 1$, $(p, q) = 1$. Because $ar^{n-1} = a(\frac{p}{q})^{n-1}$ is an integer, q^{n-1} divides a. We may take $p = q+1$, for the progression

$$100 \leq a < a\left(\frac{q+1}{q}\right) < \cdots < a\left(\frac{q+1}{q}\right)^{n-1}$$
$$\leq 1000$$

has length n and the entries are all integers (since q^{n-1} divides a) in the required range. If $q \geq 3$, then

$$1000 \geq a\left(\frac{q+1}{q}\right)^{n-1} \geq (q+1)^{n-1} \geq 4^{n-1},$$

i.e., $n \leq 5$. If $q = 1$, then

$$1000 \geq a\left(\frac{q+1}{q}\right)^{n-1} = a2^{n-1} \geq 100 \cdot 2^{n-1},$$

i.e., $n \leq 4$. If $q = 2$, then

$$1000 \geq a\left(\frac{q+1}{q}\right)^{n-1} = a\left(\frac{3}{2}\right)^{n-1}$$
$$\geq 100\left(\frac{3}{2}\right)^{n-1},$$

i.e., $n \leq 6$. Hence the longest sequence has length 6.

Problem 77.   *First solution.* The identity

$$x^n - y^n$$
$$= (x-y)(x^{n-1} + x^{n-2}y + \cdots + xy^{n-2} + y^{n-1})$$

shows that $x^n - y^n$ is divisible by $x - y$. Hence $8^n - 6^n$ is divisible by 2, $3^n - 1^n$ is divisible by 2, and $1^n + 8^n - 3^n - 6^n$ is divisible by 2.

Similarly, $8^n - 3^n$ is divisible by $8 - 3 = 5$, $6^n - 1^n$ is divisible by $6 - 1 = 5$, and $1^n + 8^n - 3^n - 6^n$ is also divisible by 5.

Thus, $1^n + 8^n - 3^n - 6^n$ is divisible by 2 and 5, and hence by 10.

Second solution. The number is obviously even. Also, mod 5,

$$1^n + 8^n - 3^n - 6^n$$
$$= 1^n + (5+3)^n - 3^n - (5+1)^n$$
$$\equiv 1^n + 3^n - 3^n - 1^n \equiv 0.$$

Thus, the number is divisible by 2 and 5 and hence by 10.

Problem 78. The three sides of each "interior" triangle determine 6 points on the circle. Conversely, every set of 6 points on the circle determines a triangle. Hence the number of triangles is $\binom{n}{6} = \dfrac{n!}{6!(n-6)!}$. (Generalize to points on a sphere and tetrahedron.)

Problem 79. Since

$$a_{n+1} - 1 = a_n(a_n - 1) = a_n a_{n-1}(a_{n-1} - 1)$$
$$= \cdots = a_n a_{n-1} \cdots a_1(a_1 - 1)$$
$$= a_n a_{n-1} \cdots a_1,$$

the result follows easily.

FIGURE 98

Problem 80. Note that

$$\frac{1}{3} + \frac{1}{4} > \frac{1}{4} + \frac{1}{4} = \frac{1}{2}$$
$$\frac{1}{5} + \frac{1}{6} + \frac{1}{7} + \frac{1}{8} > 4\left(\frac{1}{8}\right) = \frac{1}{2}$$
$$\frac{1}{9} + \frac{1}{10} + \cdots + \frac{1}{16} > 8\left(\frac{1}{16}\right) = \frac{1}{2}$$

and so on. Hence

$$1 + \frac{1}{2} + \frac{1}{3} + \frac{1}{4} > 1 + 2\left(\frac{1}{2}\right)$$
$$1 + \frac{1}{2} + \frac{1}{3} + \cdots + \frac{1}{8} > 1 + 3\left(\frac{1}{2}\right)$$
$$1 + \frac{1}{2} + \frac{1}{3} + \cdots + \frac{1}{16} > 1 + 4\left(\frac{1}{2}\right)$$

and so on. For example $1 + \frac{1}{2} + \frac{1}{3} + \cdots + \frac{1}{2^{198}} > 1 + 198\left(\frac{1}{2}\right) = 100$.

Comment. More generally, show that if a and d are positive then N can be taken so large that

$$\frac{1}{a} + \frac{1}{a+d} + \cdots + \frac{1}{a+(N-1)d} > 100.$$

Problem 81. *First solution.* By the Arithmetic-Geometric Mean Inequality

$$\sqrt[n]{\frac{a_1 a_2 \cdots a_n}{b_1 b_2 \cdots b_n}} \leq \frac{1}{n}\left(\frac{a_1}{b_1} + \frac{a_2}{b_2} + \cdots + \frac{a_n}{b_n}\right)$$

and

$$\sqrt[n]{\frac{b_1 b_2 \cdots b_n}{a_1 a_2 \cdots a_n}} \leq \frac{1}{n}\left(\frac{b_1}{a_1} + \frac{b_2}{a_2} + \cdots + \frac{b_n}{a_n}\right).$$

Since the expressions on the left are reciprocals of each other, one of them must be at least as great as 1, and the result follows.

Second solution. By Cauchy's inequality (Tool Chest, D.4)

$$\left(\sum_{i=1}^{n} \frac{a_i}{b_i}\right)\left(\sum_{i=1}^{n} \frac{b_i}{a_i}\right) \geq n^2.$$

Consequently one of the factors on the left hand side must be $\geq n$. There is equality if and only if $a_i = b_i$ for all i.

Problem 82. If $|a_0| \neq 1$, a solution is obvious. For if m is any multiple of a_0 then $f(m)$

is divisible by a_0. In fact, infinitely often $f(m)$ is a proper multiple of a_0, since $f(m)$ can assume the values a_0 and $-a_0$ only a finite number of times. To find a solution when $|a_0| = 1$, proceed as follows. Choose any integer k for which $f(k)$ is not equal to 1 or -1. (This is always possible. Why?) Let p be any prime divisor of $f(k)$. Then $f(k + rp) \equiv f(k) \pmod{p}$. Furthermore, since a polynomial can assume any value only a finite number of times, $f(k+rp)$ is a multiple of p distinct from $\pm p$ for all but finitely many values of r.

Problem 83. Let $f(n)$ denote the maximal number of regions determined by n lines. A new line l—the $n + 1$th—is intersected by the n lines, creating $n + 1$ segments on l, each of which divides a region into two, for an increase of $n + 1$ in the number of regions i.e., $f(n + 1) = f(n) + n + 1$, $n \geq 1$. Repeated use of this, and the obvious $f(1) = 2$, shows that

$$f(n + 1)$$
$$= n + 1 + f(n)$$
$$= (n + 1) + n + f(n - 1)$$
$$= (n + 1) + n + (n - 1) + f(n - 2)$$
$$\vdots$$
$$= (n + 1) + n + (n - 1) + \cdots + 2 + f(1)$$
$$= (n + 1) + n + (n - 1) + \cdots + 2 + 1 + 1$$
$$= \frac{(n + 2)(n + 1)}{2} + 1$$

or, equivalently

$$f(n) = \frac{(n + 1)n}{2} + 1, \qquad n = 1, 2, \ldots.$$

(Consider the situation in space with n planes dividing space into solid regions.)

Problem 84. *First solution.* If $f(m, n)$ denotes the number of paths from $(0, 0)$ to (m, n), it follows (see Figure 99) that

$$f(m, n) = f(m - 1, n) + f(m, n - 1),$$
$$m \geq 1, \quad n \geq 1, \qquad (\star)$$
$$f(m, 0) = f(0, n) = 1.$$

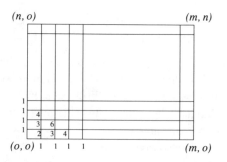

FIGURE 99

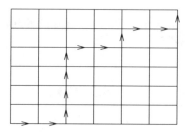

RRUUUURRURRU

FIGURE 100

Using this recurrence, we can compute $f(m, n)$ for small values of m, n. These are indicated in the diagram. Along the diagonal joining $(1, 0)$ to $(0, 1)$ the numbers are: $1 = \binom{1}{1}$, $1 = \binom{1}{0}$. Along the diagonal joining $(2, 0)$ to $(0, 2)$ they are $1 = \binom{2}{2}$, $2 = \binom{2}{1}$, $1 = \binom{2}{0}$. Along the next diagonal they are: $1 = \binom{3}{3}$, $3 = \binom{3}{2}$, $1 = \binom{3}{0}$. These suggest that $f(m, n) = \binom{m+n}{n}$, and it is easy to verify that $\binom{m+n}{n}$ does indeed satisfy the system $(\star)$.

Second solution. Every path from $(0, 0)$ to (m, n) is described by a sequence of m symbols R and n symbols U arranged in a straight line—the R corresponding to a step "right" and the U to a step "up". For example, the path in Figure 100 corresponds to RRUUUURRURRU.

The number of sequences (of m R's and n U's) is $\binom{m+n}{n}$.

Problem 85. Subtracting the second equation from the first yields

$$y - x = \frac{a^2 - b^2}{p}, \qquad \text{where } p = x + y + z;$$

hence

$$2\left(x^2 + xy + y^2\right) - (y-x)^2 = 2c^2 - \frac{\left(a^2 - b^2\right)^2}{p^2}$$

and similarly

$$2\left(y^2 + yz + z^2\right) - (z-y)^2$$
$$= 2a^2 - \frac{\left(b^2 - c^2\right)^2}{p^2},$$

$$2\left(z^2 + zx + x^2\right) - (x-z)^2$$
$$= 2b^2 - \frac{\left(c^2 - a^2\right)^2}{p^2}.$$

The sum of the last three equations reduces to

$$2p^2 = 2(a^2 + b^2 + c^2)$$
$$- \frac{\left(a^2 - b^2\right)^2 + \left(b^2 - c^2\right)^2 + \left(c^2 - a^2\right)^2}{p^2}$$

and the solutions of this quadratic equation in p^2 are

$$2p^2 = a^2 + b^2 + c^2 \pm \left\{6\left(a^2b^2 + b^2c^2 + c^2a^2\right)\right.$$
$$\left. - 3\left(a^4 + b^4 + c^4\right)\right\}^{\frac{1}{2}}. \quad (\star)$$

In a triangle ABC, with sides a, b, c, let P be any point inside and let x, y, z denote the distances from P to the vertices (Figure 101). It is well known, but not easy to prove, that if all the angles of triangle ABC are less than $120°$, then the point P for which $x + y + z$ is a minimum is located such that

$$\angle APB = \angle BPC = \angle CPA = 120°.$$

In this case (1), (2), and (3) clearly hold—they are just statements of the cosine law for triangles,

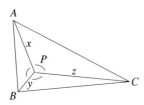

FIGURE 101

and $(\star)$ simplifies to

$$p = \left\{\frac{a^2 + b^2 + c^2 + 4\Delta\sqrt{3}}{2}\right\}^{\frac{1}{2}},$$

where Δ denotes the area of triangle ABC.

Comment. See Comment, solution to Problem 30.

Problem 86. Let v_1, $v_2, \ldots, v_6$ be the six points. Of the 5 edges joining v_1 to the other vertices, at least 3 are of the same color, say v_1 is joined to v_2, v_3, v_4 by red edges. If the triangle $v_2v_3v_4$ has all edges blue we are done. If one (or more) of the edges of triangle $v_2v_3v_4$ is red, say v_2v_3, then triangle $v_1v_2v_3$ has red edges only.

Remarks. In fact, there must be at least two triangles which are monochromatic! It can be proved that if 16 points in space are such that no three are in a line, and the 120 edges joining them in pairs are painted using 3 colors, then there is a monochromatic triangle.

Problem 87. Repeated use of the identity

$$\frac{1}{k} = \frac{1}{k+1} + \frac{1}{k(k+1)}$$

will lead to infinitely many representations, such as

$$1 = \frac{1}{2} + \frac{1}{3} + \frac{1}{6} = \frac{1}{2} + \frac{1}{3} + \frac{1}{7} + \frac{1}{42}$$
$$= \frac{1}{2} + \frac{1}{3} + \frac{1}{7} + \frac{1}{43} + \frac{1}{1806}$$
$$= \frac{1}{2} + \frac{1}{4} + \frac{1}{6} + \frac{1}{12}$$
$$= \frac{1}{2} + \frac{1}{5} + \frac{1}{6} + \frac{1}{12} + \frac{1}{20}$$

and so on. It is possible to start with any representation of 1 as the reciprocal of finitely many integers and to use the above identity to obtain a representation with distinct reciprocals.

Problem 88. The circle can be divided into any integral number of pieces of equal areas with straight edge and compasses. For simplicity, we indicate the method for $n = 4$. Divide a diameter

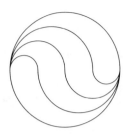

FIGURE 102

into 4 equal parts and draw semicircles as indicated in Figure 102.

Problem 89. Suppose the path determined by the listing $p_1, p_2, \ldots, p_n$ crosses itself (see Figure 103). We show that there is a shorter path. Assume that for some i and j, $1 \le i < j \le n-1$, the segments $p_i p_{i+1}$, $p_j p_{j+1}$, cross, at the point r say. Then

$$\overline{p_i p_j} + \overline{p_{i+1} p_{j+1}} < \overline{p_i r} + \overline{r p_j} + \overline{p_{i+1} r} + \overline{r p_{j+1}}$$
$$= \overline{p_i p_{i+1}} + \overline{p_j p_{j+1}}$$

so the path

$$p_1 p_2 \cdots p_i p_j p_{j-1} \cdots p_{i+2} p_{i+1} p_{j+1} p_{j+2} \cdots p_n$$

is shorter than the path

$$p_1 p_2 \cdots p_i p_{i+1} \cdots p_j p_{j+1} \cdots p_n.$$

Problem 90. We have $P(x,y) \equiv (x-y)Q(x,y)$ and $P(x,y) \equiv (y-x)Q(y,x)$, so that

$$0 \equiv P(x,y) - P(y,x)$$
$$\equiv (x-y)(Q(x,y) + Q(y,x)),$$

i.e.,

$$Q(x,y) + Q(y,x) \equiv 0,$$

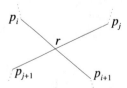

FIGURE 103

and hence $Q(x,x) \equiv 0$. Thus, the polynomial equation $Q(x,y) = 0$ has a solution $x = y$, so $x - y$ is a factor of $Q(x,y)$.

Comment. For extensions of this result see *Crux Mathematicorum* 14 (1988) 139, # 4.

Problem 91. Because Δ_8 and Δ_0 are not used as labels, triangle $P_0 P_8 P_7$ must be labelled Δ_7. Triangle $P_0 P_7 P_6$ must therefore be labelled Δ_6, triangle $P_0 P_3 P_6$ must therefore be labelled Δ_3, triangle $P_0 P_1 P_3$ must therefore be labelled Δ_1, triangle $P_1 P_2 P_3$ must then be labelled Δ_2, triangle $P_3 P_4 P_6$ must be labelled Δ_4, and triangle $P_4 P_5 P_6$ must be labelled Δ_5. This is the only labelling which satisfies the given conditions. See Figure 104.

Try to generalize this to an n-gon dissected into $n - 2$ triangles.

Problem 92. Since OT is perpendicular to TS, lines PR, OT and QS are parallel. Hence

$$\frac{\overline{RT}}{\overline{TS}} = \frac{\overline{PO}}{\overline{OQ}} = 1, \text{ or } \overline{RT} = \overline{TS}.$$

Thus, in triangles ROT and SOT, OT is a common side, $\overline{RT} = \overline{ST}$ and $\angle RTO = \angle STO$ (both are right angles). Therefore these triangles are congruent, and $\overline{OR} = \overline{OS}$. See Figure 105.

Problem 93. *First solution.* The proof is by mathematical induction. The result is valid for

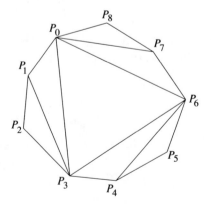

FIGURE 104

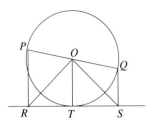

FIGURE 105

$n = 1$. Assume it is valid for $n = k$. Then

$$(1 + a_1)(1 + a_2) \cdots (1 + a_k)(1 + a_{k+1})$$
$$\geq \frac{2^k}{k+1}(1 + a_1 + \cdots + a_k)(1 + a_{k+1}).$$

We now show that

$$\frac{2^k}{k+1}(1 + s)(1 + a) \geq \frac{2^{k+1}}{k+2}(1 + s + a),$$

where $a = a_{k+1}$, $s = a_1 + \cdots + a_k$. Multiplying out (and rearranging terms) we obtain

$$2(as - k) + k(a - 1)(s - 1) \geq 0,$$

and this is valid because $a \geq 1$ for all i. Thus the result is valid for $n = k + 1$ and by induction for all n. There is equality only if $a_i = 1$ for all i.

Second solution. The given inequality is equivalent to

$$\left(\frac{1 + a_1}{2}\right)\left(\frac{1 + a_2}{2}\right) \cdots \left(\frac{1 + a_n}{2}\right)$$
$$\geq \frac{1 + a_1 + a_2 + \cdots + a_n}{n+1},$$

or, with $x_i = (a_i - 1)/2 \geq 0$,

$$(1 + x_1)(1 + x_2) \cdots (1 + x_n)$$
$$\geq 1 + \frac{2}{n+1}(x_1 + \cdots + x_n);$$

but

$$(1 + x_1)(1 + x_2) \cdots (1 + x_n)$$
$$\geq 1 + x_1 + \cdots + x_n$$
$$\geq 1 + \frac{2}{n+1}(x_1 + \cdots + x_n).$$

Problem 94. It is apparent that angle XBY is a right angle and angle AXB is constant. Hence angle APB is constant, so the locus is a circle through A and B. See Figure 106.

Problem 95. The given examples suggest the general law

$$\frac{1}{n} = \frac{1}{n+1} + \frac{1}{n(n+1)}, \qquad n = 1, 2, 3, \ldots,$$

which is easily established. After putting this in the equivalent form

$$\frac{1}{n(n+1)} = \frac{1}{n} - \frac{1}{n+1}, \qquad n = 1, 2, 3, \ldots,$$

we evaluate, for $1 \leq i < j$,

$$\frac{1}{i(i+1)} + \frac{1}{(i+1)(i+2)} + \cdots + \frac{1}{j(j+1)}$$
$$= \left(\frac{1}{i} - \frac{1}{i+1}\right) + \left(\frac{1}{i+1} - \frac{1}{i+2}\right)$$
$$+ \cdots + \left(\frac{1}{j-1} - \frac{1}{j}\right) + \left(\frac{1}{j} - \frac{1}{j+1}\right)$$
$$= \frac{1}{i} - \frac{1}{j+1}.$$

Thus we must solve

$$\frac{1}{n} = \frac{1}{i} - \frac{1}{j+1}.$$

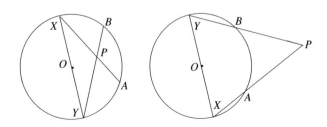

FIGURE 106

Comparing this with

$$\frac{1}{n} = \frac{1}{n-1} - \frac{1}{(n-1)n}$$

(which is the first equality with n replaced by $n-1$) we have the solution

$$i = n-1, \quad j+1 = (n-1)n.$$

Problem 96. *First solution.* Clearly

$$\tan\alpha = 1, \quad \tan\beta = \frac{1}{2}, \quad \tan\gamma = \frac{1}{3},$$

so

$$\tan(\beta+\gamma) = \frac{\tan\beta + \tan\gamma}{1 - \tan\beta\tan\gamma}$$

$$= \frac{\frac{1}{2} + \frac{1}{3}}{1 - \frac{1}{2}\cdot\frac{1}{3}} = 1 = \tan\alpha$$

and hence $\beta + \gamma = \alpha$. See Figure 107.

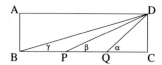

FIGURE 107

Second solution. Let $\overline{CD} = 1$. Triangles DBQ and PDQ are similar because in triangle DBQ, $\overline{BD} = \sqrt{10}$, $\overline{BQ} = 2$, $\overline{QD} = \sqrt{2}$, and in triangle PDQ, $\overline{PD} = \sqrt{5}$, $\overline{DQ} = \sqrt{2}$, $\overline{PQ} = 1$, so the sides of the former are $\sqrt{2}$ times the sides of the latter triangle. Therefore $\angle PDQ = \angle DBQ = \gamma$, and the result follows by applying the exterior angle theorem to triangle PDQ.

Problem 97. First, we make some comments. The presence of the absolute value signs makes the sum "indigestible." However, it is symmetrical with respect to the $\{x_i\}$, so that changing their order does not alter it. Thus we can arrange that the largest possible value of this sum can be obtained with a choice of $\{x_i\}$ for which $0 \le x_1 \le x_2 \le \cdots \le x_n \le 1$. Then $|x_i - x_j|$ can be written without an absolute value sign.

The sum should be rearranged to put all the x_i terms together for each i. Then those x_i with

positive coefficients can be made as large as possible and those with negative coefficients as small as possible (in keeping with $0 \le x_1 \le x_2 \le \cdots \le x_n \le 1$). To get a feeling for what is happening, consider $n = 4$, $0 \le x_1 \le x_2 \le x_3 \le x_4 \le 1$.

$$S = |x_1 - x_2| + |x_1 - x_3| + |x_1 - x_4|$$

$$+ |x_2 - x_3| + |x_2 - x_4| + |x_3 - x_4|$$

$$= (x_2 - x_1) + (x_3 - x_1) + (x_4 - x_1)$$

$$+ (x_3 - x_2) + (x_4 - x_2) + (x_4 - x_3)$$

$$= -3x_1 - x_2 + x_3 + 3x_4.$$

Clearly S is maximized by taking $x_1 = x_2 = 0$ and $x_3 = x_4 = 1$.

Now we proceed to the proof.

Assume without loss of generality that $0 \le x_1 \le x_2 \le \cdots \le x_n \le 1$. Let S denote the sum. Of course

$$|x_i - x_j| = \begin{cases} x_i - x_j & \text{if } i > j, \\ x_j - x_i & \text{if } i < j. \end{cases}$$

Hence, for $j = 1, 2, \ldots, i-1$, $|x_i - x_j| = x_i - x_j$, while for $j = i+1, \ldots, n$, $|x_i - x_j| = x_j - x_i$.

Thus

$$S = \sum_{i=1}^{n} x_i(i - 1 - (n-i))$$

$$= \sum_{i=1}^{n} (2i - n - 1)x_i.$$

If n is even, say $n = 2m$, we have

$$S = \sum_{i=m+1}^{2m} (2i - 2m - 1)x_i - \sum_{i=1}^{m}(2m + 1 - 2i)x_i$$

$$\le \sum_{i=m+1}^{2m} (2i - 2m - 1)$$

(corresponding to $x_1 = \cdots = x_m = 0$,

$$x_{m+1} = \cdots = x_{2m} = 1)$$

$$= 1 + 3 + 5 + \cdots + (2m - 1)$$

$$= \frac{m}{2}(2m) = \frac{(2m)^2}{4} = \frac{n^2}{4},$$

i.e.,

$$S \le \left(\frac{n}{2}\right)^2.$$

If n is odd, say $n = 2m + 1$,

$$S = \sum_{i=m+2}^{2m+1} (2i - 2m - 2)x_i$$

$$- \sum_{i=1}^{m+1} (2m + 2 - 2i)x_i$$

$$\leq \sum_{i=m+2}^{2m+1} (2i - 2m - 2)$$

$$= 2 + 4 + 6 + \cdots + 2m$$

$$= 2(1 + 2 + \cdots + m)$$

$$= 2\left(\frac{m}{2}\right)(m + 1) = \left(\frac{n-1}{2}\right)\left(\frac{n+1}{2}\right)$$

$$= \frac{n^2 - 1}{4},$$

$$S \leq \frac{n^2 - 1}{4} = \left[\frac{n^2}{4}\right].$$

Thus $S(n) = \left[\dfrac{n^2}{4}\right]$.

Problem 98. The suggested generalizations are

$$\frac{1}{2n-1} + \frac{1}{2n+1} = \frac{4n}{4n^2 - 1}, \quad n = 1, 2, 3, \ldots,$$

$$(4n)^2 + (4n^2 - 1)^2 = (4n^2 + 1)^2, \quad n = 1, 2, 3, \ldots,$$

and these are easily established.

Problem 99. More generally, suppose John tosses $n + 1$ coins and Mary tosses n coins. Then, either John tosses more heads than Mary does or John tosses more tails than Mary does—but not both. (Check this out!) Since these two outcomes are symmetric, each occurs with probability $\frac{1}{2}$.

Exercise. From this result obtain a binomial sum identity.

Problem 100. One can rearrange the sides of the hexagon so each pair of consecutive sides are of length a and b. Since all the angles of $ABCDEF$ are congruent, each is $120°$. By the law of cosines

$$\overline{AC} = 2R^2(1 - \cos 120°) = a^2 + b^2 - 2ab \cos 120°$$

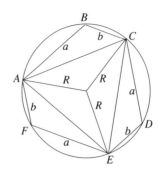

FIGURE 108

or

$$3R^2 = a^2 + ab + b^2.$$

Problem 101. Let x be the base.

(a) $10201 = x^4 + 2x^2 + 1 = (x^2 + 1)^2$.

(b) $10101 = x^4 + x^2 + 1 = (x^2 + x + 1)(x^2 - x + 1)$, and (since $x \geq 2$) both factors exceed 1.

(c) $100011 = x^5 + x + 1 = (x^2 + x + 1) \times (x^3 - x^2 + 1)$, and again both factors exceed 1.

Problem 102. Denote the n people by p_1, $p_2, \ldots, p_n$ and a call from p_i to p_j by $p_i \to p_j$. Let $f(n)$ denote the minimum number of calls which can leave everybody fully informed. The particular sequence

$$p_1 \to p_n, \ p_2 \to p_n, \ldots, p_{n-1} \to p_n,$$

$$p_n \to p_{n-1}, \ p_n \to p_{n-2}, \ldots, p_n \to p_1$$

contains $2n - 2$ calls and leaves everybody informed, showing that $f(n) \leq 2n - 2$.

Suppose we have a sequence of calls which leaves everybody fully informed. Consider the "crucial" call at the end of which the receiver (call him p) is the first to be fully informed. Clearly, each of the $n - 1$ people other than p must have placed at least one call no later than the crucial call (how else could p be fully informed?) Also, each of these $n - 1$ people (being not fully informed) must receive at least one call after the crucial one. Hence the given sequence contains at least $2(n - 1)$ calls.

Thus $f(n) = 2n - 2$.

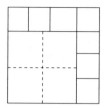

FIGURE 109

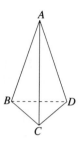

FIGURE 110

Problem 103. Let $k \geq 2$ be an integer. The square can be subdivided into $2k$ nonoverlapping squares as indicated by the "solid" lines in Figure 109 (which shows the case $k = 4$). Subdividing one of these squares into four nonoverlapping squares yields a subdivision into $2k + 3$ squares. Thus the square can be subdivided into n squares for each integer $n \geq 6$. (Can a square be subdivided into 2, 3, or 5 squares? Why?)

Problem 104. *First solution.* Let $\vec{u} = \overrightarrow{AB}$, $\vec{v} = \overrightarrow{AC}$, $\vec{w} = \overrightarrow{AD}$. Then, the given condition is

$$\vec{v}^2 + (\vec{w} - \vec{u})^2 = \vec{u}^2 + (\vec{v} - \vec{u})^2$$
$$+ (\vec{w} - \vec{v})^2 + \vec{w}^2.$$

Hence

$$\vec{v} \cdot \vec{v} + \vec{w} \cdot \vec{w} - 2\vec{w} \cdot \vec{u} + \vec{u} \cdot \vec{u}$$
$$= \vec{u} \cdot \vec{u} + \vec{v} \cdot \vec{v} - 2\vec{v} \cdot \vec{u} + \vec{u} \cdot \vec{u}$$
$$+ \vec{w} \cdot \vec{w} - 2\vec{w} \cdot \vec{v} + \vec{v} \cdot \vec{v} + \vec{w} \cdot \vec{w}.$$

This leads to

$$(\vec{u} + \vec{w} - \vec{v})^2 = \vec{u} \cdot \vec{u} + \vec{v} \cdot \vec{v}$$
$$+ \vec{w} \cdot \vec{w} - 2\vec{w} \cdot \vec{v}$$
$$- 2\vec{v} \cdot \vec{u} + 2\vec{w} \cdot \vec{u} = 0$$

or $\vec{v} = \vec{u} + \vec{w}$. Thus $ABCD$ is a parallelogram.

Second solution. Let A, B, C, D have the respective rectangular coordinates in space $(r, 0, 0)$, $(0, 0, 0)$, $(s, t, 0)$, (u, v, w). (There is no loss of generality in taking B at the origin and the plane determined by ABC as the x, y plane.) We have to show that $u = r + s$, $v = t$ and $w = 0$.

The given relation among the lengths of the sides is equivalent to

$$(s - r)^2 + t^2 + u^2 + v^2 + w^2$$
$$= r^2 + s^2 + t^2 + (u - s)^2 + (v - t)^2$$
$$+ w^2 + (u - r)^2 + v^2 + w^2$$

or

$$-2rs = (v - t)^2 + w^2 + r^2 + s^2$$
$$+ u^2 - 2us - 2ur$$

or

$$0 = (v - t)^2 + w^2 + (r + s - u)^2.$$

Hence $v - t = w = r + s - u = 0$, as required.

Problem 105. Our proof is indirect. Assume that there are F faces and that no two faces have the same number of edges. The face with the fewest edges has at least three, the next face at least four, and so on. The face with the most edges has at least $F + 2$ of them. Since there is a bordering face for each edge of this "greatest" face, there are at least $F + 3$ faces, a contradiction.

In a similar way one can show that there are at least two pairs of faces such that the faces of each pair have the same number of edges. Additionally, one can obtain a dual result by replacing face and number of edges by vertex and valence (= number of edges incident with a vertex) respectively.

Problem 106. Let BC intersect PR in T. We have

area quadrilateral $PSBT$

$$= \text{area } \Delta PSB + \text{ area } \Delta PBT$$

$$= \frac{3\sqrt{3}}{4}(\overline{SB} + \overline{BT}).$$

Since a clockwise rotation of $60°$ about P carries A onto B and S onto T, we have that $\overline{BT} = \overline{AS} = 2$. Hence $\overline{SB} + \overline{BT} = 3$ and the required area is $\frac{9\sqrt{3}}{4}$.

Problem 107. *First proof* (by induction). The result is valid for $n = 1$. Assuming its validity for $n = k$, i.e.,

$$1 - \frac{1}{2} + \frac{1}{3} - \cdots + \frac{1}{2k-1}$$

$$= \frac{1}{k} + \frac{1}{k+1} + \cdots + \frac{1}{2k-1},$$

we deduce that

$$1 - \frac{1}{2} + \frac{1}{3} - \cdots + \frac{1}{2k-1} - \frac{1}{2k} + \frac{1}{2k+1}$$

$$= \left(\frac{1}{k} + \frac{1}{k+1} + \cdots + \frac{1}{2k-1} \right)$$

$$- \frac{1}{2k} + \frac{1}{2k+1}$$

$$= \frac{1}{k+1} + \cdots + \frac{1}{2k-1}$$

$$+ \left(\frac{1}{k} - \frac{1}{2k} \right) + \frac{1}{2k+1}$$

$$= \frac{1}{k+1} + \cdots + \frac{1}{2k-1} + \frac{1}{2k} + \frac{1}{2k+1}.$$

Second proof.

$$1 - \frac{1}{2} + \frac{1}{3} - \frac{1}{4} + \cdots + \frac{1}{2n-1}$$

$$= \left(1 + \frac{1}{2} + \frac{1}{3} + \cdots + \frac{1}{2n-1} \right)$$

$$- 2 \left(\frac{1}{2} + \frac{1}{4} + \frac{1}{6} + \cdots + \frac{1}{2n-2} \right)$$

$$= \left(1 + \frac{1}{2} + \frac{1}{3} + \cdots + \frac{1}{2n-1} \right)$$

$$- \left(1 + \frac{1}{2} + \frac{1}{3} + \cdots + \frac{1}{n-1} \right)$$

$$= \frac{1}{n} + \frac{1}{n+1} + \cdots + \frac{1}{2n-1}.$$

Problem 108. *First solution* (by mathematical induction). Since

$$2 + h(1) = 2 + 1 = 2 \left(1 + \frac{1}{2} \right) = 2h(2),$$

the statement holds when $n = 2$. If

$$n + h(1) + \cdots + h(n-1) = nh(n)$$

holds for some $n \geq 2$, then, adding $1 + h(n)$ to both sides, we deduce

$$n + 1 + h(1) + h(2) + \cdots + h(n-1) + h(n)$$

$$= nh(n) + 1 + h(n) = (n+1)h(n) + 1$$

$$= (n+1) \left\{ h(n+1) - \frac{1}{n+1} \right\} + 1$$

$$= (n+1)h(n+1),$$

i.e., the statement holds for $n + 1$.

When Leo Moser was playing in the chess tournament in Toronto in 1946, a heckler was bothering the players. "Chess is a complete waste of time," he said. " It has no relation to any other branch of knowledge." "How about mathematics?" Moser asked. "I have studied mathematics for many years," he replied , "and know that chess has no relation to any of the four branches of mathematics." "What branches do you mean?" Moser asked. "You know," he replied, "addition, subtraction, multiplication and division."

Second solution.

$$n + h(1) + h(2) + \cdots + h(n-1)$$

$$= n + 1$$
$$+ 1 + \frac{1}{2}$$
$$+ 1 + \frac{1}{2} + \frac{1}{3}$$
$$\vdots$$
$$+ 1 + \frac{1}{2} + \frac{1}{3} + \cdots + \frac{1}{n-2}$$
$$+ 1 + \frac{1}{2} + \frac{1}{3} + \cdots + \frac{1}{n-2} + \frac{1}{n-1}$$

$$= n + (n-1)1 + (n-2)\frac{1}{2} + (n-3)\frac{1}{3}$$
$$+ \cdots + 2\frac{1}{n-2} + 1\frac{1}{n-1}$$

$$= n + (n-1) + \left(\frac{n}{2} - 1\right) + \left(\frac{n}{3} - 1\right)$$
$$+ \cdots + \left(\frac{n}{n-2} - 1\right) + \left(\frac{n}{n-1} - 1\right)$$

$$= n + n\left\{1 + \frac{1}{2} + \cdots + \frac{1}{n-1}\right\} - (n-1)1$$

$$= 1 + n\left\{h(n) - \frac{1}{n}\right\} = nh(n).$$

More generally, if $s_n = a_1 + a_2 + \cdots + a_n$ for $n = 1, 2, 3, \ldots$, then

$$\sum_{r=1}^{n-1} s_r = \sum_{r=1}^{n-1}\sum_{t=1}^{r} a_t = \sum_{t=1}^{n-1}\sum_{r=t}^{n-1} a_t$$

$$= \sum_{t=1}^{n-1} (n-t)a_t$$

$$= n\sum_{t=1}^{n-1} a_t - \sum_{t=1}^{n-1} ta_t$$

$$= ns_{n-1} - \sum_{t=1}^{n-1} ta_t.$$

In the present situation, $a_t = \frac{1}{t}$ and we obtain the desired result.

Problem 109. For any $k \geq 0$, the five integers $k+1$, $k+2, \ldots, k+5$ leave, on division by 5, remainders of 0, 1, 2, 3, 4 in some order. Thus,

modulo 5, the given sum is congruent to

$$0^n + 1^n + 2^n + 3^n + 4^n = 1 + 2^n + 3^n + 4^n.$$

Next, observe the following congruences modulo 5, all easily established for a nonnegative integer m:

$$2^{4m} \equiv 1, \ 2^{4m+1} \equiv 2, \ 2^{4m+2} \equiv 4, \ 2^{4m+3} \equiv 3,$$

$$3^{4m} \equiv 1, \ 3^{4m+1} \equiv 3, \ 3^{4m+2} \equiv 4, \ 3^{4m+3} \equiv 2,$$

$$4^{4m} \equiv 1, \ 4^{4m+1} \equiv 4, \ 4^{4m+2} \equiv 1, \ 4^{4m+3} \equiv 4,$$

Hence, the given sum is congruent modulo 5 to

$1 + 1 + 1 + 1 = 4$	if	$n = 4m$
$1 + 2 + 3 + 4 \equiv 0$	if	$n = 4m + 1$
$1 + 4 + 4 + 1 \equiv 0$	if	$n = 4m + 2$
$1 + 3 + 2 + 4 \equiv 0$	if	$n = 4m + 3.$

Thus the given sum is divisible by 5 for any $k \geq 0$ and any n not a multiple of 4, and not divisible by 5 for any $k \geq 0$ and any n a multiple of 4.

More compactly, the solution may be presented as follows. Modulo 5, we have

$$1^n + 2^n + 3^n + 4^n$$
$$\equiv 1^n + 2^n + (5-2)^n + (5-1)^n$$
$$\equiv 1^n + (-1)^n + 2^n + (-2)^n.$$

If n is odd, $(-1)^n = -1$ and $(-2)^n = -2^n$, so $1^n + 2^n + 3^n + 4^n \equiv 0$.

If n is even, then $n = 2m$ and $(-1)^n = 1$, $(-2)^n = 2^n$, so that

$$1^n + 2^n + 3^n + 4^n \equiv 2\left[1^{2m} + 2^{2m}\right]$$
$$= 2\left[1 + 4^m\right]$$
$$\equiv 2\left[1 + (-1)^m\right],$$

which is 0 when m is odd and nonzero when m is even.

Hence $1^n + 2^n + 3^n + 4^n$ is divisible by 5 if and only if n is not a multiple of 4.

Problem 110. Construct any triangle ABC with $\angle A = \alpha$, $\angle B = \beta$, and a line DE whose length is equal to the perimeter of ABC. Construct UV such that $\frac{\overline{UV}}{\overline{AB}} = \frac{p}{\overline{DE}}$ and triangle UVW with $\angle U = \alpha$, $\angle V = \beta$. This is the required triangle.

Problem 111. If we think of y and z as fixed and x as the variable, so that $P(x, y, z)$ is a polynomial in x, we see that: $x - (-y - z)$ is a factor of $P(x, y, z)$ if and only if $P(-y - z, y, z) = 0$. Thus, we seek values of k for which $P(-y - z, y, z) \equiv 0$. This identity is

$$(-y - z)^5 + y^5 + z^5$$
$$+ k \left((-y - z)^3 + y^3 + z^3 \right) \times$$
$$\left((-y - z)^2 + y^2 + z^2 \right) \equiv 0.$$

Simplifying, we obtain

$$-(5 + 6k)yz(y + z)\left(y^2 + yz + z^2 \right) \equiv 0,$$

so that $k = -\frac{5}{6}$.

For the second part, let x, y, z be roots of the cubic

$$t^3 - at^2 + bt - c = 0, \qquad (1)$$

where $a = x + y + z$, $b = xy + yz + zx$, $c = xyz$. Let $S_n = x^n + y^n + z^n$.

Then

$$S_1 = a \text{ and } S_2 = a^2 - 2b. \qquad (2)$$

S_3 is obtained by summing (1) over the roots, which yields

$$S_3 - aS_2 + bS_1 - 3c = 0.$$

From (2),

$$S_3 = a^3 - 3ab + 3c. \qquad (3)$$

Multiplying (1) by t and t^2 respectively, and summing over the roots produces

$$S_4 - aS_3 + bS_2 - cS_1 = 0,$$
$$S_5 - aS_4 + bS_3 - cS_2 = 0,$$

which, together with (2) and (3) yield

$$S_4 = a^4 - 4a^2b + 4ac + 2b^2,$$
$$S_5 = a^5 - 5a^3b + 5a^2c + 5ab^2 - 5bc.$$

Finally,

$$6(x^5 + y^5 + z^5) - 5(x^3 + y^3 + z^3)(x^2 + y^2 + z^2)$$
$$= 6S_5 - 5S_3S_2 = a^2(a^3 - 5ab + 15c)$$
$$= (x + y + z)^2 \{(x + y + z)^3$$
$$\quad - 5(x + y + z)(xy + yz + zx) + 15xyz\}$$
$$= (x + y + z)^2 \{x^3 + y^3 + z^3 + 6xyz$$
$$\quad - 2(x^2y + xy^2 + y^2z + yz^2 + z^2x + zx^2)\}.$$

Problem 112. From $(1 - p)^2 \geq 0$, it follows that $1 + p + p^2 \geq 3p > 0$, and similarly for q, r and s. The product of the inequalities yields the desired result.

Problem 113. Write $x = [x] + e$ where $0 \leq e < 1$. Then

$$[nx] = n[x] + [ne]$$

so that

$$\left[\frac{[nx]}{n} \right] = [x] + \left[\frac{[ne]}{n} \right] = [x],$$

since $[ne] \leq ne < n$.

Problem 114. A generalization for set A is:

$$\sum_{i=1}^{n} (-1)^{i+1} \binom{n}{i} \frac{1}{i} = \sum_{j=1}^{n} \frac{1}{j}, \quad n = 1, 2, 3, \dots.$$

This can be proved by induction. The equality is valid when $n = 1$. Assume that it holds for $n = k$:

$$\sum_{i=1}^{k} (-1)^{i+1} \binom{k}{i} \frac{1}{i} = \sum_{j=1}^{k} \frac{1}{j}.$$

Then

$$\sum_{i=1}^{k+1} (-1)^{i+1} \binom{k+1}{i} \frac{1}{i}$$

$$= \sum_{i=1}^{k+1} (-1)^{i+1} \left\{ \binom{k}{i-1} + \binom{k}{i} \right\} \frac{1}{i}$$

$$= \sum_{i=1}^{k+1} (-1)^{i+1} \binom{k}{i-1} \frac{1}{i}$$

$$\quad + \sum_{i=1}^{k} (-1)^{i+1} \binom{k}{i} \frac{1}{i}$$

$$= \sum_{i=1}^{k+1}(-1)^{i+1}\binom{k+1}{i}\frac{1}{k+1}$$

$$+ \left\{1+\frac{1}{2}+\cdots+\frac{1}{k}\right\}$$

$$= \frac{1}{k+1}\{1-(1-1)^{k+1}\} + \left\{1+\frac{1}{2}+\cdots+\frac{1}{k}\right\}$$

$$= 1+\frac{1}{2}+\frac{1}{3}+\cdots+\frac{1}{k}+\frac{1}{k+1}.$$

A generalization for set B is:

$$\sum_{i=1}^{n}(-1)^{i+1}\binom{n}{i}\left\{1+\frac{1}{2}+\cdots+\frac{1}{i}\right\} = \frac{1}{n},$$

$$n = 1,2,3,\ldots .$$

We give a direct proof. Interchanging the order of summation, we obtain, for the left side,

$$\sum_{i=1}^{n}(-1)^{i+1}\binom{n}{i}\sum_{j=1}^{i}\frac{1}{j}$$

$$= \sum_{1\le j\le i\le n}\frac{1}{j}(-1)^{i+1}\binom{n}{i}$$

$$= \sum_{j=1}^{n}\frac{1}{j}\sum_{i=j}^{n}(-1)^{i+1}\binom{n}{i}$$

$$= \sum_{j=1}^{n}\frac{1}{j}\left\{\sum_{i=j}^{n-1}(-1)^{i+1}\left[\binom{n-1}{i-1}+\binom{n-1}{i}\right]\right.$$

$$\left. +(-1)^{n+1}\right\}$$

$$= \sum_{j=1}^{n}\frac{1}{j}\left\{(-1)^{j+1}\left[\binom{n-1}{j-1}+\binom{n-1}{j}\right]\right.$$

$$\left. -\binom{n-1}{j}-\binom{n-1}{j+1}+\cdots+(-1)^{n+1}\right]\right\}$$

$$= \sum_{j=1}^{n}(-1)^{j+1}\frac{1}{j}\binom{n-1}{j-1} = \frac{1}{n}\sum_{j=1}^{n}(-1)^{j+1}\binom{n}{j}$$

$$= \frac{1}{n}\{1-(1-1)^{n}\} = \frac{1}{n}.$$

The relationship between the sets of equations A and B can be generalized. Suppose a sequence (x_n) of real numbers is given and that the sequence (y_n) is defined by

$$y_n = \sum_{i=1}^{n}(-1)^{i+1}\binom{n}{i}x_i \qquad n = 1,2,3,\ldots .$$

Then

$$x_n = \sum_{i=1}^{n}(-1)^{i+1}\binom{n}{i}y_i \qquad n = 1,2,3,\ldots .$$

We need to show that, for each n,

$$x_n = \sum_{i=1}^{n}(-1)^{i+1}\binom{n}{i}\sum_{j=1}^{i}(-1)^{j+1}\binom{i}{j}x_j.$$

Reversing the order of summation, the right-hand side becomes

$$\sum_{j=1}^{n}(-1)^{j}x_j\sum_{i=j}^{n}(-1)^{i}\binom{n}{i}\binom{i}{j}$$

$$= x_n + \sum_{j=1}^{n-1}(-1)^{j}x_j\sum_{i=j}^{n}(-1)^{i}\binom{n}{i}\binom{i}{j}.$$

But, when $1\le j\le n-1$,

$$\sum_{i=j}^{n}(-1)^{i}\binom{n}{i}\binom{i}{j}$$

$$= \sum_{i=j}^{n}(-1)^{i}\frac{n!}{i!(n-i)!}\frac{i!}{j!(i-j)!}$$

$$= \frac{n!}{(n-j)!j!}\sum_{i=j}^{n}(-1)^{i}\frac{(n-j)!}{(n-i)!(i-j)!}$$

$$= \binom{n}{j}\sum_{i=j}^{n}(-1)^{i}\binom{n-j}{i-j}$$

$$= (-1)^{j}\binom{n}{j}\sum_{r=0}^{n-j}(-1)^{r}\binom{n-j}{r}$$

$$= (-1)^{j}\binom{n}{j}(1-1)^{n-j} = 0.$$

The result follows.

This gives an alternative way of obtaining either set of equations A and B, given that the other set is valid.

Problem 115. By considering the convex hull of all the points it follows that we can always choose two of the points, say A and B, so that the remaining $2n+1$ points lie on the same side of line AB. Label these $P_1, P_2, \ldots, P_{2n+1}$ in such a way that

$$\angle AP_1B \le \angle AP_2B \le \cdots \le \angle AP_{2n+1}B.$$

In fact, these angles are all different; for, if say $\angle AP_iB = \angle AP_jB$, then the four points

A, B, P_i, P_j would all lie on a circle. Hence, the circle through AB and P_{n+1} has $P_1, P_2, \ldots, P_n$ in its exterior and $P_{n+2}, P_{n+3}, \ldots, P_{2n+1}$ in its interior.

This is a generalization of Problem 26. (Extend the result to points in space and a sphere.)

Problem 116. *First solution.* A limiting case occurs when two of the segments are parallel to the plane, e.g., $a = b = \infty$ and $c = d$, the perpendicular distance from P to the plane. Consequently, we have to show that

$$\frac{1}{a^2} + \frac{1}{b^2} + \frac{1}{c^2} = \frac{1}{d^2}.$$

Let A, B, C be the points, in the plane, at which the segments terminate. The volume of the tetrahedron $PABC$ is $\dfrac{abc}{6} = \dfrac{d\Delta}{3}$, where Δ is the area of the triangle ABC. The square of the area of any plane figure is equal to the sum of the squares of the projections of the area on three mutually perpendicular planes. Thus

$$\Delta^2 = \frac{1}{4}(a^2 b^2 + b^2 c^2 + c^2 a^2)$$

so that

$$a^2 b^2 c^2 = 4d^2 \Delta^2 = d^2(a^2 b^2 + b^2 c^2 + c^2 a^2)$$

and

$$\frac{1}{a^2} + \frac{1}{b^2} + \frac{1}{c^2} = \frac{1}{d^2}.$$

Second solution (by analytic geometry). With P as the origin and the three line segments as axes of coordinates, the equation of the plane is

$$\frac{x}{a} + \frac{y}{b} + \frac{z}{c} = 1.$$

Thus, the distance from P to the plane is given by

$$\frac{1}{\sqrt{a^{-2} + b^{-2} + c^{-2}}}.$$

Problem 117. The left inequality holds for any three positive numbers, since

$$(a + b + c)^2 - 3(bc + ca + ab)$$
$$= \frac{1}{2}[(a - b)^2 + (b - c)^2 + (c - a)^2] \geq 0.$$

To prove the second inequality, we use the conditions $c + a > b$, $a + b > c$, $b + c > a$ (satisfied by the side-lengths of a triangle) to deduce that

$$|a - b| < c, \quad |b - c| < a, \quad |c - a| < b,$$

and hence

$$4(bc + ca + ab) - (a + b + c)^2$$
$$= c^2 - (a - b)^2 + a^2 - (b - c)^2$$
$$\quad + b^2 - (c - a)^2$$
$$> 0.$$

An alternative approach to the right-hand inequality is to observe that $4(bc + ca + ab) - (a + b + c)^2$ corresponds to 16 times the square of the area of a triangle of sides $\sqrt{a}$, $\sqrt{b}$, $\sqrt{c}$. (Observe that if a, b, c are the sides of a triangle then so are $a^{1/n}$, $b^{1/n}$, $c^{1/n}$ whenever $n > 1$.)

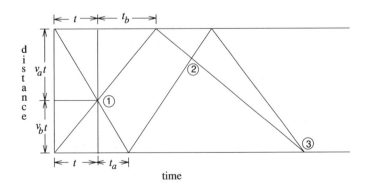

FIGURE 111

Problem 118. *First solution.* More generally, let v_a, v_b denote the speeds of Andy and Bob respectively, and let t_a, t_b denote the times it takes Andy and Bob to first reach towns B and A, respectively, after they first pass each other. The progress of the two drivers can be illustrated by the "world-line" diagram of Figure 111.

If t is the time elapsing between noon and their first passing, it follows from the diagram that

$$v_a = \frac{v_b t}{t_a}, \qquad v_b = \frac{v_a t}{t_b}$$

and hence

$$t = \sqrt{t_a t_b}, \qquad \frac{v_a}{v_b} = \sqrt{\frac{t_b}{t_a}}.$$

For the data of the problem, $t = 30$ minutes and $v_a = 60$ kph.

To find the subsequent times of meeting, just keep on extending the world lines of Andy and Bob as indicated. One then obtains a periodic pattern after five passings which is centrosymmetric about the third passing, which occurs at town B. It follows that the nth time of passing is $30 + 60(n - 1)$ minutes past noon.

Second solution. The conditions of the problem yield the equations

$$(t + 20)v_a = (t + 45)40 = (v_a + 40)t = 60d,$$

where d is the distance in kilometers between A and B, v_a is Andy's speed in kph, t is the time to first passing in minutes and the factor 60 relates the units of time. This gives $20v_a = 40t$ and $v_a t = 45 \cdot 40$, whence $v_a = 2t$ and $2t^2 = 45 \cdot 40$. Thus $t = 30$ minutes, $v_a = 60$ kph and $d = 50$ kilometers.

After the first meeting, the cars are separating at a relative velocity of 100 kph for 20 minutes until Andy arrives at B, and then approaching at a relative velocity of 20 kph for 25 minutes un-

til Bob arrives at A. During this time they have separated to a distance of $\frac{100}{3}$ kilometers, then approached to a distance of 25 kilometers. When Bob leaves A, he is approaching Andy travelling from B, with a relative velocity of 100 kph. Thus they will meet 15 minutes later. The total time between first and second meetings is 60 minutes.

A similar argument establishes that the third meeting takes place at B 60 minutes later, and, now by symmetry, subsequent meetings occur at 60 minute intervals. Thus the two pass for the nth time at $30 + 60(n - 1)$ minutes past noon.

Problem 119. Since $\angle BCG = \angle GCO = 60°$ and B, C, O, G are concyclic, it follows that $\angle BOG = \angle GBO = 60°$ and hence triangle BGO is equilateral. Let X be the center of the larger hexagon . A counterclockwise rotation of $60°$ about G maps B and C onto O and X respectively. Hence $\overline{BC} = \overline{OX}$, and

$$\overline{FO} = \overline{FC} + \overline{CO} = 2\overline{BC} + \overline{CO}$$
$$= 2\overline{OX} + \overline{CO} = \overline{OX} + \overline{CX} = \overline{OJ}.$$

Problem 120. *First solution.* The given condition is equivalent to

$$(1 + x)^n \equiv 1 + x^n \qquad (\text{mod } 2). \qquad (\star)$$

Now observe that

$$(1 + x)^2 \equiv 1 + x^2 \qquad (\text{mod } 2),$$
$$(1 + x)^4 \equiv (1 + x^2)^2 \equiv 1 + x^4 \qquad (\text{mod } 2),$$

and, using mathematical induction, we can prove that

$$(1 + x)^{2^k} \equiv 1 + x^{2^k} \qquad (\text{mod } 2),$$

i.e., $(\star)$ holds if n is a power of 2. If n is not a power of 2, then

$$n = 2^{k_1} + 2^{k_2} + \cdots$$

An elegant solution is generally considered to be one characterized by clarity, conciseness, logic and surprise.

Charles W. Trigg, *Mathematical Quickies* (Dover, 1985), p. vii

with at least two distinct k_i's. Then

$$(1+x)^n = (1+x)^{2^{k_1}}(1+x)^{2^{k_2}}\cdots$$

$$\equiv \left(1+x^{2^{k_1}}\right)\left(1+x^{2^{k_2}}\right)\cdots \pmod 2$$

and $(\star)$ is not satisfied. Hence $\binom{n}{1},\ldots,\binom{n}{n-1}$ are all even integers if and only if n is a power of 2.

Second solution. To motivate the approach, consider why $495 = \binom{12}{4}$ has to be odd. We can write

$$\binom{12}{4} = \frac{12\cdot 11\cdot 10\cdot 9}{4\cdot 3\cdot 2\cdot 1}.$$

Look at how often 2 occurs as a factor in both the numerator and the denominator. The highest power of 2 dividing 12 is 4, the same as the highest power dividing 4 in the denominator. Therefore, necessarily, 2 is the highest power of 2 dividing the next even number 10 or 2 in the two products respectively. Thus every factor 2 in the numerator will cancel out every factor 2 in the denominator. An extension of this argument can be applied to $n = 2^k p$, where p is odd, to show that $\binom{n}{2^k}$ is odd.

Consider

$$\binom{n}{2^k} = \frac{n(n-1)\cdots(n-2^k+1)}{2^k(2^k-1)\cdots 1}$$

$$= \prod_{i=0}^{2^k-1} \frac{n-i}{2^k-i}.$$

For each i with $0 \le i \le 2^k - 1$, write $i = 2^r q$ where q is odd and $r < k$. Then $2^k - i \equiv n - i \equiv -i$ modulo 2^r and modulo 2^{r+1}. Hence 2 divides $n-i$ and $2^k - i$ to exactly the same power. Thus $\binom{n}{2^k}$ is odd. If n is not a power of 2, then $p > 1$ and $1 \le 2^k \le n - 1$, so that not all $\binom{n}{j}$, $1 \le j \le n-1$, are even. If n is a power of 2, a similar argument shows that the numerator of $\binom{n}{j}$, $1 \le j \le n - 1$, is divisible by a higher power of 2 than the denominator.

Problem 121. $x - y$ is a divisor of $x^n - y^n$, $n = 1, 2, \ldots$ Since $2141 - 1863 = 1770 - 1492 = 278$, the given expression is divisible by 278. Similarly, $2141 - 1770 = 1863 - 1492 = 371$, which is relatively prime to 278, and also divides the given

expression. Hence $(278)(371) = (53)(1946)$ is a divisor.

Problem 122. Consider the polynomial $Q(x) = xP(x) - 1$. Using the properties of $P(x)$, we have that $Q(x)$ is a polynomial of degree $n + 1$ with zeros at $x = 1, 2, \ldots, n + 1$. Since also $Q(0) = -1$, it follows that

$$Q(x) = \frac{(x-1)(x-2)\ldots(x-n-1)}{(-1)^n(n+1)!}.$$

Hence $Q(n+2) = (-1)^n$ and so

$$P(n+2) = \frac{1+(-1)^n}{n+2} = \begin{cases} 0 & n \text{ odd,} \\ \frac{2}{n+2} & n \text{ even.} \end{cases}$$

Problem 123. Suppose that $\log x = \dfrac{f(x)}{g(x)}$ and let n be any positive integer. Then

$$\frac{f(x^n)}{g(x^n)} = \log x^n = n\log x = \frac{nf(x)}{g(x)}.$$

Hence $f(x^n)g(x) = nf(x)g(x^n)$ for each n.

If $f(x) = a_p x^p + \cdots + a_0$ and $g(x) = b_q x^q + \cdots + b_0$, then $f(x^n)g(x) = a_p b_q x^{np+q} + \cdots + a_0 b_0$ while $nf(x)g(x^n) = na_p b_q x^{p+nq} + \cdots + na_0 b_0$. Comparison of leading coefficients gives $a_p b_q = na_p b_q$ for each n, which is impossible.

Remark. More generally, it can be shown that $\log x$ is not algebraic, i.e., $y = \log x$ does not satisfy any equation of the form

$$P_n(x)y^n + P_{n-1}(x)y^{n-1} + \cdots + P_1(x)y$$

$$+ P_0(x) = 0$$

where each $P_i(x)$ is a polynomial. This can be deduced from the fact that $\lim\limits_{x\to\infty} \dfrac{\log x}{x} = 0$. Or, one may assume that the above is the equation of least degree; obtain a contradiction by replacing x by x^m and deriving an equation of degree one lower.

Problem 124. Let the time that the driver looked at his watch be m minutes past h ($0 \le h \le 11$). Since the minute hand makes 12 revolutions for 1 revolution of the hour hand, we have

$$5h + \frac{m}{60}5 = m$$

so that

$$m = \frac{60}{11}h.$$

But the journey took $14\frac{6}{11}$ minutes, and the train left the station at a whole number of minutes past the hour. Hence $m - \frac{6}{11}$ must be an integer, i.e.,

$\frac{60}{11}h - \frac{6}{11}$ is an integer or $60h - 6 \equiv 0 \pmod{11}$

or

$10h \equiv 1 \pmod{11}$ or $h = 10$ (since $0 \le h \le 11$).

Thus $m = 54\frac{6}{11}$, and the train left the station at 10:40 (A.M. or P.M.).

Problem 125. Suppose the sides have the given lengths a, b, c, d with AB parallel to CD, and $c > a$, as in Figure 112. If X is on DC with $\overline{DX} = a$, the triangle XBC has sides of length d, b and $c - a$. Thus, we construct first a triangle XBC with these sides, then produce CX to D so that $\overline{CD} = c$, then draw the line through B parallel to CD, and finally locate A so that $\overline{BA} = a$.

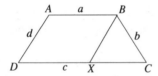

FIGURE 112

Problem 126. Assume that the sides of the tiles are of unit length. Observe that the number of them required to form an equilateral triangular array of sidelength x is

$$1 + 3 + \cdots + (2x - 1) = x^2.$$

The circumscribing triangle for any hexagonal array of the desired type must be an equilateral triangle, whose side-length is l, say (see Figure 113). The hexagon is formed by removing equilateral triangular corners of sidelengths a, b, c where $a \ge b \ge c > 0$ and $l > a + b$.

Hence n is possible if and only if

$$n = l^2 - a^2 - b^2 - c^2$$

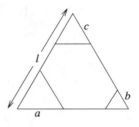

FIGURE 113

where $a \ge b \ge c > 0$ and $l > a + b$.

Rider. If $L = 2l - (a + b + c)$, $A = l - (b + c)$, $B = l - (a + c)$, $C = l - (a + b)$, deduce the identity $L^2 - A^2 - B^2 - C^2 = l^2 - a^2 - b^2 - c^2$.

Problem 127. Because the polynomial $P(x)$ has integral coefficients and a, b, c are distinct integers, it follows from the Factor Theorem that

$$\frac{P(a) - P(b)}{a - b}, \quad \frac{P(b) - P(c)}{b - c}, \quad \frac{P(c) - P(a)}{c - a}$$

are integers. Suppose that $P(a) = b$, $P(b) = c$ and $P(c) = a$. Then

$$\frac{b - c}{a - b}, \quad \frac{c - a}{b - c}, \quad \frac{a - b}{c - a}$$

are integers whose product is obviously 1. Hence the absolute value of each of them is 1, so, consequently,

$$|a - b| = |b - c| = |c - a|.$$

But this is impossible, because a, b, c are distinct.

By similar arguments, it can be shown that if $a_1, a_2, \ldots, a_n$ are distinct integers, then not all of the following can hold:

$$P(a_1) = a_2,$$
$$P(a_2) = a_3, \ldots, P(a_{n-1}) = a_n,$$
$$P(a_n) = a_1.$$

Problem 128. Since $a_1 = \sum_{i=1}^{n} r_i$ and $a_2 = \sum_{1 \le i < j \le n} r_i r_j$, it follows that

$$a_1^2 = \sum_{i=1}^{n} r_i^2 + 2a_2.$$

Thus, the given inequality is equivalent to

$$(n-1)\left\{\sum_{i=1}^{n} r_i^2 + 2a_2\right\} \geq 2na_2$$

or

$$(n-1)\sum_{i=1}^{n} r_i^2 \geq 2 \sum_{1\leq i<j\ leqn} r_i r_j$$

or

$$\sum_{1\leq i<j\leq n} (r_i - r_j)^2 \geq 0.$$

The last inequality is obvious, and shows that equality occurs exactly when $r_1 = r_2 = \cdots = r_n$.

Problem 129. For any positive value of n (not necessarily integral), the degenerate (or limiting) case $A = \pi$, $B = C = 0$ shows that

$$(k(n) + \sin 0) + (k(n) + \sin 0) \geq k(n) + \sin \frac{\pi}{n}$$

so

$$k(n) \geq \sin \frac{\pi}{n}.$$

When $n = 1$, we have in particular that $k(1) \geq 0$. On the other hand it is well known that (by the Law of Sines) $\sin A$, $\sin B$ and $\sin C$ are the sides of a triangle similar to $\triangle ABC$. Hence $k(1) = 0$.

Suppose $n \geq 2$. Since $\sin \theta$ is a positive increasing function of θ for $0 \leq \theta \leq \frac{\pi}{2}$ and since $\frac{A}{n} \leq \frac{\pi}{n} \leq \frac{\pi}{2}$ it follows that $\sin(\frac{A}{n}) \leq \sin(\frac{\pi}{n})$. Hence

$$\left(\sin \frac{\pi}{n} + \sin \frac{B}{n}\right) + \left(\sin \frac{\pi}{n} + \sin \frac{C}{n}\right)$$
$$\geq 2\sin \frac{\pi}{n} \geq \sin \frac{\pi}{n} + \sin \frac{A}{n}.$$

Similar inequalities hold for any permutation of A, B, C. Hence $k(n) = \sin \frac{\pi}{n}$.

Remark. This result can be extended to any real $n \geq 1$. If $u + \sin \frac{A}{n}$, $u + \sin \frac{B}{n}$, $u + \sin \frac{C}{n}$ are sides of a triangle, then the fact that the sum of the lengths of two sides exceeds the length of the third implies that

$$u \geq \sin \frac{A}{n} - \sin \frac{B}{n} - \sin \frac{C}{n}$$

with symmetric inequalities for A, B, C permuted in any order. $k(n)$ being the smallest such u, we

must have

$$k(n) = \max_{A+B+C=\pi} \left\{\sin \frac{A}{n} - \sin \frac{B}{n} - \sin \frac{C}{n}\right\}$$
$$= \max \left\{\sin \frac{A}{n} - 2\sin \frac{B+C}{2n} \cos \frac{B-C}{2n}\right\}.$$

For each fixed A, $B + C$ is fixed and the maximum is attained when

$$\cos \frac{B-C}{2n} = \cos \frac{(\pi - A) - 2B}{2n}$$

is minimum; this occurs either when $B = 0$ or $B = \pi - A$, i.e., $C = 0$. Then

$$k(n) = \max_{A+B=\pi} \left\{2\sin \frac{A-B}{2n} \cos \frac{A+B}{2n}\right\}$$
$$= 2\sin \frac{\pi}{2n} \cos \frac{\pi}{2n} = \sin \frac{\pi}{n}.$$

Similarly, it can be shown that

$$\min_{A+B+C=\pi} \left\{\sin \frac{A}{n} - \sin \frac{B}{n} - \sin \frac{C}{n}\right\}$$
$$= -2\sin \frac{\pi}{2n}.$$

The situation for $0 < n < 1$ is apparently not so easily analyzed.

Problem 130. Repeated use of the inequality $\left(\frac{a+b}{2}\right)^2 \geq ab$, for $a \geq 0$, $b \geq 0$, yields

$$\left(\frac{n+1}{2}\right)^2 \geq n \cdot 1$$

$$\left(\frac{n+1}{2}\right)^2 = \left(\frac{n-1+2}{2}\right)^2 \geq (n-1)\cdot 2$$

$$\left(\frac{n+1}{2}\right)^2 = \left(\frac{n-2+3}{2}\right)^2 \geq (n-2)\cdot 3$$

$$\vdots$$

$$\left(\frac{n+1}{2}\right)^2 = \left(\frac{2+n-1}{2}\right)^2 \geq 2\cdot(n-1)$$

$$\left(\frac{n+1}{2}\right)^2 = \left(\frac{1+n}{2}\right)^2 \geq 1\cdot n.$$

Multiplying all these inequalities shows that

$$\left(\frac{n+1}{2}\right)^{2n} \geq (n!)^2 \quad \text{or} \quad (n+1)^n \geq 2^n n!. \quad \text{(i)}$$

From this we deduce

$$6^n(n!)^2 \le 6^n \left\{ \left(\frac{n+1}{2} \right)^n \right\}^2$$

$$= 2^n \cdot \left(\frac{n+1}{2} \right)^n \cdot 3^n \left(\frac{n+1}{2} \right)^n$$

$$= (n+1)^n \left(\frac{3n+3}{2} \right)^n$$

$$\le (n+1)^n \left(\frac{4n+2}{2} \right)^n$$

$$= (n+1)^n (2n+1)^n. \qquad \text{(ii)}$$

This last inequality follows from the obvious $3n + 3 \le 4n + 2$ for $n = 1, 2, 3, \ldots$.

Equality occurs in both (i) and (ii) only if $n = 1$.

Inequality (i) can also be proved by induction. Assuming its truth for n, then, since $\left(1 + \frac{1}{n+1} \right)^{n+1} \ge 2$ (from the Binomial Theorem), we have

$$(n+2)^{n+1} \ge 2(n+1)^{n+1} \ge 2(n+1)2^n \cdot n!$$

$$= 2^{n+1}(n+1)!.$$

Problem 131. *First solution.* Let $a = z_1 + z_2 + z_3$, $b = z_1 z_2 + z_2 z_3 + z_3 z_1$, $c = z_1 z_2 z_3$ so that z_1, z_2, z_3 are the roots of the cubic $z^3 - az^2 + bz - c$. The given conditions imply that $c = 1$ and $a = b$. Thus the cubic reduces to $z^3 - az^2 + az - 1$, and this obviously has 1 as a root.

Second solution. Because of (1), (2) becomes

$$z_2 + z_3 + \frac{1}{z_2 z_3} = \frac{1}{z_2} + \frac{1}{z_3} + z_2 z_3,$$

so that $(1 - z_2)(1 - z_3) = \left(1 - \frac{1}{z_2} \right) \left(1 - \frac{1}{z_3} \right)$. Upon multiplying by $z_2 z_3$, we get $(z_2 z_3 - 1)(1 - z_2)(1 - z_3) = 0$. Thus $z_2 = 1$, $z_3 = 1$ or $z_2 z_3 = 1$, in which case $z_1 = 1$.

Problem 132. It suffices to prove that the median from each vertex is at least as long as the angle bisector from the same vertex. Accordingly, let A be a vertex of $\triangle ABC$, AM be the median and AT an angle bisector. Assume without loss of generality that $\overline{AB} \le \overline{AC}$.

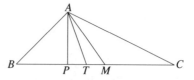

FIGURE 114

Since $\frac{\overline{BT}}{\overline{TC}} = \frac{\overline{AB}}{\overline{AC}}$, it follows that $\overline{BT} \le \overline{BM}$.

Consequently $\overline{AT} \le \overline{AM}$.

(To see this, note that if AP is an altitude then $\angle PAB \le \angle PAC$, so that $\angle PAB \le \angle TAB = \angle TAC \le \angle PAC$. Thus T lies between P and M.)

Problem 133. In the special case $n = 5$, the values of S corresponding to $r = 0, 1, \ldots, 5$ are, respectively, 1, -4, 6, -4, 1, 0. Consequently, we guess that $S = (-1)^r \binom{n-1}{r}$.

To prove this, note that for any function f,

$$\sum_{k=1}^{r} \{ f(k) - f(k-1) \} = f(r) - f(0).$$

Taking $f(k) = (-1)^k \binom{n-1}{k}$ yields

$$\sum_{k=1}^{r} \left\{ (-1)^k \binom{n-1}{k} - (-1)^{k-1} \binom{n-1}{k-1} \right\}$$

$$= (-1)^r \binom{n-1}{r} - 1.$$

> Discovery consists in seeing what everybody has seen and thinking what nobody has thought.
>
> Albert Szent-Györgyi

Thus

$$(-1)^r \binom{n-1}{r}$$

$$= 1 + \sum_{k=1}^{r} (-1)^k \left\{ \binom{n-1}{k} + \binom{n-1}{k-1} \right\}$$

$$= \sum_{k=0}^{r} (-1)^k \binom{n}{k}.$$

Problem 134. An easy manipulation shows that the given inequality is equivalent to

$$\left(\frac{x}{y} - \frac{y}{z} \right)^2 + \left(\frac{y}{z} - \frac{z}{x} \right)^2 + \left(\frac{z}{x} - \frac{x}{y} \right)^2 \geq 0.$$

This obviously holds; equality occurs only when $x = y = z$.

It also follows from

$$\frac{1}{3} \left(\frac{x^2}{y^2} + \frac{y^2}{z^2} + \frac{z^2}{x^2} \right) \geq \left\{ \frac{1}{3} \left(\frac{x}{y} + \frac{y}{z} + \frac{z}{x} \right) \right\}^2$$

$$\geq \frac{1}{3} \left(\frac{x}{y} + \frac{y}{z} + \frac{z}{x} \right), \quad (\star)$$

that

$$\frac{x^2}{y^2} + \frac{y^2}{z^2} + \frac{z^2}{x^2} \geq \max \left\{ \frac{x}{y} + \frac{y}{z} + \frac{z}{x}, \frac{y}{x} + \frac{z}{y} + \frac{x}{z} \right\}.$$

(Justify the steps in $(\star)$.)

Problem 135. *First solution.* If all the chords are concurrent, then, by symmetry, we would expect the common point to lie on the axis of the parabola, the x-axis. Thus we should check whether the x-intercept of the chords is independent of their endpoints.

Suppose that the endpoints of a chord subtending a right angle at the vertex $(0,0)$ are

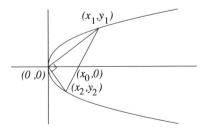

FIGURE 115

(x_1, y_1) and (x_2, y_2), and that its x-intercept is $(x_0, 0)$. Of course

$$x_1^2 + y_1^2 + x_2^2 + y_2^2 = (x_1 - x_2)^2 + (y_1 - y_2)^2,$$

and hence

$$x_1 x_2 + y_1 y_2 = 0. \qquad (1)$$

Because the endpoints lie on the parabola,

$$y_1^2 = 4ax_1, \qquad y_2^2 = 4ax_2. \qquad (2)$$

From (1) and (2) we have

$$(y_1 y_2)^2 = 16a^2 x_1 x_2 = -16a^2 y_1 y_2$$

and hence

$$4a(x_2 y_1 - x_1 y_2) = y_1 y_2 (y_2 - y_1)$$
$$= -16a^2 (y_2 - y_1).$$

From $\dfrac{y_1}{y_1 - y_2} = \dfrac{x_1 - x_0}{x_1 - x_2}$, we obtain

$$x_0 = \frac{x_2 y_1 - x_1 y_2}{y_1 - y_2} = \frac{4a(y_1 - y_2)}{y_1 - y_2} = 4a.$$

Hence all chords contain the point $(4a, 0)$.

Second solution. Let (x_1, y_1) and (x_2, y_2) be the endpoint of the chord, as above. Since the slopes of the lines from the origin to these points are negative reciprocals,

$$x_1 x_2 + y_1 y_2 = 0. \qquad (3)$$

If the equation of the chord is $y = mx + b$ then x_1 and x_2 are the roots of the quadratic equation $(mx+b)^2 = 4ax$, i.e., of $m^2 x^2 + (2bm - 4a)x + b^2 = 0$. Thus

$$x_1 x_2 = \frac{b^2}{m^2},$$

$$y_1 y_2 = (mx_1 + b)(mx_2 + b)$$
$$= m^2 x_1 x_2 + mb(x_1 + x_2) + b^2 = \frac{4ab}{m}.$$

Hence, from (3), $\dfrac{b^2}{m^2} + \dfrac{4ab}{m} = 0$, or $b = -4am$. Thus every chord which subtends a right angle at the vertex has an equation of the form $y = mx - 4am = m(x - 4a)$, and this line clearly passes through the point $(4a, 0)$.

Problem 136. Let a, m, b be the lengths of the altitudes to DC of the triangles ADN, MDC and BNC. Suppose that $\overline{AM} = r\overline{AB}$. Then $\overline{NC} = r\overline{DC}$ and $m = (1-r)a + rb$. (If $AB \parallel DC$, the latter equation is obvious. Otherwise, it may be obtained by considering three similar right-angled triangles each with hypotenuse along AB and an adjacent side along DC.) Then (with $[ABC]$ denoting the area of triangle ABC)

$$[ADN] = \frac{1}{2}a \cdot \overline{DN} = \frac{1}{2}a(1-r) \cdot \overline{DC}$$

$$[MDC] = \frac{1}{2}m \cdot \overline{DC}$$

$$[BNC] = \frac{1}{2}b \cdot \overline{NC} = \frac{1}{2}br \cdot \overline{DC}.$$

Hence $[MDC] = [ADN] + [BNC]$. Subtracting the sum of the areas of triangles PDN and QNC from both sides yields the desired result.

Problem 137. Choose any four of the points. In the given situation, there is a uniquely determined sphere containing them; let its center be C and its radius be r. If d is the distance between C and the fifth point, then the sphere with center C and radius $\frac{1}{2}(r + d)$ (if, say, the fifth point is outside of the sphere) is equidistant from the five points. Thus there are at least five spheres with the desired property.

Depending on the configuration of the points, there may be finitely or infinitely many other solutions. The locus of the center of a sphere containing three of the points is a line; the locus of the center of a sphere containing the other two is a plane. This line may be parallel to the plane, intersect the plane in one point or lie in the plane. If O is a point of intersection of line and plane, then, in a manner similar to that described in the first paragraph, a sphere with center O equidistant from the five points (having three of them on one side and two on the other) can be found.

Problem 138. We have

$$5^{2n+1} + 11^{2n+1} + 17^{2n+1}$$
$$\equiv 2^{2n+1} + 2^{2n+1} + 2^{2n+1}$$
$$\equiv 0 \pmod 3$$

and

$$5^{2n+1} + 11^{2n+1} + 17^{2n+1}$$
$$= (11 - 6)^{2n+1} + 11^{2n+1} + (11 + 6)^{2n+1}$$
$$\equiv (-6)^{2n+1} + 6^{2n+1}$$
$$\equiv 6^{2n+1}(-1 + 1)$$
$$\equiv 0 \pmod{11}.$$

Thus, the given sum is divisible by the product of 3 and 11.

Problem 139. *First solution.* Setting $x = 1$ in $(x + 1)^n = \sum_{r=0}^{n} \binom{n}{r}x^r$ yields

$$2^n = \sum_{r=0}^{n} \binom{n}{r}.$$

Let

$$Q(x) = \binom{x}{0} + \binom{x}{1} + \cdots + \binom{x}{n}.$$

Since, for $0 \le r \le n$,

$$\binom{x}{r} = \frac{x(x-1)\cdots(x-r+1)}{r!},$$

$Q(x)$ is a polynomial of degree n. Furthermore $P(x) = Q(x)$ for $x = 0, 1, 2, \ldots, n$. Hence the polynomials $P(x)$ and $Q(x)$ must be identical, and

$$P(n + 1)$$
$$= Q(n + 1)$$
$$= \binom{n+1}{0} + \binom{n+1}{1} + \cdots + \binom{n+1}{n}$$
$$= 2^{n+1} - \binom{n+1}{n+1} = 2^{n+1} - 1.$$

A high school student wrote (in an essay on famous musicians): *Bach was the most famous composer in the world, and so was Handel. Handel was half German, half Italian and half English.*

Second solution. For each fixed n, the polynomial is uniquely determined by its values at $0, 1, \ldots, n$. Write P_n for the polynomial, to show its dependence on n. For $n \geq 1$ and $x = 0, 1, 2, \ldots, n-1$, we have

$$P_{n-1}(x) = P_n(x+1) - P_n(x). \qquad (1)$$

Since both sides of (1) are of degree $n-1$, the equation (1) must hold for all values of x. In particular, for $n \geq 1$,

$$P_{n-1}(n) = P_n(n+1) - P_n(n).$$

Hence

$$\begin{aligned}
P_n(n+1) &= P_{n-1}(n) + P_n(n) \\
&= P_{n-1}(n) + 2^n \\
&= P_{n-2}(n-1) + 2^{n-1} + 2^n \\
&\;\;\vdots \\
&= P_0(1) + 2 + 2^2 + \cdots + 2^n \\
&= 1 + 2 + 2^2 + \cdots + 2^n \\
&= 2^{n+1} - 1.
\end{aligned}$$

Rider. Redo the problem with 2^k replaced by 3^k.

Problem 140 (by induction). The result is clearly true for $n = 1$. For $n = 2$, we have to prove $2(1 + x_1 x_2) \geq (1 + x_1)(1 + x_2)$.

This is equivalent to $(1 - x_1)(1 - x_2) \geq 0$, which is valid (with equality if and only if either of x_1 and x_2 equals 1).

Suppose the result holds for all values of n up to $k \geq 2$, with equality occurring under the stated condition. Then, given $0 \leq x_i \leq 1$, $i = 1, 2, \ldots, k+1$,

$$\begin{aligned}
2^k(1 &+ x_1 x_2 \cdots x_k x_{k+1}) \\
&= 2^{k-1}\{2(1 + \overbrace{x_1 x_2 \cdots x_k}\, x_{k+1})\} \\
&\geq 2^{k-1}(1 + x_1 x_2 \cdots x_k)(1 + x_{k+1}) \\
&\geq (1 + x_1)(1 + x_2) \cdots (1 + x_k)(1 + x_{k+1}),
\end{aligned}$$

using the result for $n = 2$ and $n = k$. If equality occurs, at least $k - 1$ of the quantities $x_1, x_2, \ldots, x_k$ are 1. If only $k - 1$ of these quantities are 1, then x_{k+1} must equal 1 as well.

Problem 141. If you did you would lose more often than you won. The number of combinations of three cards possible is $\frac{52 \times 51 \times 50}{1 \times 2 \times 3} = 22100$. The number of combinations of spot cards is only $\frac{40 \times 39 \times 38}{1 \times 2 \times 3} = 9880$. Therefore the chance of picking 3 spot cards is $\frac{9880}{22100} = \frac{38}{85}$.

Problem 142. Multiplying the four equations yields

$$xyzw = abcd\lambda,$$

where λ is a root of the polynomial $t^7 - 1$. Hence

$$x = \frac{(abcd\lambda)^2}{a^7}, \quad y = \frac{(abcd\lambda)^2}{b^7},$$

$$z = \frac{(abcd\lambda)^2}{c^7}, \quad w = \frac{(abcd\lambda)^2}{d^7}.$$

Problem 143. Let c be the longest chord of the solid. Every cross-section of the solid which contains c must be a circle with diameter c. Hence the solid is a solid of revolution about c, and thus it is a sphere.

Problem 144. The number of ways of stacking the coins without restriction is $n!$. The number of ways of stacking the coins when the particular two are kept together is $2 \cdot (n-1)!$. (The factor 2 arises since the order of the two coins can be reversed.) Hence the number of permutations in which the two coins are kept apart is

$$n! - 2(n-1)! = (n-2)(n-1)!$$

If we wish to distinguish stackings obtained by turning over some of the coins, the count would be $2^n(n-2)(n-1)!$.

Problem 145. To get some understanding of the situation, consider the degenerate case when the variable circle is a straight line. This line would be a common tangent to the circles, having both circles either on the same side or on opposite sides. Thus, one might expect the fixed points to show up as intersections of common tangents. In addition, by symmetry considerations, one might expect the points to lie on the line of centers of the two circles.

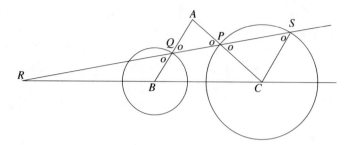

FIGURE 116

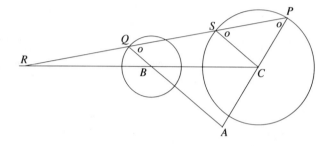

FIGURE 117

Case 1a: Both circles are touched externally.

Let B, C be the centers of the two fixed disjoint circles (as in Figure 116), and let the circle of center A touch these circles externally at Q and P, respectively. Then AQB and APC are each collinear and $\overline{AQ} = \overline{AP}$.

Let QP meet BC in R and the circle with center C again in S. Then

$$\angle BQR = \angle AQP = \angle APQ$$

$$= \angle CPS = \angle CSP,$$

so that triangles RQB and RSC are similar, and $\dfrac{\overline{RB}}{\overline{RC}} = \dfrac{\overline{QB}}{\overline{SC}}$, a constant ratio. Hence R is independent of the touching circle.

Case 1b: Both circles are touched internally.

With A, B, C as centers of the three circles (as in Figure 117), we again have $\overline{AQ} = \overline{AP}$ and triangles RQB, RSC are similar, so that $\dfrac{\overline{RB}}{\overline{RC}}$ is a constant ratio.

Case 2: One circle is touched internally, one externally.

Let A, B, C be the centers of the three circles, such that the variable circle of center A touches the circle of center B externally and the circle of center C internally (as in Figure 118). Since $\overline{AP} = \overline{AQ}$,

$$\angle APQ = \angle AQP = \angle BQS = \angle BSQ,$$

and triangles RBS, RCP are similar. Hence $\dfrac{\overline{RB}}{\overline{RC}} = \dfrac{\overline{BS}}{\overline{CP}}$, a constant ratio. Thus R is fixed.

Riders. (a) Verify that the point R in Cases 1a and 1b is the intersection of the common tangents having both circles on the same side, while in Case 2, R is the intersection of the other two common tangents.

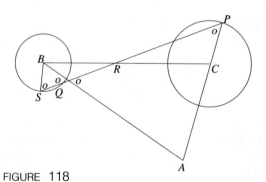

FIGURE 118

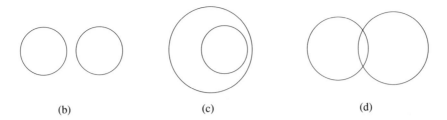

FIGURE 119

(b) What happens if the two circles are congruent?

(c) What happens if the two circles are nested?

(d) What happens if the two circles are intersecting?

Remark. The same result can be obtained from the converse of Menelaus' Theorem. Referring to Figure 116 (Case 1a), we have: if three points R, Q, P are taken in the sides CB, BA, AC respectively of a triangle such that

$$\frac{\overline{AQ}}{\overline{QB}} \cdot \frac{\overline{BR}}{\overline{RC}} \cdot \frac{\overline{CP}}{\overline{PA}} = -1$$

then R, P, Q are collinear. Here

$$\frac{r}{r_1} \cdot \frac{\overline{BR}}{\overline{RC}} \cdot \frac{r_2}{r} = -1,$$

where r, r_1, r_2 are the radii of circles with centers A, B, C respectively. Thus, if R is taken on BC such that $r_2\overline{BR} = -r_1\overline{RC}$, then R is the fixed point. The other cases can be similarly analyzed.

Problem 146. *First solution.* By the Cauchy Inequality (Tool Chest, D. 4),

$$n^2 \le \left(\sum_{i=1}^{n} \frac{S}{S-x_i} \right) \left(\sum_{i=1}^{n} \frac{S-x_i}{S} \right)$$

$$= \left(\sum_{i=1}^{n} \frac{S}{S-x_i} \right) \left(\sum_{i=1}^{n} \left(1 - \frac{x_i}{S} \right) \right)$$

$$= \left(\sum_{i=1}^{n} \frac{S}{S-x_i} \right) \left(n - \frac{\sum x_i}{S} \right)$$

$$= \left(\sum_{i=1}^{n} \frac{S}{S-x_i} \right) (n-1)$$

with equality if and only if $\sqrt{\dfrac{S}{S-x_i}}$ is a constant multiple of $\sqrt{\dfrac{S-x_i}{S}}$, i.e., when $\dfrac{S}{S-x_i}$ is constant.

Second solution. When $a_1, a_2, \ldots, a_n$ are positive, the Harmonic-Arithmetic Means Inequality asserts that

$$\frac{n}{a_1^{-1} + a_2^{-1} + \cdots + a_n^{-1}}$$

$$= \left(\frac{a_1^{-1} + a_2^{-1} + \cdots + a_n^{-1}}{n} \right)^{-1}$$

$$\le \frac{a_1 + a_2 + \cdots + a_n}{n},$$

with equality if and only if all the a_i are equal. Set $a_i = \dfrac{S}{S-x_i}$ so that $\sum_{i=1}^{n} a_i = n-1$. Then the result is immediate.

Problem 147. Let $P(a,b,c)$ denote the expression. Since $P(a,a,c) = 0$, $a-b$ is a factor. Similarly, $b-c$ and $c-a$ are factors. Dividing through, we obtain

$$P(a,b,c)$$
$$= (b-c)(c-a)(a-b)\times$$
$$\left\{ abc + (a^3 + b^3 + c^3) \right.$$
$$\left. + (a^2b + a^2c + b^2a + b^2c + c^2a + c^2b) \right\}$$

or, more briefly,

$$P(a,b,c) = (b-c)(c-a)(a-b)\times$$
$$\left\{ abc + \sum_{\text{symm}} a^3 + \sum_{\text{symm}} a^2b \right\}.$$

	Score	5	4	3	2	1	0	Number of ways 7 problems can be scored
Number of		6	0	0	0	0	1	$7 = 7$
problems		5	1	0	0	1	0	$7 \cdot 6 = 42$
given score		5	0	1	1	0	0	$7 \cdot 6 = 42$
		4	2	0	1	0	0	$\binom{7}{4}\binom{3}{2} = 105$
		4	1	2	0	0	0	$\binom{7}{4}\binom{3}{1} = 105$
		3	3	1	0	0	0	$\binom{7}{3}\binom{4}{3} = 140$
		2	5	0	0	0	0	$\binom{7}{2} = 21$

Total number of possibilities: 462

TABLE 2

Problem 148. Table 2 lists all the possibilities.

Problem 149. A generalization is

$$(\underbrace{55\ldots5}_{n}6)^2 - (\underbrace{44\ldots4}_{n}5)^2 = \underbrace{11\ldots1}_{2n+2}$$

$(n = 0, 1, 2, \ldots)$. Indeed, since $a^2 - b^2 = (a - b)(a + b)$, the left side becomes

$$(\underbrace{11\ldots1}_{n+1})(1\underbrace{00\ldots0}_{n}1)$$

which, upon multiplication, yields the right side.

Problem 150. The given equations are equivalent to the following:

$$(y - a)(z - a) = a^2 + r$$
$$(z - a)(x - a) = a^2 + s$$
$$(x - a)(y - a) = a^2 + t$$

from which

$$(x - a)(y - a)(z - a)$$
$$= \pm\sqrt{(a^2 + r)(a^2 + s)(a^2 + t)}.$$

Hence

$$x - a = \frac{(x - a)(y - a)(z - a)}{a^2 + r}$$
$$= \pm\sqrt{\frac{(a^2 + s)(a^2 + t)}{a^2 + r}}$$

from which x can be found. Similarly, y and z can be found. (The signs before the radical must be chosen to be compatible with the given equation.)

Problem 151. Let $a = \overline{AB} = \overline{BC} = \overline{CA}$; then the area of $\triangle ABC$ is $\left(\frac{\sqrt{3}}{4}\right)a^2$. This area is also the sum of the areas of triangles PAB, PBC, and

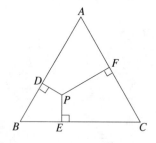

FIGURE 120

PCA. Hence

$$\frac{\sqrt{3}a^2}{4} = a\frac{\overline{PD}}{2} + a\frac{\overline{PE}}{2} + a\frac{\overline{PF}}{2}$$

$$= \frac{a(\overline{PD} + \overline{PE} + \overline{PF})}{2},$$

and the desired equality follows.

Rider. Show, more generally, that for any triangle and any interior point P,

$$\min\,(h_1, h_2, h_3) \le r_1 + r_2 + r_3$$

$$\le \max\,(h_1, h_2, h_3),$$

where h_1, h_2, h_3 are the altitudes to, and r_1, r_2, r_3 are the distances from P to, the three sides whose lengths are a, b, c, respectively.

Problem 152. Cubing both sides yields

$$(13x + 37) - 3\sqrt[3]{13x + 37} \cdot \sqrt[3]{13x - 37}\times$$

$$(\sqrt[3]{13x + 37} - \sqrt[3]{13x - 37}) - (13x - 37)$$

$$= 2.$$

Hence $24 = \sqrt[3]{2}\sqrt[3]{13^2x^2 - 37^2}$, and cubing again yields $4 \cdot 12^3 = 13^2x^2 - 37^2$, so that $x = \pm 7$. Both values satisfy the given equation.

Problem 153. Since $A = 10^{a+1} - 10^a = 9 \cdot 10^a$ and $B = 10^b - 10^{b-1} = 9 \cdot 10^{b-1}$, $\log A - \log B = a - (b-1)$. Hence $(\log A - a) - (\log B - b) = 1$.

Problem 154. Let $h = r\cos\theta$, $k = r\sin\theta$, $x - h = 2ru$, $y - k = 2rv$. Then the two equations become

$$u^2 + v^2 = 1,$$

$$(2u + \cos\theta)(2v + \sin\theta) = \sin\theta\cos\theta.$$

Let $z = u + iv$, so that $\bar{z} = u - iv$, $2u = z + \bar{z}$, $2v = \dfrac{z - \bar{z}}{i}$,

$$z\bar{z} = 1 \qquad (1)$$

and (with $\operatorname{cis}\theta = \cos\theta + i\sin\theta$)

$$z(z + \operatorname{cis}\theta) = \bar{z}(\bar{z} + \operatorname{cis}(-\theta)). \qquad (2)$$

Multiplying (2) by z^2 and using (1) yields

$$z^3(z + \operatorname{cis}\theta) = 1 + z\operatorname{cis}(-\theta),$$

which can be manipulated to give

$$(z^3 - \operatorname{cis}(-\theta))(z + \operatorname{cis}\theta) = 0.$$

The root $z = -\operatorname{cis}\theta$ corresponds to $x = h + 2ru = h - 2r\cos\theta = -h$ and $y = k + 2rv = k - 2r\sin\theta = -k$, the coordinates of the point $(-h, -k)$. The other factor, $z^3 - \operatorname{cis}(-\theta)$, has three roots:

$$z = \operatorname{cis}\frac{-\theta}{3}, \qquad \text{i.e., } u = \cos\frac{\theta}{3}, \quad v = -\sin\frac{\theta}{3};$$

$$z = \operatorname{cis}\frac{-\theta + 2\pi}{3},$$

$$\text{i.e., } u = \cos\frac{\theta - 2\pi}{3}, \quad v = -\sin\frac{\theta - 2\pi}{3};$$

$$z = \operatorname{cis}\frac{-\theta - 2\pi}{3},$$

$$\text{i.e., } u = \cos\frac{\theta + 2\pi}{3}, \quad v = -\sin\frac{\theta + 2\pi}{3}.$$

These roots give the points of intersection

$$\left(h + 2r\cos\frac{\theta}{3},\ k - 2r\sin\frac{\theta}{3}\right),$$

$$\left(h + 2r\cos\frac{\theta - 2\pi}{3},\ k - 2r\sin\frac{\theta - 2\pi}{3}\right),$$

$$\left(h + 2r\cos\frac{\theta + 2\pi}{3},\ k - 2r\sin\frac{\theta + 2\pi}{3}\right),$$

which are the vertices of an equilateral triangle.

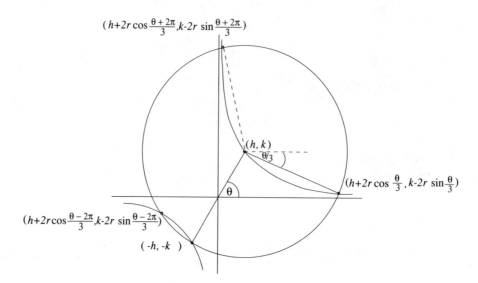

FIGURE 121

For a geometrical interpretation, take any point (h, k) on an equilateral hyperbola $xy = a^2$. Construct a circle with center (h, k) and passing through $(-h, -k)$. This circle will intersect the hyperbola in three other points which are the vertices of an equilateral triangle.

Problem 155. Let

$$I_m = (\sqrt{3} + 1)^{2m} + (\sqrt{3} - 1)^{2m}.$$

If we expand I_m we find that all surd terms cancel, so that I_m is an integer. Since $(\sqrt{3} - 1) < 1$, $(\sqrt{3} - 1)^{2m} < 1$. Hence I_m must be the smallest integer exceeding $(\sqrt{3} + 1)^{2m}$.

Note that $I_m = (4 + 2\sqrt{3})^m + (4 - 2\sqrt{3})^m$; also $4 + 2\sqrt{3}$ and $4 - 2\sqrt{3}$ are the roots of the quadratic equation $x^2 = 8x - 4$. Hence $4 + 2\sqrt{3}$ and $4 - 2\sqrt{3}$ satisfy all the equations

$$x^{m+2} = 8x^{m+1} - 4x^m \quad m = 0, 1, \dots,$$

and we have

$$I_{m+2} = 8I_{m+1} - 4I_m.$$

Since $I_1 = 8$ and $I_2 = 56$, 2^{m+1} divides I_m for $m = 1, 2$. Suppose 2^{m+1} divides I_m for $m = 1, 2, \dots, k$. Then $4I_{k-1}$ and $8I_k$ are each multiples of 2^{k+2}. Hence 2^{k+2} divides I_{k+1}. The result follows by induction.

Problem 156. We have that

$$\left| \frac{r+2}{r+1} - \sqrt{2} \right| = \left| \frac{(r+2) - \sqrt{2}(r+1)}{r+1} \right|$$

$$= \frac{(\sqrt{2} - 1)|r - \sqrt{2}|}{1 + r}$$

$$\leq (\sqrt{2} - 1)|r - \sqrt{2}| < |r - \sqrt{2}|.$$

Rider. If r denotes a nonnegative rational approximation to (a) $\sqrt{3}$, (b) $\sqrt[3]{2}$, find an always-better rational approximation.

Problem 157. Since $\cos 0 = 1$, $\cos \frac{\pi}{3} = \frac{1}{2}$ and $\cos \frac{\pi}{2} = 0$, three values of k for which $\cos k\pi$ is rational are $0, \frac{1}{3}$ and $\frac{1}{2}$. Suppose $\cos \theta = \frac{p}{q}$ with $q \geq 3$, $p < q$ and p and q coprime. Then

$$\cos 2\theta = 2\cos^2 \theta - 1 = \frac{2p^2 - q^2}{q^2}.$$

Any common divisor of numerator and denominator of this latter fraction must divide both $2p^2$ and q^2; since the only common divisor of p and q is 1, the greatest common divisor of $2p^2 - q^2$ and q^2 is either 1 or 2. Thus, when $\frac{2p^2 - q^2}{q^2}$ is written in its lowest terms, the denominator is at least $\frac{q^2}{2} > q$, and hence $\cos 2\theta \neq \cos \theta$.

Continuing on, we prove that if $\cos \theta$ is rational (in lowest terms, with denominator exceeding

2), $\cos\theta$, $\cos 2\theta$, $\cos 4\theta$, $\cos 8\theta$, $\cos 16\theta$, ... are all rational and the denominators form an increasing sequence. Hence $\cos\theta$, $\cos 2\theta$, $\cos 4\theta$, $\cos 8\theta$, ... are all distinct.

Suppose $\theta = 2^i(\frac{u}{v})\pi$ where u and v are odd and coprime. Since v is odd, there is a positive integer $w > |i|$ such that $2^w - 1$ is divisible by v. (Tool Chest, B 8.) Hence

$$2^{2w-i+1}\theta - 2^{w-i+1}\theta = \frac{2^w - 1}{v}2^w u(2\pi)$$

is an integral multiple of 2π, so that $\cos(2^{2w-i+1}\theta) = \cos(2^{w-i+1}\theta)$. By what was shown earlier, $\cos\theta$ cannot be rational with denominator exceeding 2 when $\theta = 2^i(\frac{u}{v})\pi$.

Hence $\cos k\pi$ is rational only if $k = 0, \frac{1}{3}, \frac{1}{2}$.

Problem 158. We form a cubic polynomial whose roots are x, y, z and then find its roots by an alternative method; this approach is suggested by the symmetric role played by the three variables in the given system and by the fact that the coefficients of any polynomial are symmetric functions of the roots.

The cubic polynomial with roots x, y, z is

$$(t - x)(t - y)(t - z) = t^3 - pt^2 + qt - r,$$

where $p = x + y + z$, $q = xy + yz + zx$, $r = xyz$. Since $p = 0$, $6ab = x^2 + y^2 + z^2 = (x + y + z)^2 - 2q = -2q$ and (using the fact that x, y, z are each roots of the cubic),

$$(x^3 + y^3 + z^3) - p(x^2 + y^2 + z^2)$$
$$+ q(x + y + z) - 3r = 0,$$

we have that $q = -3ab$ and $r = a^3 + b^3$.

Thus, x, y, z are the roots of the cubic

$$t^3 - 3abt - (a^3 + b^3).$$

By inspection we find that one root is $a+b$. Hence the cubic is divisible by $t - (a + b)$:

$$t^3 - 3abt - (a^3 + b^3)$$
$$= [t - (a + b)][t^2 + (a + b)t + (a^2 - ab + b^2)].$$

Hence the remaining roots are $\frac{1}{2}\{-(a + b) \pm i\sqrt{3}(a - b)\}$.

Problem 159. Let $ABCD$ be the tetrahedron, and let P be the foot of the perpendicular dropped from A to the face BCD. Then, comparing areas and using $[ABC]$ to denote the area of triangle ABC,

$$[ABC] > [PBC],$$
$$[ACD] > [PCD],$$
$$[ADB] > [PDB].$$

If P falls within triangle BCD,

$$[BCD] = [PBC] + [PCD] + [PDB].$$

On the other hand, if P falls outside of BCD, then

$$[BCD] < [PBC] + [PCD] + [PDB].$$

In either case,

$$[ABC] + [ACD] + [ADB] > [BCD].$$

The other choices of three faces can be handled similarly.

Alternatively, the result is equivalent to showing that

$$|\overrightarrow{BA} \times \overrightarrow{CA}| + |\overrightarrow{CA} \times \overrightarrow{DA}| + |\overrightarrow{DA} \times \overrightarrow{BA}|$$
$$\geq |\overrightarrow{BA} \times \overrightarrow{CA} + \overrightarrow{CA} \times \overrightarrow{DA} + \overrightarrow{DA} \times \overrightarrow{BA}|,$$

and this inequality follows from the triangle inequality.

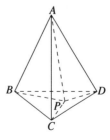

FIGURE 122

Problem 160. The desired inequality is equivalent to

$$a + b \leq \sqrt{2}\sqrt{a^2 + b^2},$$

which is equivalent to

$$a^2 + 2ab + b^2 \le 2(a^2 + b^2)$$

or $(a - b)^2 \ge 0$.

Remark. More generally, for any triangle with sides a, b, c, $a + b \le c \csc \frac{C}{2}$, where C is the angle opposite side c, with equality if and only if $a = b$.

Problem 161. Let $0 \le \theta \le \frac{\pi}{2}$. Then $0 \le \sin^5 \theta \le \sin^2 \theta$, $0 \le \cos^5 \theta \le \cos^2 \theta$, so that $\sin^5 \theta + \cos^5 \theta \le \sin^2 \theta + \cos^2 \theta = 1$, with equality if and only if θ is equal to 0 or $\frac{\pi}{2}$.

Problem 162. If a is the first term and d the common difference of the arithmetic progression then $p = a + (q - 1)d$ and $q = a + (p - 1)d$; hence, $d = -1$ and $a = p + q - 1$. Therefore the nth term is

$$(p + q - 1) - (n - 1) = p + q - n,$$

so that the $p + q$th term is 0.

Problem 163. *First solution.* With $[ABC]$ denoting the area of triangle ABC and $x = \overline{CD}$,

$$[BCD] = \frac{1}{2} ax \sin \frac{C}{2},$$
$$[ACD] = \frac{1}{2} bx \sin \frac{C}{2},$$
$$[ABC] = \frac{1}{2} ab \sin C = ab \sin \frac{C}{2} \cos \frac{C}{2}.$$

Hence

$$\frac{1}{2} ax \sin \frac{C}{2} + \frac{1}{2} bx \sin \frac{C}{2} = ab \sin \frac{C}{2} \cos \frac{C}{2},$$

and the result follows.

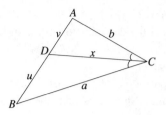

FIGURE 123

Second solution. Using the fact that $\frac{b}{a} = \frac{v}{u}$, we obtain

$$\frac{b^2}{a^2} = \frac{v^2}{u^2} = \frac{b^2 + x^2 - 2bx \cos \frac{C}{2}}{a^2 + x^2 - 2ax \cos \frac{C}{2}}.$$

If $a \ne b$, cross multiplying and cancelling $a^2 b^2$ leads to $(b^2 - a^2)x^2 = 2ab(b - a)x \cos \frac{C}{2}$, which yields the result upon division by $(b - a)x$. If $a = b$, then it can be seen directly from the diagram for an isosceles triangle that

$$x = a \cos \frac{C}{2} = \frac{2ab \cos \frac{C}{2}}{a + b}.$$

Third solution. Through B draw the line parallel to DC, and let it meet AC produced in a point E. Then

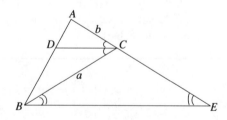

FIGURE 124

Setting the questions is not a simple matter. They must not be too easy but also not so difficult as to induce a feeling of frustration. It is sometimes rather hard to decide whether a question is easy. It may appear simple enough when the solution is known.

L. J. Mordell, "The Putnam Competition," *Amer. Math. Monthly* 70 (1963) p. 483.

$$\overline{CE} = \overline{CB} = a, \qquad \overline{BE} = 2a\cos\frac{C}{2}$$

and $\qquad \dfrac{\overline{CD}}{\overline{BE}} = \dfrac{\overline{AC}}{\overline{AE}}.$

Thus

$$\frac{\overline{CD}}{2a\cos\frac{C}{2}} = \frac{b}{a+b}.$$

Problem 164. $(1367631)_b = 1 + 3b + 6b^2 + 7b^3 + 6b^4 + 3b^5 + b^6 = (1 + b + b^2)^3$. Thus the given number is perfect cube in all bases greater than 7.

Problem 165. The inequality is equivalent to

$$1 + x + x^2 + \cdots + x^{2n} \geq (2n+1)x^n$$

which is the same as

$$(x^n + x^{-n} - 2) + (x^{n-1} + x^{1-n} - 2)$$
$$+ \cdots + (x + x^{-1} - 2) \geq 0.$$

The latter inequality holds since

$$t + \frac{1}{t} - 2 = \left(\sqrt{t} - \frac{1}{\sqrt{t}}\right)^2 \geq 0 \quad \text{for } t > 0.$$

Rider. Show that the given inequality is valid for all $n \geq 0$ and $x \geq 0$, $x \neq 1$.

Problem 166. *First solution.* If a is the initial term and d is the common difference, $q = a + (p-1)d$, $r = a + (q-1)d$, $p = a + (r-1)d$. Elimimation of a and d yields

$$q(q-r) + r(r-p) + p(p-q) = 0.$$

Hence $(p-q)^2 + (q-r)^2 + (r-p)^2 = 0$ so that $p = q = r$ and the required difference is zero.

Second solution. An arithmetic progression, unless constant, is either continually increasing or continually decreasing. In the former case, if $p \leq q \leq r$, then $q \leq r \leq p$, so $p = q = r$. The remaining possibilities also lead to $p = q = r$, so that inevitably, the required difference is 0.

Problem 167. That the required construction exists can be seen from the following argument. In

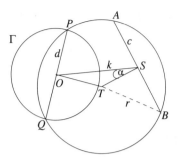

FIGURE 125

Figure 125 the circle to be constructed is $PQBA$; PQ is a diameter and O is the center of the given circle Γ. The center T of the required circle is at the intersection of the right bisectors OT and ST of PQ and AB respectively.

Let $\overline{OP} = d$, $\overline{AS} = c$, $\overline{OS} = k$, $\angle OST = \alpha$ and let r be the radius $\overline{AT} = \overline{BT} = \overline{PT} = \overline{QT}$ of the required circle. Then c, d, k, α are known while r is unknown. Since $\overline{ST} = \sqrt{r^2 - c^2}$, $\overline{OT} = \sqrt{r^2 - d^2}$, the Law of Cosines applied to $\triangle OST$ yields

$$r^2 - d^2 = r^2 - c^2 + k^2 - 2k\sqrt{r^2 - c^2}\cos\alpha,$$

or

$$\sqrt{r^2 - c^2} = \frac{k^2 + d^2 - c^2}{2k\cos\alpha}.$$

A rectangle whose area is $k^2 + d^2 - c^2$ can be constructed with one side having length $2k\cos\alpha$. An adjacent side has length $\sqrt{r^2 - c^2} = \overline{ST}$. Hence the position of T is determined.

Comment. For a generalization, see *Crux Mathematicorum* 7 (1981) 207, Problem #562.

Problem 168. *First solution.* Let $x = t^3$. Then, from the given polynomial equation, we have

$$-(x + c) = \sqrt[3]{x}(b + a\sqrt[3]{x}).$$

Cubing both sides yields

$$-(x + c)^3 = x(b^3 + a^3 x + 3ab\sqrt[3]{x}(b + a\sqrt[3]{x}))$$

or

$$-(x + c)^3 = x(b^3 + a^3 x - 3ab(x + c)).$$

Second solution. If m, n, p are the roots of the given polynomial equation, we have $m + n + p = -a$, $mn + mp + np = b$ and $mnp = -c$. The roots of the required equation are m^3, n^3, p^3. Therefore

$m^3 + n^3 + p^3$

$\quad = (m + n + p)^3$

$\qquad - 3(m + n + p)(mn + mp + np) + 3mnp$

$\quad = -a^3 + 3ab - 3c,$

$m^3n^3 + m^3p^3 + n^3p^3$

$\quad = 3m^2n^2p^2 + (mn + mp + np)^3$

$\qquad - 3(m + n + p)(mn + mp + np)mnp$

$\quad = 3c^2 + b^3 - 3abc,$

$m^3n^3p^3 = -c^3.$

The equation sought is

$x^3 + (a^3 - 3ab + 3c)x^2 + (3c^2 + b^3 - 3abc)x + c^3 = 0.$

Problem 169. The inequality

$$\frac{a^2 + b^2 + c^2}{a + b + c} \geq \frac{a + b + c}{3}$$

is seen to be equivalent to $(a - b)^2 + (b - c)^2 + (c - a)^2 \geq 0$, which is clearly true. Applying this to each term of the left side gives the result.

Rider. Show more generally that if

$$S_2 = a_1^2 + a_2^2 + \cdots + a_n^2,$$

$$S_1 = a_1 + a_2 + \cdots + a_n, \qquad a_i > 0,$$

then

$$\sum_{i=1}^{n} \frac{S_2 - a_i^2}{S_1 - a_i} \geq S_1.$$

Problem 170. A positive integer with initial digit 6 is of the form $6 \cdot 10^n + m$, where $0 \leq m \leq 10^n - 1$.

(a) Here the condition is $25m = 6 \cdot 10^n + m$, which simplifies to $m = 2^{n-2} \cdot 5^n$. Thus the numbers are of the form

$$6 \cdot 10^n + 2^{n-2} \cdot 5^n = 625 \cdot 10^{n-2},$$

namely, $625, 6250, 62500, \ldots.$

(b) Suppose that the first digit is d, so that $35m = d \cdot 10^m + m$ or $17m = d \cdot 2^{n-1} \cdot 5^n$. Since $1 \leq d \leq 9$, this is impossible.

Problem 171. Let the angles of the polygon be $\alpha_1, \alpha_2, \ldots, \alpha_n$ and the corresponding exterior angles be $\beta_1, \beta_2, \ldots, \beta_n$. Then $\alpha_i + \beta_i = 180°$ for $i = 1, 2, \ldots, n$. If $n \geq 3$ and $\alpha_n = \alpha_{n-1} = \alpha_{n-2} = 60°$ then $\beta_n = \beta_{n-1} = \beta_{n-2} = 120°$, so $\beta_n + \beta_{n-1} + \beta_{n-2} = 360°$. But, it is known that the sum of *all* exterior angles of a convex polygon add up to $360°$. It follows that $n = 3$.

If a convex polygon has four of its angles as right angles, then the polygon must be a rectangle. More generally, if a convex polygon has r of its angles adding up to $(r - 2)180°$, then the polygon must be an r-gon with the given angles.

Problem 172. Since $1 + y^4 \geq 2y^2$, $1 + z^4 \geq 2z^2$, $1 + x^4 \geq 2x^2$ (with equality if and only if $x^2 = y^2 = z^2 = 1$), it is enough to show that

$$x^4y^2 + y^4z^2 + z^4x^2 \geq 3x^2y^2z^2.$$

But this is an immediate consequence of the arithmetic—geometric mean inequality. There is equality if and only if $x = y = z = 0$ or $x^2 = y^2 = z^2 = 1$.

Problem 173. The number of kth powers between 1 and 10^{30} inclusive is equal to the number of integers between 1 and $10^{\frac{30}{k}}$ inclusive, for $k = 1, 2, \ldots$. By the Principle of Inclusion and Exclusion the answer is

$$10^{30} - 10^{\frac{30}{2}} - 10^{\frac{30}{3}} - 10^{\frac{30}{5}} + 10^{\frac{30}{2 \cdot 3}}$$

$$+ 10^{\frac{30}{2 \cdot 5}} + 10^{\frac{30}{3 \cdot 5}} - 10^{\frac{30}{2 \cdot 3 \cdot 5}}$$

$$= 10^{30} - 10^{15} - 10^{10} - 10^6$$

$$+ 10^5 + 10^3 + 10^2 - 10.$$

Problem 174. The first few values of $16^n + 10n - 1$ are $25, 275, 4125, \ldots$. Suspecting that the answer is 25, we try to isolate a factor 25 by manipulation. For each positive integer n,

$$16^n + 10n - 1 = (1 + 15)^n + 10n - 1$$
$$= 1 + 15n + 15^2k + 10n - 1$$
$$= 25(n + 9k)$$

where k is some integer (because each term from the third on in the binomial expansion of $(1+15)^n$ contains the factor 15^2). It follows that the greatest common divisor is 25.

Problem 175. *First solution* (by induction). The result is valid for $n = 1$ and for $n = 2$. Assume it holds for $n = k \geq 2$. Then

$$a_1 a_2 \cdots a_k \geq 1 - k + a_1 + a_2 + \cdots + a_k.$$

Multiplying by a_{k+1} yields

$$a_1 a_2 \cdots a_k a_{k+1}$$
$$\geq a_{k+1}(1 - k + a_1 + a_2 + \cdots + a_k)$$
$$\geq a_{k+1} + (1 - k + a_1 + a_2 + \cdots + a_k) - 1$$

(from the $n = 2$ case)

$$= a_1 + a_2 + \cdots + a_k + a_{k+1} - k.$$

Equality occurs if and only if either $a_{k+1} = 1$ and all but at most one of $a_1, a_2, \ldots, a_k$ equals unity or $1 - k + a_1 + a_2 + \cdots + a_k = 1$. In either case, the condition is that all the a_i are unity with at most one exception.

Second solution. Let $a_i = 1 + t_i$ ($i = 1, 2, \ldots, n$). Then $t_i \geq 0$ and the inequality to be established is

$$n + (1 + t_1)(1 + t_2) \cdots (1 + t_n)$$
$$\geq 1 + n + t_1 + \cdots + t_n.$$

But

$$(1 + t_1)(1 + t_2) \cdots (1 + t_n)$$
$$= 1 + t_1 + t_2 + \cdots + t_n + p,$$

where $p = \prod_i (1+t_i) - 1 - \sum_i t_i$ is a nonnegative quantity which vanishes if and only if at least $n-1$ of the t_i's are 0.

Problem 176. By the Remainder Theorem, we have $p(a) = a$, $p(b) = b$, $p(c) = c$. Assuming that

a, b, c are distinct, we can write

$$p(x) = (x - a)(x - b)(x - c)q(x) + r(x),$$

where $r(x)$ is a polynomial of degree at most 2. We must have $r(a) = a$, $r(b) = b$, $r(c) = c$. Hence the polynomial $r(x) - x$ has degree at most 2 and roots a, b, c, and must therefore be the zero polynomial. Thus, x is the remainder.

Rider. Show that if $a_1, a_2, \ldots, a_n$ are all distinct and a_i is the remainder when $p(x)$ is divided by $x - a_i$ ($i = 1, 2, \ldots, n$), then $p(x) = q(x) \prod (x - a_i) + x$ for some polynomial $q(x)$.

Problem 177. Replacing x by $1 - x$ gives

$$(1 - x)^2 F(1 - x) + F(x) = 2(1 - x) - (1 - x)^4.$$

Solving this and the given equation simultaneously for $F(x)$ gives $F(x) = 1 - x^2$.

Problem 178. Considering the equation modulo 4, we can see that it is impossible for there to be exactly one, two or three odd numbers in the set $\{a, b, c\}$. Hence all three must be even, and for some integers a_1, b_1, c_1, we have $a = 2a_1, b = 2b_1, c = 2c_1$ whence $a_1^2 + b_1^2 + c_1^2 = 4a_1^2 b_1^2 \equiv 0 \pmod 4$. But we can see that a_1, b_1, c_1 also have to be all even, so $a_1 = 2a_2, b_1 = 2b_2, c_1 = 2c_2$ for some integers a_2, b_2, c_2. We can continue the argument to show that these are even, and so on. But this cannot continue indefinitely. Consequently, there are no positive solutions.

Comment. For extensions, see M.S. Klamkin, *USA Mathematical Olympiads 1972–1986*, Math. Assoc. Amer., Washington, D.C. 1988, pp. 32–33.

Problem 179. *First solution.* Since $a_{n+2} - 1 = a_1 a_2 \cdots a_{n+1}$, it follows that $a_{n+1} = \dfrac{a_{n+2} - 1}{a_{n+1} - 1}$. Hence

$$\frac{1}{a_{n+2} - 1} = \frac{1}{(a_{n+1} - 1)a_{n+1}}$$
$$= \frac{1}{a_{n+1} - 1} - \frac{1}{a_{n+1}},$$

for $n = 1, 2, 3, \ldots$. Therefore

$$1 + \sum_{n=1}^{\infty} \frac{1}{a_{n+1}} = 1 + \sum_{i=1}^{\infty} \left\{ \frac{1}{a_{n+1}-1} - \frac{1}{a_{n+2}-1} \right\}$$

$$= 1 + \frac{1}{a_2 - 1} = 2.$$

Second solution. By induction, it can be shown that, for each positive integer n,

$$\sum_{i=1}^{n} \frac{1}{a_i} = 2 - \frac{1}{a_1 a_2 \cdots a_n}$$

from which the result follows.

Problem 180. Draw and extend lines l_1, l_2 from B, which are close to A. Draw arbitrary lines PA, PC, PD as shown in Figure 126. G and F are then determined by the intersections of AE and AD with PC. Consider triangles GHE and FID. By Desargue's Theorem (see Tool Chest, E. 37), HG and IF will intersect in a point on AB close to A. Step by step, AB can now be drawn.

Problem 181. The plane determined by the two centers and the midpoint of the common chord (or, as appropriate, the point of tangency), also contains the two lines perpendicular to the planes of the two circles and passing through the two centers. However, a line perpendicular to the plane of a circle which passes through its center is the locus of points equidistant from points on its circumference. Hence, the two lines must intersect in a point equidistant from the points on both circumferences, including in particular the endpoints of the common chord (or the point of tangency).

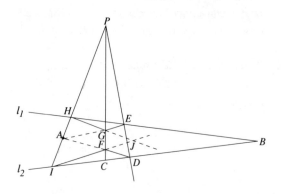

FIGURE 126

This point is the center of a sphere containing both circles.

Problem 182. Assume that there are distinct primes p, q, r whose cube roots are in arithmetic progression, i.e., for real numbers a and d and integers b and c,

$$\sqrt[3]{p} = a, \quad , \sqrt[3]{q} = a + bd, \quad \sqrt[3]{r} = a + cd.$$

Since $d = \dfrac{\sqrt[3]{q} - \sqrt[3]{p}}{b} = \dfrac{\sqrt[3]{r} - \sqrt[3]{p}}{c}$, we obtain

$$u\sqrt[3]{p} = v\sqrt[3]{q} + w\sqrt[3]{r},$$

where $u = c - b$, $v = c$, $w = -b$ are integers. Cubing this equation gives

$$u^3 p = v^3 q + w^3 r + 3vw\sqrt[3]{qr}(v\sqrt[3]{q} + w\sqrt[3]{r})$$

$$= v^3 q + w^3 r + 3uvw\sqrt[3]{pqr}.$$

But this would imply that $\sqrt[3]{pqr}$ is a rational number, viz., $\dfrac{u^3 p - v^3 q - w^3 r}{3uvw}$, which is not true. Hence primes p, q, r with the stated property do not exist.

Rider. Show that the result is still valid if "cube roots" is changed to "nth roots" ($n \geq 2$), or if geometric progressions, instead of arithmetic progressions, are considered.

Problem 183. Let ED be a line perpendicular to plane BOC. Then ED is parallel to AO, and hence lies in the plane AOD. Now, BC is perpendicular to both AD and ED, and is therefore perpendicular to the plane AOD and to every line in that plane. In particular, BC is perpendicular to OD.

Problem 184. *First solution.* Let A be at the origin of a 3-dimensional space, and let $\overrightarrow{x} = \overrightarrow{AB}$, $\overrightarrow{y} = \overrightarrow{AC}$, $\overrightarrow{z} = \overrightarrow{AD}$. Note that a quadrilateral is a parallelogram if and only if its diagonals bisect each other.

Case 1: $\{\overrightarrow{x}, \overrightarrow{y}, \overrightarrow{z}\}$ is a linearly independent set. Suppose that the four vertices P, Q, R, S are $t\overrightarrow{x}$, $(1-u)\overrightarrow{x} + u\overrightarrow{y}$, $(1-v)\overrightarrow{y} + v\overrightarrow{z}$, $w\overrightarrow{z}$, respectively, where each of t, u, v, w lies between

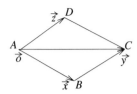

FIGURE 127

0 and 1 inclusive. Then $PQRS$ is a parallelogram if and only if

$$\frac{1}{2}\{t\overrightarrow{x} + (1-v)\overrightarrow{y} + v\overrightarrow{z}\}$$
$$= \frac{1}{2}\{(1-u)\overrightarrow{x} + u\overrightarrow{y} + w\overrightarrow{z}\}.$$

Since $\overrightarrow{x}$, $\overrightarrow{y}$, $\overrightarrow{z}$ are linearly independent, $u = 1 - t$, $v = t$, $w = t$ and the locus of the center is

$$\left\{\frac{1}{2}(t\overrightarrow{x} + (1-t)\overrightarrow{y} + t\overrightarrow{z}) : 0 \le t \le 1\right\}$$
$$= \left\{t\left(\frac{\overrightarrow{x} + \overrightarrow{z}}{2}\right) + (1-t)\frac{\overrightarrow{y}}{2} : 0 \le t \le 1\right\},$$

i.e., a line segment joining the midpoints of the diagonals.

Case 2: $\{\overrightarrow{x}, \overrightarrow{y}, \overrightarrow{z}\}$ is a linearly dependent set (i.e., A, B, C, D are coplanar). In this case, for some real numbers r and s, $\overrightarrow{y} = r\overrightarrow{x} + s\overrightarrow{z}$. Then P, Q, R, S are, respectively

$$t\overrightarrow{x},$$
$$(1-u)\overrightarrow{x} + u\overrightarrow{y} = (1-u+ur)\overrightarrow{x} + us\overrightarrow{z},$$
$$(1-v)\overrightarrow{y} + v\overrightarrow{z} = (1-v)r\overrightarrow{x} + [(1-v)s+v]\overrightarrow{z},$$
$$w\overrightarrow{z}.$$

The condition that the diagonals bisect each other gives

$$(1-u+ur)\overrightarrow{x} + (us+w)\overrightarrow{z}$$
$$= [t + (1-v)r]\overrightarrow{x} + [(1-v)s+v]\overrightarrow{z},$$

which leads to

$$1-u+ur = t+r-vr \quad \text{and} \quad us+w = s-vs+v.$$

If $r = 1$, we have $1 = t + 1 - v$, or $t = v$, and the midpoint is

$$\frac{1}{2}\left[\overrightarrow{x} + (us+w)\overrightarrow{z}\right]$$
$$= \frac{1}{2}\left[\overrightarrow{x} + (s - ts + t)\overrightarrow{z}\right]$$
$$= (1-t)\left(\frac{\overrightarrow{x} + s\overrightarrow{z}}{2}\right) + t\left(\frac{\overrightarrow{x} + \overrightarrow{z}}{2}\right),$$

where $0 \le t \le 1$.

If $s = 1$, we have $u+w = 1$ and the midpoint is

$$\frac{1}{2}\left[(1-u+ur)\overrightarrow{x} + \overrightarrow{z}\right]$$
$$= (1-u)\left(\frac{\overrightarrow{x} + \overrightarrow{z}}{2}\right) + u\left(\frac{r\overrightarrow{x} + \overrightarrow{z}}{2}\right).$$

In general, the midpoint is

$$\frac{1}{2}\{[(1-u) + ur]\overrightarrow{x} + [(1-v)s + v]\overrightarrow{z}\},$$
$$0 \le u, \ v \le 1.$$

The locus is shaded in Figure 128. It consists of a parallelogram with sides parallel to the vectors $\overrightarrow{x}$ and $\overrightarrow{z}$ and with its four vertices located at

$$\frac{\overrightarrow{x} + \overrightarrow{z}}{2}, \qquad \frac{\overrightarrow{x} + s\overrightarrow{z}}{2},$$
$$\frac{r\overrightarrow{x} + s\overrightarrow{z}}{2}, \qquad \frac{r\overrightarrow{x} + \overrightarrow{z}}{2}.$$

Second solution. (when A, B, C, D are non-coplanar). First, observe that $PQRS$ is coplanar and that its planar extension is either parallel to AC or meets AC produced in a single point, say T. In the latter case, T would have to lie on both PQ and RS produced, which would contradict the

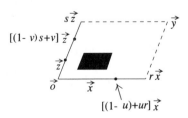

FIGURE 128

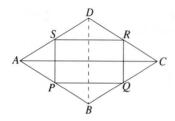

FIGURE 129

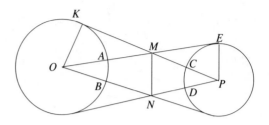

FIGURE 130

fact that they are parallel. Hence AC is parallel to $PQRS$. Since PQ lies in the plane ABC, PQ and AC cannot be skew. Hence $AC \parallel PQ \parallel RS$. Similarly $BD \parallel PS \parallel QR$. See Figure 129.

Let $\overrightarrow{AP} = k\overrightarrow{AB}$. Then $\overrightarrow{AS} = k\overrightarrow{AD}$, $\overrightarrow{AQ} = \overrightarrow{AB} + (1-k)\overrightarrow{BC}$ and $\overrightarrow{AR} = \overrightarrow{AD} + (1-k)\overrightarrow{DC}$. The center M of $PQRS$ is given by

$$\overrightarrow{AM} = \frac{1}{4}\{\overrightarrow{AP} + \overrightarrow{AQ} + \overrightarrow{AR} + \overrightarrow{AS}\}$$

$$= \frac{1}{4}\{(1+k)\overrightarrow{AB} + (1+k)\overrightarrow{AD}$$

$$+ (1-k)\overrightarrow{BC} + (1-k)\overrightarrow{DC}\}$$

$$= (1-k)\left\{\frac{\overrightarrow{AB}+\overrightarrow{BC}}{4} + \frac{\overrightarrow{AD}+\overrightarrow{DC}}{4}\right\}$$

$$+ k\left\{\frac{\overrightarrow{AB}+\overrightarrow{AD}}{2}\right\}$$

$$= (1-k)2\frac{\overrightarrow{AC}}{4} + k\left\{\frac{\overrightarrow{AB}+\overrightarrow{AD}}{2}\right\}$$

$$= (1-k)\frac{\overrightarrow{AC}}{2} + k\frac{\overrightarrow{AB}+\overrightarrow{AD}}{2}.$$

The value k ranges between 0 and 1 inclusive, so that $\overrightarrow{AM}$ ranges over the segment joining the midpoints of the diagonals AC and BD.

Problem 185. *First solution.* From $\triangle OKM \sim \triangle PEM$ follows $\dfrac{\overline{OK}}{\overline{OM}} = \dfrac{\overline{PE}}{\overline{PM}}$. Since $\overline{OK} = \overline{OA}$ and $\overline{PE} = \overline{PC}$ we have

$$\frac{\overline{OA}}{\overline{OM}} = \frac{\overline{PC}}{\overline{PM}}. \tag{1}$$

From $\triangle OAB \sim \triangle OMN$ and $\triangle PCD \sim \triangle PMN$ we have respectively

$$\frac{\overline{OA}}{\overline{OM}} = \frac{\overline{AB}}{\overline{MN}} \quad \text{and} \quad \frac{\overline{PC}}{\overline{PM}} = \frac{\overline{CD}}{\overline{MN}}. \tag{2}$$

Now (1) and (2) combine to give $\dfrac{\overline{AB}}{\overline{MN}} = \dfrac{\overline{CD}}{\overline{MN}}$, so $\overline{AB} = \overline{CD}$.

Second solution. Let $O = (0,0)$, $P = (d,0)$ and let the radii of the circles with centers O and P be r and R respectively. Let $y = mx$ be the equation of OE. The distance of P from OE is $R = \dfrac{|md|}{\sqrt{1+m^2}}$. Thus, $\dfrac{m^2}{1+m^2} = \dfrac{R^2}{d^2}$. The line $y = mx$ intersects the circle $x^2 + y^2 = r^2$ in the point whose ordinate is given by $y^2(1+m^2) = m^2r^2$, i.e., $y^2 = \dfrac{R^2r^2}{d^2}$. Hence, the length of AB is $\dfrac{2Rr}{d}$. Since the expression for $\overline{AB}$ is symmetrical in the radii of the circles, $\overline{CD}$ must be given by the same quantity. Hence $\overline{AB} = \overline{CD}$.

Problem 186. Suppose to begin with, that $PQ \parallel AC$. Since A, C, R, S and A, P, Q, C determine two planes intersecting along AC, and since PQ and RS could therefore intersect in a point on AC, it follows that $PQ \parallel SR$. Hence $SR \parallel AC$.

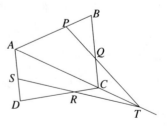

FIGURE 131

It is then clear that $P'Q' \parallel A'C'$ and $S'R' \parallel A'C'$, so that $P'Q' \parallel S'R'$ and P', Q', R', S' are coplanar. We now assume that neither PQ nor SR is parallel to AC. Then PQ and SR both intersect AC in T, the point of intersection of the line AC produced and the plane $PQRS$. By Menelaus' Theorem (Tool Chest, E. 20),

$$\frac{\overline{AT}}{\overline{TC}} = -\frac{\overline{QB}}{\overline{CQ}} \cdot \frac{\overline{PA}}{\overline{BP}} = -\frac{\overline{RD}}{\overline{CR}} \cdot \frac{\overline{SA}}{\overline{DS}}.$$

Since the figure $APBQC$ is congruent to $A'P'B'Q'C'$, $P'Q'$ must meet $A'C'$ in a point T' such that

$$\frac{\overline{A'T'}}{\overline{T'C'}} = -\frac{\overline{Q'B'}}{\overline{C'Q'}} \cdot \frac{\overline{P'A'}}{\overline{B'P'}} = -\frac{\overline{QB}}{\overline{CQ}} \cdot \frac{\overline{PA}}{\overline{BP}}.$$

Similarly, $S'R'$ meets $A'C'$ in a point T'', such that

$$\frac{\overline{A'T''}}{\overline{T''C'}} = -\frac{\overline{R'D'}}{\overline{C'R'}} \cdot \frac{\overline{S'A'}}{\overline{D'S'}} = -\frac{\overline{RD}}{\overline{CR}} \cdot \frac{\overline{SA}}{\overline{DS}}$$

$$= -\frac{\overline{QB}}{\overline{CQ}} \cdot \frac{\overline{PA}}{\overline{BP}} = \frac{\overline{A'T'}}{\overline{T'C'}}.$$

Hence $T' = T''$, so $P'Q'$ and $S'R'$ intersect in T'. Hence P', Q', R', S' are coplanar.

Problem 187. By the Law of Sines,

$$\frac{a}{\sin A} = \frac{b}{\sin B} = \frac{c}{\sin C},$$

so that (a) reduces to

$$\sin A + \sin B = k \sin C = k \sin(\pi - A - B)$$

$$= k \sin(A + B)$$

or

$$2k \sin \frac{A+B}{2} \cos \frac{A+B}{2}$$

$$= 2 \sin \frac{A+B}{2} \cos \frac{A-B}{2}.$$

Since $0 < \frac{A+B}{2} < \frac{\pi}{2}$, we can cancel $2 \sin \frac{A+B}{2}$ to get

$$k \cos \frac{A+B}{2} = \cos \frac{A-B}{2}. \qquad (1)$$

By (b),

$$\frac{\cos \frac{A}{2}}{\sin \frac{A}{2}} + \frac{\cos \frac{B}{2}}{\sin \frac{B}{2}} = k \frac{\sin \frac{A+B}{2}}{\cos \frac{A+B}{2}}$$

so that

$$\cos \frac{A+B}{2} \left\{ \cos \frac{A}{2} \sin \frac{B}{2} + \cos \frac{B}{2} \sin \frac{A}{2} \right\}$$

$$= k \sin \frac{A+B}{2} \sin \frac{A}{2} \sin \frac{B}{2}.$$

Hence

$$\cos \frac{A+B}{2} \sin \frac{A+B}{2}$$

$$= \frac{1}{2} k \sin \frac{A+B}{2} \left\{ \cos \frac{A-B}{2} - \cos \frac{A+B}{2} \right\}.$$

Thus

$$2 \cos \frac{A+B}{2}$$

$$= k \left\{ \cos \frac{A-B}{2} - \cos \frac{A+B}{2} \right\}. \qquad (2)$$

Now (1) and (2) together yield $k^2 = 2 + k$ or

$$(k - 2)(k + 1) = 0.$$

Since $k = -1$ is inadmissible, $k = 2$. For example, when $k = 2$, (a) and (b) hold for equilateral triangles.

Problem 188. *First solution.* Assume A, B, C, D are not coplanar. Consider the tetrahedron $ABCD$. Since the sum of any two face angles of a trihedral angle is greater than the third face angle, we have

$$\angle CAB + \angle CAD > \angle DAB = 90°,$$

$$\angle DBC + \angle DBA > \angle ABC = 90°,$$

$$\angle ACD + \angle ACB > \angle BCD = 90°,$$

$$\angle BDA + \angle BDC > \angle CDA = 90°.$$

Adding the four inequalities to the equality $\angle DAB + \angle ABC + \angle BCD + \angle CDA = 360°$ gives that the sum of all the face angles of the

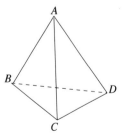

FIGURE 132

four triangular faces of the tetrahedron exceeds $720° = 4(180°)$. Since this is false, A, B, C, D must be coplanar.

Second solution. The point D lies both on a plane through A perpendicular to AB and on a plane through C perpendicular to BC. Hence D lies on the line l of intersecton of these two planes. Since the normals AB and BC of the two planes are perpendicular, the planes themselves are perpendicular and the line l is perpendicular to the plane ABC.

Suppose l intersects ABC in D'; then $ABCD'$ is a planar rectangle. Thus AC subtends right angles at D and D', so that both D and D' lie on the intersection of l and a sphere with diameter AC. But l is perpendicular to a radius of the sphere (joining D to the midpoint of AC) and so is tangent to the sphere. Thus D and D' coincide, and A, B, C, D are coplanar.

Third solution. Determine the point D' as in the second solution. We have

$$\overline{AD}^2 + \overline{DC}^2$$
$$= \overline{AC}^2 = \overline{AD'}^2 + \overline{D'C}^2$$
$$= (\overline{AD}^2 + \overline{DD'}^2) + (\overline{DD'}^2 + \overline{DC}^2)$$
$$= \overline{AD}^2 + \overline{DC}^2 + 2\overline{DD'}^2$$

(by Pythagoras' Theorem).

Hence $\overline{DD'} = 0$, and $D = D'$.

Fourth solution. Use 3-dimensional coordinate geometry. Let B lie at the origin, A on the x-axis and C on the y-axis, so that, for some nonzero a and c, $A = (a, 0, 0)$, $B = (0, 0, 0)$, $C = (0, c, 0)$. Suppose $D = (x, y, z)$.

$\angle BAD = 90°$

$\Rightarrow\; 0 = (a, 0, 0) \cdot (x - a, y, z) = a(x - a)$

$\Rightarrow\; x = a;$

$\angle BCD = 90°$

$\Rightarrow\; 0 = (0, c, 0) \cdot (x, y - c, z) = c(y - c)$

$\Rightarrow\; y = c;$

$\angle ADC = 90°$

$\Rightarrow\; 0 = (x - a, y, z) \cdot (x, y - c, z)$

$\qquad = x(x - a) + y(y - c) + z^2 = z^2.$

Hence $D = (a, c, 0)$ and the result follows.

Problem 189. Since $\dfrac{a}{\sin A} = \dfrac{b}{\sin B} = 2R$, we have to show that

$$2(\sin A - \sin B) = 1,$$

i.e., that $2(\cos 36° - \cos 72°) = 2(\sin 126° - \sin 18°) = 1$.

Set $\theta = 72°$. Then $\cos 3\theta = \cos(360° - 2\theta) = \cos 2\theta$. Because $\cos 3\theta = 4\cos^3 \theta - 3\cos\theta$ and $\cos 2\theta = 2\cos^2 \theta - 1$, we find that $x = \cos\theta$ is a positive number satisfying the equation

$$4x^3 - 2x^2 - 3x + 1 = 0$$

or

$$(x - 1)(4x^2 + 2x - 1) = 0.$$

Since $x \neq 1$, $4x^2 + 2x - 1 = 0$ and

$$\cos 72° = x = \frac{\sqrt{5} - 1}{4}, \quad \cos 36° = \frac{\sqrt{5} + 1}{4};$$

thus $2(\cos 36° - \cos 72°) = 1$, as required.

Problem 190. Let P be the point of intersection of AB produced and DC produced. Noting that $ABCD$ is concyclic, we have that

$$\angle PDA = 180° - \angle ABC = \angle PBC,$$
$$\angle PAD = 180° - \angle BCD = \angle PCB.$$

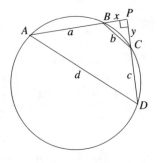

FIGURE 133

Hence $\triangle PAD \sim \triangle PCB$. Thus, if $y = \overline{PC}$, $x = \overline{PB}$,

$$\frac{y}{b} = \frac{a+x}{d} \quad \text{and} \quad \frac{x}{b} = \frac{c+y}{d}.$$

Solving for x and y we obtain

$$(d^2 - b^2)y = b(ad + bc)$$

and

$$(d^2 - b^2)x = b(ab + cd).$$

Since $x^2 + y^2 = b^2$, the desired equality follows.

Problem 191. Since PS and QR intersect, P, S, Q, R are coplanar. Either PQ and RS intersect or $PQ \parallel RS$. Suppose PQ and RS intersect in a point T. Since T lies on PQ produced, T lies in the plane ABC. Similarly T lies in the plane ACD. Hence, T lies on the intersection of these two planes, namely AC produced. Thus, in this case, PQ, RS and AC are concurrent.

Suppose PQ and RS do not intersect when produced. Then $PQ \parallel RS$. We show that AC produced does not meet the plane of $PQRS$. Suppose, on the contrary, they meet at a point V. V lies on both planes, $PQRS$ and ABC, hence on their line of intersection PQ produced. Similarly, V lies on RS produced, and we get a contradiction. Thus AC is parallel to

$PQRS$. Hence AC and PQ do not meet. Since AC and PQ are coplanar, they must be parallel. Thus $PQ \parallel RS \parallel AC$.

Problem 192. Let P be the foot of the perpendiculars from C and D to AB. See Figure 135. Then $\angle CPD = \theta$. Take the length of AC to be 1. Then $CP = \frac{\sqrt{3}}{2}$. Let $\angle CAD = \alpha$. By the Law of Cosines,

$$\overline{CD}^2 = 1 + 1 - 2\cos\alpha$$

and

$$\overline{CD}^2 = \frac{3}{4} + \frac{3}{4} - \frac{9}{8}\cos\theta.$$

Hence $2\cos\alpha = \frac{1}{2} + \frac{9}{8}\cos\theta$, or

$$\alpha = \arccos\left(\frac{4 + 9\cos\theta}{16}\right).$$

Problem 193. The Law of Cosines easily leads to $\cos A = \frac{3}{4}$ and $\cos C = \frac{1}{8}$. Hence

$$\cos C = \frac{1}{8} = 2\left(\frac{3}{4}\right)^2 - 1$$
$$= 2\cos^2 A - 1 = \cos 2A.$$

Since both $2A$ and C lie between 0 and π, $2A = C$.

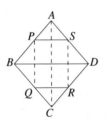

FIGURE 134

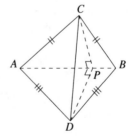

FIGURE 135

My preferences are for problems that are not technical and that can be easily understood by the general reader. There should be a certain elegance about the problems; the best problems are elegant in statement (short and sweet), elegant in result, and elegant in solution. Such problems are not easy to come by.

Murray S. Klamkin, Problem Corner, *Math. Intelligencer*, vol. 5, #1, (1983), p. 59.

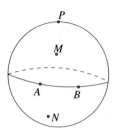

FIGURE 136

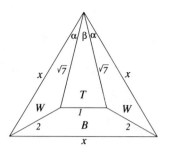

FIGURE 137

Problem 194. Draw an arbitrary circle on the sphere; let its center be P. From two arbitrary points A, B on the circle, construct equal arcs intersecting at M and N. The circle through PMN is a great circle. (P, M, N are equidistant from A and B; hence they lie on a plane right-bisecting AB. This plane passes through the center of the sphere and intersects the surface of the sphere in a great circle.) With the compasses, transfer the chordal distances PM, MN and NP to the plane and construct the circumcircle of triangle PMN (as transferred to the plane). The radius of this circumcircle will equal the radius of the sphere.

Problem 195. If we have n points, $A, B, C, \ldots$, then the left side can be interpreted as the number of pairs of segments formed by the n points. Now add an extra point Q and consider each combination of four points. The combination Q, A, B, C gives rise to three pairs of segments of the original set, i.e., AB, AC; AB, BC; BC, AC. The combination A, B, C, D also gives rise to three pairs of segments, i.e., AB, CD; AC, BD; AD, BC. Hence the number of pairs of segments is equal to three times the number of ways of choosing 4 of the $n + 1$ points, i.e., $3\binom{n+1}{4}$.

Problem 196. By Euler's Formula, $E + 2 = F + V$, where E, F, V denote the number of edges, faces and vertices, respectively. Also, since each face has at least three edges and each edge belongs to two faces, $2E \geq 3F$. Similarly, since each vertex has at least three edges emanating from it and since each edge emanates from two vertices $2E \geq 3V$.

(a) In this case, $14 \geq 3F$ and $14 \geq 3V$ so that $F < 4$, $V < 4$. But then $E + 2 < 8$, which contradicts $E = 7$.

(b) Here, $2E = 6F$, so that, by Euler's Formula $3V = 3E + 6 - 3F = 2E + 6$. But this contradicts $2E \geq 3V$.

Problem 197. *First solution.* (See Figure 137.) By the Law of Cosines, $1 = 7 + 7 - 14\cos\beta$, whence $\cos\beta = \frac{13}{14}$, $\sin\beta = \frac{\sqrt{27}}{14}$ and

$$2\cos^2\alpha = 1 + \cos 2\alpha = 1 + \cos\left(\frac{\pi}{3} - \beta\right)$$

$$= 1 + \frac{1}{2} \cdot \frac{13}{14} + \frac{\sqrt{3}}{2} \cdot \frac{\sqrt{27}}{14}$$

$$= 1 + \frac{11}{14} = \frac{25}{14}.$$

Thus $\cos\alpha = \dfrac{5}{2\sqrt{7}}$.

Again, by the Law of Cosines, $4 = 7 + x^2 - 2x\sqrt{7}\cos\alpha = 7 + x^2 - 5x$, so that $x^2 - 5x + 3 = 0$. Since $x > 1$, we must have $x = \frac{1}{2}(5 + \sqrt{13})$.

Second solution. Referring to the diagram and using Heron's rule (Tool Chest, E. 44) for calculating triangular areas, we find that triangle T has area $\dfrac{3\sqrt{3}}{4}$ and height $\dfrac{3\sqrt{3}}{2}$. The height of the trapezoid B is $\dfrac{(x-3)\sqrt{3}}{2}$, so that its area is $\dfrac{(x^2 - 2x - 3)\sqrt{3}}{4}$. Now

$2(\text{area } W)$

$= (\text{area of large triangle}) - \text{area } T - \text{area } B$

$= \dfrac{\sqrt{3}x^2}{4} - \dfrac{3\sqrt{3}}{4} - \dfrac{\sqrt{3}(x^2 - 2x - 3)}{4} = \dfrac{\sqrt{3}x}{2}.$

But Heron's rule yields

$$2(\text{area } W) = 2\sqrt{\frac{x+2+\sqrt{7}}{2} \cdot \frac{x+2-\sqrt{7}}{2} \cdot \frac{\sqrt{7}+(x-2)}{2} \cdot \frac{\sqrt{7}-(x-2)}{2}}$$

$$= \frac{1}{2}\sqrt{\{(x+2)^2 - 7\}\{7 - (x-2)^2\}}$$

$$= \frac{1}{2}\sqrt{\{4x + (x^2 - 3)\}\{4x - (x^2 - 3)\}}$$

$$= \frac{1}{2}\sqrt{22x^2 - x^4 - 9}.$$

Hence $\sqrt{3}x = \sqrt{22x^2 - x^4 - 9}$, so that $x^4 - 19x^2 + 9 = 0$. The relevant root satisfies $x^2 = \frac{19 + 5\sqrt{13}}{2}$, so $x = \frac{5 + \sqrt{13}}{2}$.

Problem 198. Let $ABCD$ be any quadrilateral. Since $\overrightarrow{AB} + \overrightarrow{BC} + \overrightarrow{CD} + \overrightarrow{DA} = \overrightarrow{O}$,

$$\overrightarrow{DA}^2 = (\overrightarrow{AB} + \overrightarrow{BC} + \overrightarrow{CD})^2.$$

Using the notation a, b, c, d for the lengths of the sides as in Figure 138, we have

$$d^2 = a^2 + b^2 + c^2 + 2\overrightarrow{AB} \cdot \overrightarrow{BC}$$
$$+ 2\overrightarrow{AB} \cdot \overrightarrow{CD} + 2\overrightarrow{BC} \cdot \overrightarrow{CD}.$$

This is equivalent to

$$b^2 + d^2$$
$$= a^2 + c^2 + 2(\overrightarrow{AB} + \overrightarrow{BC}) \cdot (\overrightarrow{BC} + \overrightarrow{CD})$$
$$= a^2 + c^2 + 2\overrightarrow{AC} \cdot \overrightarrow{BD}.$$

From this, we deduce that a necessary and sufficient condition that the diagonals of a quadrilateral $ABCD$ are orthogonal is that $b^2 + d^2 = a^2 + c^2$.

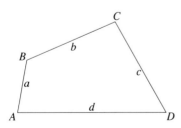

FIGURE 138

In the situation of the problem, if a, b, c, d are the side lengths of $ABCD$ and a', b', c', d' are the side lengths of $A'B'C'D'$, we have $a = a'$, $b = b'$, $c = c'$, $d = d'$. Hence,

$$AC \perp BD \Rightarrow b^2 + d^2 = a^2 + c^2 \Rightarrow b'^2 + d'^2$$
$$= a'^2 + c'^2 \Rightarrow A'C' \perp B'D'.$$

Problem 199. More generally, we show that if $P_0(x), P_1(x), \ldots, P_{n-2}(x)$ $(n \geq 2)$ and $S(x)$ are polynomials such that

$$P_0(x^n) + xP_1(x^n) + \cdots + x^{n-2}P_{n-2}(x^n)$$
$$= (x^{n-1} + x^{n-2} + \cdots + x + 1)S(x),$$

then $x - 1$ is a factor of $P_i(x)$ for all i.

Let $\omega_i, i = 1, 2, \ldots, n-1$ denote the complex nth roots of unity other than 1. Since

$$x^n - 1 = (x - 1)(1 + x + \cdots + x^{n-1}),$$

it follows that

$$1 + \omega_i + \omega_i^2 + \cdots + \omega_i^{n-1} = 0$$

for all i. If we now substitute ω_i in the given identity above, we find that

$$P_0(1) + \omega_i P_1(1) + \cdots + \omega_i^{n-2}P_{n-2}(1) = 0$$

for all i. Since the $(n - 2)$th degree polynomial $\sum_{k=0}^{n-2} P_k(1)x^k$ has $n-1$ distinct roots ω_i, it must vanish identically. Thus,

$$P_0(1) = P_1(1) = \ldots = P_{n-2}(1) = 0.$$

Finally, by the Factor Theorem, $x - 1$ is a factor of each of $P_0(x), P_1(x), \ldots, P_{n-2}(x)$.

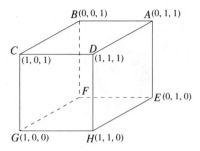

FIGURE 139

Problem 200. Set up a rectangular coordinate system, as in Figure 139, with the origin at F and the x-, y-, z-axes along FG, FE, FB respectively. Take $\overline{AB} = 1$, so that the coordinates of P, Q, R are given by

$$P = (0, c, 1) \quad Q = (1, 0, a) \quad R = (b, 1, 0),$$

where $0 \le a, b, c \le 1$. The problem is to determine a, b, c with $0 \le a, b, c \le 1$, so that

$$L = \sqrt{a^2 + (1-b)^2 + 1}$$
$$+ \sqrt{b^2 + (1-c)^2 + 1} + \sqrt{c^2 + (1-a)^2 + 1}$$

is a minimum.

By Minkowski's Inequality

$$\sqrt{x_1^2 + x_2^2 + x_3^2} + \sqrt{y_1^2 + y_2^2 + y_3^2}$$
$$+ \sqrt{z_1^2 + z_2^2 + z_3^2}$$
$$\ge \sqrt{\begin{array}{c}(x_1 + y_1 + z_1)^2 + (x_2 + y_2 + z_2)^2 \\ + (x_3 + y_3 + z_3)^2\end{array}}.$$

With $(x_1, x_2, x_3) = (1, a, 1 - b)$, $(y_1, y_2, y_3) = (1, b, 1 - c)$, $(z_1, z_2, z_3) = (1, c, 1 - a)$, this yields

$$L^2 \ge 9 + s^2 + (3 - s)^2 = 2(s - \tfrac{3}{2})^2 + \tfrac{27}{2},$$

where $s = a + b + c$.

Since $2(s - \tfrac{3}{2})^2 + \tfrac{27}{2}$ attains its minimum value when $s = \tfrac{3}{2}$, we have $L^2 \ge \tfrac{27}{2}$. However, when $a = b = c = \tfrac{1}{2}$, $s = \tfrac{3}{2}$ and it turns out that $L = 3\sqrt{\tfrac{3}{2}} = \sqrt{\tfrac{27}{2}}$. Thus the minimum perimeter is $3\sqrt{\tfrac{3}{2}}$.

Remark. Minkowski's Inequality has equality if and only if (x_1, x_2, x_3), (y_1, y_2, y_3), (z_1, z_2, z_3)

are proportional. In the present situation this leads to $L^2 \ge 9 + s^2 + (3 - s)^2$, with equality if and only if $a = b = c = \tfrac{1}{2}$.

Problem 201. He cannot get to work on time! Since the time required to travel 2 km. at 12 kph equals the time required to travel 6 km. at 36 kph, he has used up all his time covering the first two kilometers.

Problem 202. *First solution.* Pick any face for January. There are $\binom{11}{5}$ ways of choosing the months to go into the ring of five faces adjacent to January, and 4! essentially different ways of arranging them. There is a second ring of five faces, each adjacent to two of January's neighbors; the months for these can be chosen in $\binom{6}{5}$ ways, and there are 5! essentially different ways of arranging them relative to the first ring. Finally, the month for the face antipodal to January's face is now determined. Hence the number of essentially different ways of making the calendar is

$$\binom{11}{5} 4! \binom{6}{5} 5! = \frac{11!}{5}.$$

Second solution. If the faces were distinguishable, there would be 12! ways of placing the months. However, one of these arrangements can be carried onto various other ones by means of a rigid transformation in space of the dodecahedron onto itself. Such a rigid transformation must carry vertices to vertices, and is uniquely determined by specifying the images of two adjacent faces (reflections, not realizable in three-dimensional space, are not counted). The first face has twelve possible images; once this is specified, its neighbor has five possible images. Thus there are $60 = 12 \times 5$ symmetries of the dodecahedron, so each of the 12! arrangements is essentially the same as 60 arrangements (including itself). Hence the number of essentially different arrangements is $\frac{12!}{60} = \frac{11!}{5}$.

Problem 203. (a) Let n be a number equal to the sum of the squares of its digits. If n has k digits, we must have $10^{k-1} \le n \le 9^2 k = 81k$. Now, for $p \ge 4$, the assumption that $81p < 10^{p-1}$ implies that $81(p+1) < 10^{p-1}+81 < 10^p$. Since $81\times4 < 10^3$, it follows by induction that $81p < 10^{p-1}$ for $p \geq 4$. Thus $k \leq 3$ and $n = a_2 10^2 + a_1 10 + a_0$ for digits a_0, a_1, a_2. Since $a_2 10^2 + a_1 10 + a_0 = a_2^2 + a_1^2 + a_0^2$, it follows that

$$(10^2 - a_2)a_2 + (10 - a_1)a_1 = a_0^2 - a_0.$$

Both sides of this equality are nonnegative, and $1 \leq a_0 \leq 9$ implies that $a_0^2 - a_0$ is at most 72. Since $10^2 - a_2 \geq 91$, we must have $a_2 = 0$. Hence $(10 - a_1)a_1 = a_0^2 - a_0$.

Since the possible values of $(10 - a_1)a_1$ are 0, 9, 16, 21, 24, 25 and the possible values of $a_0^2 - a_0$ are 0, 2, 6, 12, 20, 30, 42, 56, 72, the result follows.

(b) Let n be equal to the sum of the cubes of its digits. If n has k digits, then $10^{k-1} \leq 729k$. As in (a), it can be argued that $k \leq 4$.

If $k = 4$ and $n = a_3 10^3 + a_2 10^2 + a_1 10 + a_0 = a_3^3 + a_2^3 + a_1^3 + a_0^3$ it follows that

$$a_3(919) = a_3(10^3 - 9^2) \leq a_3(10^3 - a_3^2)$$

$$= a_3 10^3 - a_3^3$$

$$= a_2(a_2^2 - 10^2) + a_1(a_1^2 - 10)$$

$$+ a_0(a_0^2 - 1)$$

$$\leq 0 + 9 \cdot 71 + 9 \cdot 80 = 1359.$$

Hence $a_3 \leq 1$. It is not hard to see that $a_3 \neq 1$, so that $k \leq 3$.

Since 28, 35, 65, 72 and 91 are the only numbers of two digits which are the sum of two cubes, it is clear that $k \neq 2$. The only possibility for $k = 1$ is $n = 1$.

Hence we may suppose that $k = 3$ and

$$n = a_2 10^2 + a_1 10 + a_0 = a_2^3 + a_1^3 + a_0^3$$

with $a_2 \neq 0$. This implies that

$$a_2(100 - a_2^2) + a_1(10 - a_1^2) - a_0(a_0^2 - 1) = 0.$$

The possible choices of a_2, a_1, a_0 can be found from the following table:

x	$x(100 - x^2)$	$x(10 - x^2)$	$-x(x^2 - 1)$
0	0	0	0
1	99	9	0
2	192	12	-6
3	273	3	-24
4	336	-24	-60
5	375	-75	-120
6	384	-156	-210
7	357	-273	-336
8	288	-432	-504
9	171	-639	-720

Clearly, a_1 and a_0 cannot be either 8 or 9. By looking at the last digits, the possibilities are quickly narrowed down to 153, 370, 371 and 407.

Problem 204. There are 6 essentially different ways as shown in Figure 140.

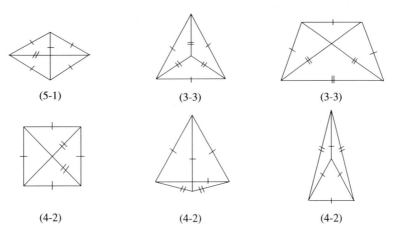

(5-1)   (3-3)   (3-3)

(4-2)   (4-2)   (4-2)

FIGURE 140

Problem 205. It is best to work backwards. At each stage, there is essentially only one way in which the money can be distributed. Observe that the total amount held by all three men is always $72.

After 3 games, amounts are $24, $24, $24.

After 2 games, amounts are $12, $12, $48.

After 1 game, amounts are $6, $42, $24.

Originally, the amounts were $39, $21, $12.

Problem 206. (a) By the triangle inequality, $2\overline{BC} = \overline{AB} < \overline{AC} + \overline{BC}$, so that $\overline{BC} < \overline{AC}$. Hence BC is the shortest side.

Since $24 = \overline{AB} + \overline{BC} + \overline{AC} = 3\overline{BC} + \overline{AC} > 4\overline{BC}$, we have $\overline{BC} < 6$.

Since $24 = \overline{AB} + \overline{BC} + \overline{AC} < \overline{AB} + \overline{BC} + (\overline{AB} + \overline{BC}) = 6\overline{BC}$, we have $\overline{BC} > 4$.

(b) Suppose the triangle is ABC and $3\overline{BC} = \overline{AB}$. Then, as in (a), it can be argued that $2\overline{BC} < \overline{AC}$, and BC is the shortest side.

Thus $24 = 4\overline{BC} + \overline{AC} > 6\overline{BC}$, so that $\overline{BC} < 4$.

Also $24 < 2(\overline{AB} + \overline{BC}) = 8\overline{BC}$, so that $\overline{BC} > 3$.

Problem 207. Equality holds in (a) for $k = 1, 2$. Strict inequality occurs for $k = 0$. When $k \geq 3$,

$$1^{2k} + 2^{2k} + 3^{2k} > 3^{2k} = 9^k > 2 \cdot 7^k.$$

As for (b), we have equality for $k = 0, 1$. When $k \geq 2$,

$$1^{2k+1} + 2^{2k+1} + 3^{2k+1} > 3^{2k+1}$$
$$= 3^k 3^{k+1}$$
$$> 2 \cdot 2^k \cdot 3^{k+1}$$
$$= 6^{k+1}.$$

Problem 208. Multiplying out, we have

$$x^3 + 6x^2 + 11x + 6 = x^3 + 6x^2 - 7x - 60,$$

which leads to $x = -\frac{11}{3}$.

Problem 209. If M is the number, $M + 1$ must be the least common multiple of $2, 3, \ldots, 10$, namely $2^3 \cdot 3^2 \cdot 5 \cdot 7$. Hence $M = 2519$.

Problem 210. The situation can be illustrated graphically as in Figure 141. Suppose units of time are chosen so that the faster car covers the distance from A to B in 4 units of time, and the slower car in 5 units of time. Let A be at milestone a and B at milestone b. The polygonal lines in the graph illustrate the progress of the cars; x is the time of the second passing and u the time of the third. From similar triangles, we find that

$$\frac{b - 145}{145 - a} = \frac{10 - x}{x - 5} = \frac{x - 4}{8 - x},$$

whence $x = \frac{20}{3}$ and $2a + b = 435$. Similarly,

$$\frac{b - 201}{201 - a} = \frac{12 - u}{u - 8} = \frac{u - 10}{15 - u},$$

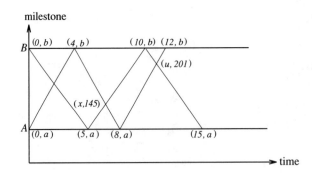

FIGURE 141

whence $u = \frac{100}{9}$ and $2a + 7b = 1809$. It is now determined that $a = 103$ and $b = 229$. Hence A is at the 103rd milestone and B is at the 229th milestone.

Problem 211. Observe that $7^2 = 49$, $7^3 = 343$ and $7^4 = 2401 = 24 \cdot 100 + 1$. This suggests that we isolate factor 7^4 as far as possible. Thus

$$7^{9999} = 7^3 \cdot 7^{9996} = 343 \cdot (7^4)^{2499}$$

$$= (343)(2401)^{2499}$$

$$= (343)(1 + 2400)^{2499}$$

$$\equiv (343)(1 + 2499 \cdot 2400) \pmod{1000}$$

(by the Binomial Theorem)

$$= (343)(1 + (2500 - 1)(2400))$$

$$\equiv (343)(1 - 2400) \pmod{1000}$$

$$\equiv (343)(601) \pmod{1000}$$

$$\equiv 143 \pmod{1000}.$$

Hence the last three digits are 143.

Problem 212. *First solution.* Draw a line through Y parallel to BC to meet AB at Z. Let CZ and BY intersect at D; let CZ and XY intersect at E. Join D, X. See Figure 142.

Since $\angle ZCB = \angle YBC = 60°$, $\triangle DBC$ is equilateral. Hence $\angle ZDY = 60°$, so that the isosceles triangle ZDY is in fact equilateral.

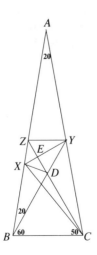

FIGURE 142

Consider $\triangle BXC$. Since $\angle BXC = 50° = \angle BCX$, $\overline{BX} = \overline{BC} = \overline{BD}$. Hence $\angle BXD = \angle BDX = 80°$ so that $\angle ZXD = 100°$.

Now

$$\angle XZD = 180° - \angle AZY - \angle YZD = 40°,$$

$$\angle XDZ = 180° - \angle BDC - \angle BDX = 40°.$$

Hence $\overline{XZ} = \overline{XD}$. Since also $\overline{YZ} = \overline{YD}$, $\triangle XZY \cong \triangle XDY$. Thus

$$\angle ZXY = \angle YXD = \frac{1}{2}\angle ZXD = 50°.$$

Second solution. We may consider A as the center of a regular 18-gon and BC as one edge. Let D, B, C, E, F, G, H be seven vertices of this 18-gon in order and join D, H. Let DH intersect AB, AC and AE in X, Y and W respectively. We show that the points X, Y obtained in this way are exactly the two points specified in the problem.

Consider the heptagon $DBCEFGH$. The sum of its angles is $900°$. Since $\angle HDB = \angle DHG$ and each of the other five angles is $160°$ we must have that $\angle HDB = 50°$. Since $\triangle DXB \cong \triangle CXB$, $\angle XCB = \angle XDB = 50°$.

Since $\angle DAE = 60°$ and $\overline{AD} = \overline{AE}$, $\triangle ADE$ is equilateral. Similarly $\triangle AHE$ is equilateral. Therefore

$$\overline{AD} = \overline{AE} = \overline{DE} = \overline{AH} = \overline{HE},$$

so $ADEH$ is a rhombus and segments DH and AE right-bisect each other. Hence $\overline{AY} = \overline{YE} = \overline{YB}$, so that $\angle ABY = \angle BAY = 20°$. Since $\angle ABC = 80°$, it follows that $\angle YBC = 60°$.

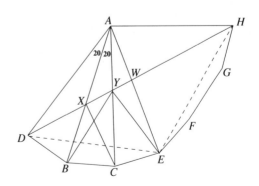

FIGURE 143

Having thus identified X and Y, we observe that, in $\triangle XDB$, $\angle XBD = 80°$ and $\angle XDB = 50°$ so $\angle DXB = 50°$. Hence $\angle AXY = 50°$.

Remark. This problem is treated in the book *Mathematical Gems II*, pp. 16–18, by Ross Honsberger, published by the Mathematical Association of America, 1976.

Problem 213. The volume of a tetrahedron is proportional to the area of the base (one of its triangular faces) and the height (perpendicular distance from the remaining vertex to the face produced). Therefore, we can solve the problem by showing that each of those ingredients remains unchanged. Clearly, it is enough to show the result when one of the segments is held fixed and the other is permitted to move.

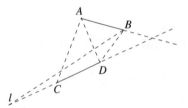

FIGURE 144

Let AB be the fixed segment and let CD vary along the line l. Let π be the plane containing A and the line l. Then the area of $\triangle ACD$ remains constant for all positions of CD. The distance from B to the face ACD produced is the distance from B to π, which is constant. The result follows.

Remark. It can also be shown that the volume of a tetrahedron is equal to one-sixth the product of: the lengths of a pair of opposite edges, the shortest distance between these edges and the sine of the angle between these edges. (See Tool Chest, E 41, 42.)

Problem 214. (a) Let u be the speed of the men and x the distance travelled by the inspecting officer. The officer's speed is ux. While he is moving forward, the speed of the officer relative to the column is $u(x - 1)$, so the time it takes him to reach the front of the column is $\frac{1}{u(x-1)}$. On return to the rear, his relative speed is $u(x + 1)$, so the time required is $\frac{1}{u(x+1)}$. Since the time required for the column to travel the kilometer is $\frac{1}{u}$, we obtain (after multiplying by u),

$$\frac{1}{x-1} + \frac{1}{x+1} = 1.$$

Thus $x^2 - 2x - 1 = 0$. The negative root is not applicable, so $x = 1 + \sqrt{2}$. Hence the officer travels $1 + \sqrt{2}$ kilometers.

Second solution. Let v and u denote the speeds of the officer and men respectively. Also, let t and s denote the time taken for the officer to get to the front and back again respectively. Then $(v-u)t = 1$, $(v + u)s = 1$ and $u(t + s) = 1$. Hence

$$u\left(\frac{1}{v-u} + \frac{1}{v+u}\right) = 1,$$

or

$$v^2 - 2vu - u^2 = 0,$$

so that $v = u(1 + \sqrt{2})$. The distance travelled by the officer is therefore

$$v(t + s) = u(1 + \sqrt{2})\frac{1}{u} = 1 + \sqrt{2}$$

kilometers.

(b) As in (a), the speed of the officer relative to the phalanx on the forward and backward journeys are $u(x - 1)$ and $u(x + 1)$ respectively. While crossing the front or rear of the phalanx, his velocity has a component u in the direction of march and v in the perpendicular direction, where $u^2 + v^2 = u^2x^2$. Hence $v = u\sqrt{x^2 - 1}$. Hence the total time for the circuit is

$$\frac{1}{u(x-1)} + \frac{1}{u(x+1)} + \frac{2}{u\sqrt{x^2-1}} = \frac{1}{u}.$$

This yields

$$\frac{2}{\sqrt{x^2-1}} = 1 - \left(\frac{1}{x-1} + \frac{1}{x+1}\right)$$

$$= 1 - \frac{2x}{x^2-1}$$

or

$$2\sqrt{x^2-1} = x^2 - 2x - 1.$$

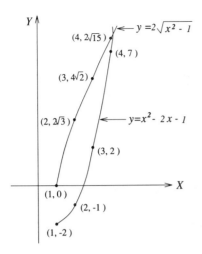

FIGURE 145

The conditions of the problem require that $x > 1$. (Indeed, he travels at least one kilometer to reach the front, and two more kilometers as he rides across the front and the rear, so $x > 3$.)

A graphical solution for x is given in Figure 145: The root sought satisfies $4 < x < 5$. Removing the surd leads to $x^4 - 4x^3 - 2x^2 + 4x + 5 = 0$. An approximate solution is $x = 4.18$.

Problem 215. Let v be the volume of the tetrahedron and t the common area of the four faces. Then the lengths of the four perpendiculars from the vertices to their opposite faces are each h, where $3v = th$. Since $\overline{OA} + \overline{OL} \geq h$,

$$t(\overline{OA} + \overline{OL}) \geq 3v = t(\overline{OL} + \overline{OM} + \overline{ON} + \overline{OP}),$$

so that $\overline{OA} \geq \overline{OM} + \overline{ON} + \overline{OP}$. Adding this inequality to similar inequalities involving OB, OC and OD yields the result.

Rider. Does the same inequality hold when the faces are not necessarily of equal area?

Problem 216. Let x be the amount of water present when the pumping begins, y the amount leaking in per hour and z the amount each man can remove per hour. Suppose $h(n)$ is the amount of time in hours needed by n men to pump the boat dry. Then

$$x + h(n)y = nh(n)z. \tag{$*$}$$

In particular,

$$x + 3y = 12 \cdot 3 \cdot z = 36z,$$

and

$$x + 10y = 5 \cdot 10 \cdot z = 50z,$$

whence $y = 2z$ and $x = 30z$. Thus $(*)$ becomes

$$30 + 2h(n) = nh(n) \quad \text{or} \quad h(n)(n-2) = 30.$$

When $h(n) = 2$, $n = 17$ and 17 men are needed to do the job in 2 hours.

Problem 217. Color the squares of the checkerboard "alternately" black and white so that any pair of adjacent squares are colored differently. Any pair of squares a knight's move apart have different colors. Since any knight's tour visits mn squares, and mn is odd, the last square occupied has the same color as the first, and hence cannot be a knight's move apart.

Rider. Is there necessarily a non-closed knight's tour?

Problem 218. Suppose, by the watch, x minutes elapse between successive eclipses of the hands. While the minute hand covers x units, the hour hand covers $\frac{x}{12}$ units. Hence $x - 60 = \frac{x}{12}$, so that $x = 60\left(\frac{12}{11}\right) = 60 + \frac{60}{11} = 65 + \frac{5}{11}$. Thus my watch gains $\frac{5}{11}$ minutes every 65 minutes.

In order for my watch to gain an hour, the amount of time that must elapse is

$$\frac{1}{60} \cdot \frac{60}{5/11} \cdot 65 = 143 \quad \text{hours.}$$

Little Jack Horner
Sits in a corner
Extracting square roots to infinity
An occupation for boys
That will minimize noise
And produce a more peaceful vicinity.

H. Winson

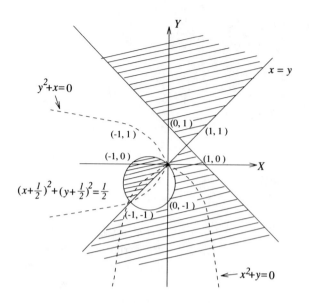

FIGURE 146

Problem 219. If $x^2 + y \geq 0$ and $y^2 + x \geq 0$, the inequality is $(y - x)(x + y - 1) \geq 0$.

If $x^2 + y \leq 0$ and $y^2 + x \leq 0$, the inequality is $(x - y)(x + y - 1) \geq 0$.

If $x^2 + y \leq 0$ and $y^2 + x \geq 0$, the inequality is $-(x^2 + y) \leq y^2 + x$ or $x^2 + y^2 + x + y \geq 0$, or $(x + \frac{1}{2})^2 + (y + \frac{1}{2})^2 \geq \frac{1}{2}$.

If $x^2 + y \geq 0$ and $y^2 + x \leq 0$, the inequality is $(x + \frac{1}{2})^2 + (y + \frac{1}{2})^2 \leq \frac{1}{2}$.

As a check, observe that if $y = 0$, the inequality reduces to $|x^2| \leq |x|$ or $|x| \leq 1$, while if $x = 0$, the inequality reduces to $y = 0$ or $|y| \geq 1$. Every point for which $x = y$ and for which $x^2 + y = 0$ lies in the locus, while the only points on the parabola $y^2 + x = 0$ on the locus are $(-1, -1)$ and $(0, 0)$, i.e., those for which also $x^2 + y = 0$.

There was a mathematician named Moser[1],
Well-known as a problem proposer.
He gave some that were silly
To his brother named Willy.
Did that stump him; the answer is No Sir.

---

[1] Leo Moser

Problem 220. The idea is to get an expression for the left side involving sums and products of squares, which are positive. By the Factor Theorem, $a - b$ divides $3a^4 - 4a^3b + b^4$. Thus

$$3a^4 - 4a^3b + b^4$$
$$= 3a^4 - 3a^3b - a^3b + b^4$$
$$= (a - b)3a^3 - b(a^3 - b^3)$$
$$= (a - b)3a^3 - b(a - b)(a^2 + ab + b^2)$$
$$= (a - b)(3a^3 - a^2b - ab^2 - b^3)$$
$$= (a - b)(a^3 - b^3 + a^3 - a^2b + a^3 - ab^2)$$
$$= (a - b)^2(a^2 + ab + b^2 + a^2 + a^2 + ab)$$
$$= (a - b)^2(2a^2 + (a + b)^2) \geq 0.$$

Alternative solution. By the Arithmetic-Geometric Mean Inequality,

$$\frac{a^4 + a^4 + a^4 + b^4}{4} \geq \sqrt[4]{a^{12}b^4} = a^3b.$$

Problem 221. Let the side lengths be a, b, c with angles of magnitudes α and 2α opposite sides of length b and c respectively. By the Sine Law,

$$2\cos\alpha = \frac{\sin 2\alpha}{\sin\alpha} = \frac{c}{b}.$$

By the Cosine Law,

$$b^2 = a^2 + c^2 - 2ac\cos\alpha = a^2 + c^2 - \frac{ac^2}{b},$$

so that

$$b(b^2 - a^2) = c^2(b - a).$$

If $b = a$, then the angles are α, α, 2α, which leads to $\alpha = 45°$, in which case a, b, c cannot all be integers. Hence $b - a \neq 0$ so that $b(b + a) = c^2$.

By dividing out a common factor if necessary, we may assume that the greatest common divisor of a, b, and c is 1. Then b and $b + a$ must be perfect squares. Hence, for some coprime pair r, s, we have $a = s^2 - r^2$, $b = r^2$, $c = rs$. Since $b < c < 2b$, we have $r < s < 2r$.

In general the sides of the triangle must be

$$\left(k(s^2 - r^2), kr^2, krs\right),$$

where k, r, s are natural numbers with $r < s < 2r$ and greatest common divisor of r and s equal to 1.

We now show that, in fact, every set (k, r, s) satisfying the above conditions gives rise to a triangle of the type specified. For, if $a = s^2 - r^2$, $b = r^2$, $c = rs$, then

$$(a + c) - b = s(s + r) - 2r^2 > 0,$$

$$(a + b) - c = s(s - r) > 0,$$

$$(b + c) - a = 2r^2 + rs - s^2$$

$$= (2r - s)(r + s) > 0,$$

so that a, b, c are sides of a triangle. If $\angle B$ and $\angle C$ are the angles opposite b and c respectively, then, using $b(b + a) = c^2$, we have

$$\cos C = \frac{-c^2 + a^2 + b^2}{2ab} = \frac{a^2 - ab}{2ab} = \frac{a - b}{2b},$$

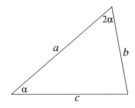

FIGURE 147

$$\cos B = \frac{-b^2 + a^2 + c^2}{2ac} = \frac{a(a + b)}{2ac} = \frac{c}{2b},$$

and

$$\cos 2B = 2\left(\frac{c^2}{4b^2}\right) - 1 = \frac{a - b}{2b} = \cos C;$$

hence $2\angle B = \angle C$.

Problem 222. (a) A number is triangular if and only if it is of the form $\frac{1}{2}k(k + 1)$. If $n = \frac{1}{2}k(k + 1)$, then $9n + 1 = \frac{1}{2}(3k + 1)(3k + 2)$, which is triangular.

(b) Suppose, for each n, n is triangular implies that $an + b$ is triangular. If $n = \frac{1}{2}k(k + 1)$ and $an + b = \frac{1}{2}r(r + 1)$, then

$$\frac{ak(k + 1)}{2} + b = \frac{r(r + 1)}{2},$$

so that

$$r^2 + r - [ak(k + 1) + 2b] = 0.$$

For each value of k, this quadratic equation must have an integer solution r. This will happen if and only if a and b are such that, for each k, the discriminant is a perfect square. If $1 + 4ak(k + 1) + 8b = (ku + v)^2$ for some u and v independent of k then, by comparison of coefficients, $4a = u^2 = 2uv$, so $2v = u$. Hence $b = \frac{v^2 - 1}{8}$ and $a = v^2$. If v is odd then b will be an integer.

Thus, if v is any odd number, $a = v^2$ and $b = \frac{v^2 - 1}{8}$ and $n = \frac{k(k + 1)}{2}$ is a triangular number, then

$$an + b = \frac{v^2 k(k + 1)}{2} + \frac{v^2 - 1}{8}$$

$$= \frac{1}{2}\left[vk + \frac{v - 1}{2}\right]\left[vk + \frac{v + 1}{2}\right]$$

is also triangular.

Problem 223. By squaring, it is easy to verify that

$$\sqrt{4n + 1} < \sqrt{n} + \sqrt{n + 1}$$

$$< \sqrt{4n + 3} \qquad (n = 1, 2, 3, \ldots).$$

Neither $4n + 2$ nor $4n + 3$ are squares, so

$$\left[\sqrt{4n + 1}\right] = \left[\sqrt{4n + 2}\right] = \left[\sqrt{4n + 3}\right]$$

and the result follows.

Problem 224. First, we should characterize the position of P which maximizes $\angle APB$. Draw the

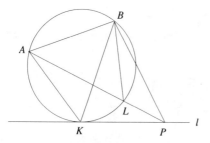

FIGURE 148

circle through A and B which is tangent to l, at K say (Figure 148). Then, for any point P on l other than K,

$$\angle AKB = \angle ALB > \angle APB.$$

Knowing this, we can easily construct a configuration (e.g., Figure 149) for which the statement is false.

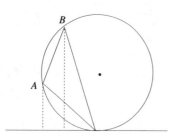

FIGURE 149

Problem 225. Since

$$\sin C \le 1 \quad \text{and} \quad \sin A \sin B \ge 0,$$

$$1 \ge \cos(A - B)$$

$$= \cos A \cos B + \sin A \sin B$$

$$= 1 + \sin A \sin B(1 - \sin C) \ge 1.$$

Hence $\cos(A - B) = 1$, $\sin C = 1$, so $\angle C = 90°$, $\angle A = \angle B = 45°$. Thus ABC is a right-angled isosceles triangle.

Problem 226. One obvious generalization is

$$a_n = (n + 1)^2 + (n + 2)^2 + ((n + 1)(n + 2))^2$$

$$= (n^2 + 3n + 3)^2.$$

Problem 227. *First solution.* If $xyz = 0$, say $z = 0$, $x \ge 0$, $y \ge 0$, the inequality to be proved is

$$8(x^3 + y^3)^2 \ge 9x^2y^2xy,$$

and this is equivalent to the obvious inequality

$$8x^6 + 7x^3y^3 + 8y^6 \ge 0.$$

If $xyz \ne 0$, i.e., $x > 0$, $y > 0$, $z > 0$ then, taking

$$a = \frac{x^2}{yz}, \quad b = \frac{y^2}{zx}, \quad c = \frac{z^2}{xy},$$

the inequality to be proved is equivalent to

$$8(a + b + c)^2$$

$$-9\left(1 + \frac{1}{a}\right)\left(1 + \frac{1}{b}\right)\left(1 + \frac{1}{c}\right) \ge 0,$$

$a, b, c > 0$, $abc = 1$. Now the left side is seen to be equal to

$$8(a^2 + b^2 + c^2) + 16(ab + bc + ca) - 18$$

$$-9\left(\frac{1}{a} + \frac{1}{b} + \frac{1}{c}\right) - 9\left(\frac{1}{ab} + \frac{1}{bc} + \frac{1}{ca}\right)$$

$$= 8(a^2 + b^2 + c^2) + 7\left(\frac{1}{a} + \frac{1}{b} + \frac{1}{c}\right)$$

$$-9(a + b + c) - 18$$

$$= \frac{1}{a}(8a^3 - 9a^2 - 6a + 7)$$

$$+ \frac{1}{b}(8b^3 - 9b^2 - 6b + 7)$$

$$+ \frac{1}{c}(8c^3 - 9c^2 - 6c + 7)$$

$$= \frac{(a - 1)^2(8a + 7)}{a} + \frac{(b - 1)^2(8b + 7)}{b}$$

$$+ \frac{(c - 1)^2(8c + 7)}{c},$$

and this expression is certainly not negative.

Second solution. By the Arithmetic-Geometric Mean Inequality

$$(x^2 + yz)(y^2 + zx)(z^2 + xy)$$

$$\le \left(\frac{x^2 + y^2 + z^2 + yz + zx + xy}{3}\right)^3.$$

Since

$$0 \le (x - y)^2 + (y - z)^2 + (z - x)^2$$

$$= 2(x^2 + y^2 + z^2 - yz - zx - xy),$$

it follows that $yz + zx + xy \le x^2 + y^2 + z^2$ and hence

$$(x^2+yz)(y^2+zx)(z^2+xy) \le \frac{8}{27}(x^2+y^2+z^2)^3.$$

Since

$$\left(\frac{x^m + y^m + z^m}{3}\right)^{1/m}$$

is an increasing function of m (see Tool Chest D7),

$$\left(\frac{x^2 + y^2 + z^2}{3}\right)^{\frac{1}{2}} \le \left(\frac{x^3 + y^3 + z^3}{3}\right)^{\frac{1}{3}}$$

so that $(x^2 + y^2 + z^2)^3 \le 3(x^3 + y^3 + z^3)^2$. Use of this now gives the result.

Remark. Prove the generalization: for $x_1, x_2, \ldots, x_n \ge 0$,

$$2^n \left\{\frac{x_1^n + x_2^n + \cdots + x_n^n}{n}\right\}^{n-1}$$

$$\ge \prod_{i=1}^{n} \left(x_i^{n-1} + \frac{x_1 x_2 \cdots x_n}{x_i}\right).$$

Problem 228. Let S be the set of people who have made an even number of handshakes and T the set of those who have made an odd number of handshakes. For any person k, let h_k be the number of handshakes he made.

Since each handshake involves exactly two people, $\sum_{x \in S} h_x + \sum_{y \in T} h_y$ is double the total number of handshakes and is therefore even. Since for $x \in S$, h_x is even, it follows that $\sum_{x \in S} h_x$ is even. Hence $\sum_{y \in T} h_y$ is also even. But h_y is odd for each $y \in T$. Therefore T must contain an even number of people.

Problem 229. Define $p = x + y + z$. Subtracting (2) from (1) yields

$$p(x - y) = a^2 - b^2. \qquad (4)$$

Similarly,

$$p(y - z) = b^2 - c^2, \qquad (5)$$

$$p(z - x) = c^2 - a^2. \qquad (6)$$

Consider first the case that $a^2 = b^2 = c^2$. Then either $x = y = z$, which can occur only when $a = b = c = 0$, or else $p = x + y + z = 0$. In the latter situation, each of the three equations (1), (2), (3) yields

$$x^2 + xy + y^2 = a^2.$$

Thus x, say, can be chosen arbitrarily and the remaining variables then determined.

Now suppose a^2, b^2, c^2 are not all equal. In particular, $a^2 + b^2 + c^2 \ne 0$. It follows from (4), (5), (6) that $p \ne 0$. Since $py = px - (a^2 - b^2)$ and $pz = px + (c^2 - a^2)$,

$$p^2 = p(x + y + z) = 3px + (b^2 + c^2 - 2a^2)$$

or

$$3px = p^2 + (2a^2 - b^2 - c^2). \qquad (7)$$

Similarly,

$$3py = p^2 + (2b^2 - a^2 - c^2), \qquad (8)$$

$$3pz = p^2 + (2c^2 - a^2 - b^2). \qquad (9)$$

If we can separately determine p, the equations will be solved. Adding (1), (2), (3) gives

$$(x - y)^2 + (y - x)^2 + (z - x)^2 = 2(a^2 + b^2 + c^2).$$

Using (4), (5), (6) with this gives

$$\begin{aligned}
p^2 &= \frac{(a^2 - b^2)^2 + (b^2 - c^2)^2 + (c^2 - a^2)^2}{2(a^2 + b^2 + c^2)} \\
&= \frac{a^4 + b^4 + c^4 - a^2 b^2 - a^2 c^2 - b^2 c^2}{a^2 + b^2 + c^2} \\
&= \frac{a^6 + b^6 + c^6 - 3a^2 b^2 c^2}{(a^2 + b^2 + c^2)^2}.
\end{aligned}$$

Leo Moser had signed a contract with a publisher to write a book (a collection of problems). Every year the publisher's representative asked Leo, "When will you have the manuscript completed?" On the seventh such request, Leo replied, " I'll publish the book posthumously," to which the representative replied, "Well, O K, but make it soon!"

From (7)

$$x = \frac{1}{3p}\{p^2 + (2a^2 - b^2 - c^2)\}$$

$$= \frac{(a^2 + b^2 + c^2)\{(a^4 + b^4 + c^4 - a^2b^2 - a^2c^2 - b^2c^2) + (a^2 + b^2 + c^2)(2a^2 - b^2 - c^2)\}}{3\sqrt{a^6 + b^6 + c^6 - 3a^2b^2c^2}\,(a^2 + b^2 + c^2)}$$

$$= \frac{3a^4 - 3b^2c^2}{3\sqrt{a^6 + b^6 + c^6 - 3a^2b^2c^2}}$$

Thus, we eventually obtain

$$\frac{x}{a^4 - b^2c^2} = \frac{y}{b^4 - c^2a^2} = \frac{z}{c^4 - a^2b^2}$$

$$= \frac{1}{\sqrt{a^6 + b^6 + c^6 - 3a^2b^2c^2}}.$$

(Note that $a^6 + b^6 + c^6 - 3a^2b^2c^2$ does not vanish, since $p \neq 0$, and therefore it has two square roots. The vanishing of $a^4 - b^2c^2$ is equivalent to the vanishing of x, with a similar comment for y and z.)

Problem 230. When n is even , we have

$$(1^2 - 2^2) + (3^2 - 4^2) + \cdots + ((n-1)^2 - n^2))$$

$$= (1 - 2)(1 + 2) + (3 - 4)(3 + 4)$$

$$+ \cdots + (\overline{n-1} - n)(\overline{n-1} + n)$$

$$= -(1 + 2) - (3 + 4) - \cdots - (\overline{n-1} + n)$$

$$= -(1 + 2 + 3 + 4 + \cdots + n)$$

$$= (-1)^{n+1}(1 + 2 + 3 + \cdots + n).$$

When n is odd, we have

$$1^2 - (2^2 - 3^2) - (4^2 - 5^2)$$

$$- \cdots - ((n-1)^2 - n^2)$$

$$= 1 - (2 - 3)(2 + 3) - (4 - 5)(4 + 5)$$

$$- \cdots - (\overline{n-1} - n)(\overline{n-1} + n)$$

$$= 1 + 2 + 3 + \cdots + \overline{n-1} + n$$

$$= (-1)^{n+1}(1 + 2 + 3 + \cdots + n).$$

Remarks. This result may also be proved by induction or by writing the left side as

$$(1^2 + 2^2 + 3^2 + 4^2 + \cdots) - 2 \cdot 2^2(1^2 + 2^2 + 3^2 + \cdots)$$

and using the formula for the sum of squares.

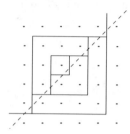

FIGURE 150

The equation can also be seen diagrammatically, as Figure 150 for the case $n = 6$ shows.

Add 6^2 dots, remove 5^2 dots, add 4^2 dots, remove 3^2 dots, etc., as indicated; the dots remaining lie below the "dotted" line.

Problem 231. Since $(a + b - c)(a - b + c) = a^2 - (b - c)^2 \leq a^2$ we have

$$(a + b - c)(a - b + c) \leq a^2,$$

$$(b + c - a)(b - c + a) \leq b^2, \qquad (*)$$

$$(c + a - b)(c - a + b) \leq c^2.$$

Since $a + b - c > 0$, $b + c - a > 0$, and $c + a - b > 0$, we may multiply the three inequalities in $(*)$ and then take the square root to obtain the desired inequality.

Comment. For other proofs and generalizations see *Crux Mathematicorum* 10 (1984) 46–48.

Problem 232. Let O be the center of the region and AB be any chord. Let $A'B'$ be the centrosymmetric image of AB. See Figure 151. Then

$$\overline{AB} \leq \overline{AO} + \overline{OB} \leq 2\max(\overline{AO}, \overline{OB})$$

$$= \max(2\overline{AO}, 2\overline{OB}) = \max(\overline{AA'}, \overline{BB'}).$$

The result follows.

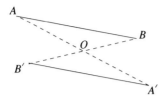

FIGURE 151

Problem 233. It is possible to find a succession of moves whereby all coins end up in the same sector if and only if k is odd.

Suppose $k = 2m + 1$. Number the sectors consecutively

$$-m, -m+1, \ldots, -2, -1, 0, 1, 2, \ldots, m-1, m.$$

A sequence of moves of the type $(-i \to -i + 1, i \to i - 1)$ will eventually move all coins to the sector numbered 0.

On the other hand, suppose $k = 2m$. Number the sectors consecutively $1, 2, 3, \ldots, 2m$. For each ranging of the coins, let n_i be the number in ith sector and form the sum

$$s = n_1 + 2n_2 + 3n_3 + \cdots + kn_k.$$

The sum s is not changed (modulo k) by each move.

For the initial position of the coins,

$$s = 1+2+\cdots+2m = m(2m+1) \not\equiv 0 \pmod{2m}.$$

For all coins in sector i, $s = 2mi \equiv 0 \pmod{2m}$. Hence, it is not possible for a sequence of moves to put all coins in the same sector.

Problem 234. The given equations are equivalent to

$$2 \sin \frac{x+y}{2} \cos \frac{x-y}{2} = a, \tag{1}$$

$$2 \cos \frac{x+y}{2} \cos \frac{x-y}{2} = b. \tag{2}$$

Dividing (1) by (2) yields

$$\tan \frac{x+y}{2} = \frac{a}{b}. \tag{3}$$

Squaring both equations (1) and (2) and adding gives

$$4 \cos^2 \frac{x-y}{2} = a^2 + b^2. \tag{4}$$

From (3) and (4) we have

$$\cos \frac{x+y}{2} = \frac{b}{\sqrt{a^2+b^2}}$$

and

$$\cos \frac{x-y}{2} = \frac{\sqrt{a^2+b^2}}{2}.$$

(We might also have taken a minus sign in front of *both* expressions; the final result will be the same.)

Hence

$$2 \cos \frac{x}{2} \cos \frac{y}{2} = \frac{b}{\sqrt{a^2+b^2}} + \frac{\sqrt{a^2+b^2}}{2}$$
$$= \frac{a^2+b^2+2b}{2\sqrt{a^2+b^2}},$$

$$2 \sin \frac{x}{2} \sin \frac{y}{2} = \frac{\sqrt{a^2+b^2}}{2} - \frac{b}{\sqrt{a^2+b^2}}$$
$$= \frac{a^2+b^2-2b}{2\sqrt{a^2+b^2}},$$

so that

$$\tan \frac{x}{2} \tan \frac{y}{2} = \frac{a^2+b^2-2b}{a^2+b^2+2b}. \tag{5}$$

Therefore

$$\tan \frac{x}{2} + \tan \frac{y}{2} = \tan \frac{x+y}{2} \left(1 - \tan \frac{x}{2} \tan \frac{y}{2}\right)$$
$$= \frac{a}{b} \left(\frac{4b}{a^2+b^2+2b}\right)$$
$$= \frac{4a}{a^2+b^2+2b}. \tag{6}$$

From (5) and (6), we can identify $\tan \frac{x}{2}$ and $\tan \frac{y}{2}$ as roots of the quadratic
$(a^2+b^2+2b)t^2 - 4at + (a^2+b^2-2b).$

Problem 235. Referring to Figure 152, we see that, since $\triangle BMN$ is isosceles, $\angle BNM = \frac{1}{2}\angle BMA$, which is a constant angle. Hence N must lie on a circle passing through points A, B. The center of this circle will be at that point O for which $\overline{AO} = \overline{BO} = \overline{OQ}$. However, the locus

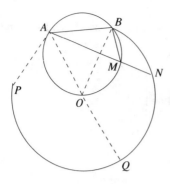

FIGURE 152

will include only part of the circle, namely the arc from B to the point P such that $\overline{AP} = \overline{AB}$ and AP is tangent to the circle AOB. For if N lies on the arc, then $\angle BNA = \frac{1}{2}\angle AOB = \frac{1}{2}\angle AMB$ whence $\angle BNM = \angle MBN$ and $\overline{BM} = \overline{BN}$.

Taking into account the possibility that M is on either side of AB, we see that the locus of N consists of two circular arcs, both emanating from B, both part of circles through A and B with centers at the ends of the diameter bisecting AB and both lying on the same side of the tangent to the given circle through A.

Question. How would you interpret the remainder of the circles?

Problem 236. The equation is

$$x^6 + x^5 + x^4 - 28x^3 + x^2 + x + 1 = 0.$$

Obviously $x = 0$ is not a solution, so the equation is equivalent to

$$x^3 + x^2 + x - 28 + \frac{1}{x} + \frac{1}{x^2} + \frac{1}{x^3} = 0.$$

Let $z = x + \frac{1}{x}$. Then $z^2 = x^2 + 2 + \frac{1}{x^2}$ and $z^3 = x^3 + 3(x + \frac{1}{x}) + \frac{1}{x^3}$; hence $x^2 + \frac{1}{x^2} = z^2 - 2$ and $x^3 + \frac{1}{x^3} = z^3 - 3z$. Thus, we obtain the equation

$$z^3 + z^2 - 2z - 30 = 0.$$

By trial we find that $z = 3$ is a root. Then, by the Factor Theorem,

$$(z - 3)(z^2 + 4z + 10) = 0,$$

whence $z = 3$ is the only real solution.

Since x real implies z real, real solutions x must satisfy $3 = x + \frac{1}{x}$ or $x^2 - 3x + 1 = 0$, i.e.,

$$x = \frac{3 + \sqrt{5}}{2} \qquad \text{or} \qquad x = \frac{3 - \sqrt{5}}{2}.$$

Remark. By Descartes' Rule of Signs, one can see at the outset that the equation has at most two positive solutions.

Problem 237. *First solution.* Let n portions, $b_1, b_2, \ldots, b_n$, of gas be placed at positions $P_1, P_2, \ldots, P_n$, listed clockwise in order around the track. Suppose the amount of gas required to get from position P_i to position P_{i+1} is a_i ($i = 1, 2, \ldots, n - 1$) and the amount to get from position P_n to P_1 is a_n.

Let $x_i = b_i - a_i$ and $s_i = x_1 + x_2 + \ldots + x_i$ ($i = 1, 2, \ldots, n$). Note that $s_n = 0$. Choose r such that $s_r \le s_i$ for each i. Then, for each i,

$$x_{r+1} = s_{r+1} - s_r \ge 0,$$

$$x_{r+1} + x_{r+2} \ge s_{r+2} - s_r \ge 0,$$

$$\vdots$$

$$x_{r+1} + x_{r+2} + \cdots + x_i = s_i - s_r \ge 0.$$

Hence, if we start the car at position $r + 1$, since $b_{r+1} \ge a_{r+1}$, $b_{r+1} + b_{r+2} \ge a_{r+1} + a_{r+2}$, and so on, the car will always have enough gas to get from each position to the next.

Second solution (by induction). We show that a clockwise circuit is always possible. The result clearly holds when all the gas is placed at one spot.

Suppose it holds for any division of the gas into $n - 1$ portions. Now, let amounts $b_1, b_2, \ldots, b_n$ be placed at positions $P_1, P_2, \ldots, P_n$ (listed clockwise). For at least one position, there must be enough gas to move clockwise around to the next position; without loss of generality, we may assume that b_1 is enough gas to go from P_1 to P_2.

Put the gas at P_2 with that at P_1. By the induction hypothesis, the circuit can be completed, starting at some point r with gas pickups of $b_1 + b_2$ at P_1 and b_i at P_i ($i = 3, 4, \ldots, n$). Since b_1 is enough gas at P_1 for the car to get to P_2, the same circuit can be completed if the amount b_2 is restored to P_2.

Problem 238. Because of periodicity, we need consider only values of x belonging to some interval of length 2π. Thus, assume that $-\frac{1}{2}\pi \le x < \frac{3}{2}\pi$.

Suppose that $\frac{1}{2}\pi < x < \frac{3}{2}\pi$. Then $-1 \le \cos x < 0$ so that $\sin(\cos x) < 0$, while $-\frac{1}{2}\pi < -1 \le \sin x \le 1 < \frac{1}{2}\pi$ insures that $\cos(\sin x) \ge 0$. Hence $\cos(\sin x) > \sin(\cos x)$ for $\frac{1}{2}\pi < x < \frac{3}{2}\pi$.

Consider the case that $-\frac{1}{2}\pi \le x \le \frac{1}{2}\pi$. Clearly, the inequality holds for $x = 0$, while for $x < 0$,

$$\cos(\sin(-x)) = \cos(-\sin x) = \cos(\sin x)$$

and $\sin(\cos(-x)) = \sin(\cos x)$. Therefore it suffices to show the result for $0 < x \le \frac{1}{2}\pi$. Now $0 < \sin x < x$ and since $\cos x$ is decreasing for $0 < x \le \frac{1}{2}\pi$,

$$\cos(\sin x) > \cos x > \sin(\cos x).$$

Rider. Investigate the relationship between $\sin(\sin x)$ and $\cos(\cos x)$.

Problem 239. The given equation is equivalent to

$$(x + y + 1)(x + y - m - 1) = 0.$$

Since $x + y + 1 > 0$ for each applicable (x, y), $x + y = m + 1$ which clearly has the m solutions $(x, y) = (i, m + 1 - i)$ $(i = 1, 2, \ldots, m)$.

Problem 240. Choose P, Q, S to be the midpoints of AB, BC, DA respectively. Since the areas of $PQRS$ and $PQR'S$ are equal,

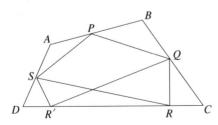

FIGURE 153

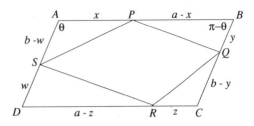

FIGURE 154

Area SQR = Area SQR', whence $SQ \parallel DC$. So must $SQ \parallel AB$; hence $AB \parallel CD$. Similarly, $AD \parallel BC$. Thus $ABCD$ must be a parallelogram.

We now show that the area equality result holds for an arbitrary convex quadrilateral $PQRS$ inscribed in any parallelogram $ABCD$. Referring to Figure 154 for notation, we see that

$$2(\text{Area } ABCD - \text{Area } PQRS)$$
$$= (\sin\theta)\{x(b - w) + y(a - x)$$
$$\quad + z(b - y) + w(a - z)\}$$
$$= (\sin\theta)\{(a - x)w + (b - y)x$$
$$\quad + (a - z)y + (b - w)z\}$$
$$= 2(\text{Area } ABCD - \text{Area} P'Q'R'S').$$

Hence Area $PQRS$ = Area $P'Q'R'S'$.

Problem 241. By the given equations, the fourth degree polynomial in t,

$$t^4 - wt^3 - zt^2 - yt - x$$

has roots a, b, c, d. Hence it is equal to $(t - a) \times (t - b)(t - c)(t - d)$. Expanding and comparing coefficients gives

$$w = a + b + c + d,$$
$$z = -(ab + bc + cd + da + ac + bd),$$
$$y = abc + abd + acd + bcd,$$
$$x = -abcd.$$

Rider. What can be said if a, b, c, d are not all distinct?

Problem 242. *First solution.* Since a and b must be of the form $2mn$ and $m^2 - n^2$ (not necessarily respectively), we have to show that

$$P \equiv mn(m-n)(m+n) \times$$

$$(m^2 - 2mn - n^2)(m^2 + 2mn - n^2)$$

is divisible by $2 \cdot 3 \cdot 7$. It suffices to consider the case that m and n are relatively prime.

By congruences, modulo 2 and 3, we see that $mn(m-n)(m+n)$ is divisible by 6.

Take congruences, modulo 7. If $m \equiv 0$, $n \equiv 0$, $m \equiv 0$, $m \equiv n$ or $m \equiv -n$ (mod 7), then $P \equiv 0$. The remaining cases are

$$(m,n) \equiv (1,2), \ (1,3), \ (1,4), \ (1,5),$$

$$(2,3), \ (2,4), \ (2,6), \ (3,5),$$

$$(3,6), \ (4,5), \ (4,6), \ (5,6).$$

For each of these,

$$(m^2 - 2mn - n^2)(m^2 + 2mn - n^2)$$

$$\equiv 0 \pmod 7.$$

Second solution. It suffices to prove the result when the greatest common divisor of a, b, c is 1. Since c and one of a, b, say a, is odd, $b^2 = c^2 - a^2$ is divisible by 8. Hence $b \equiv 0 \pmod 4$, so $ab \equiv 0 \pmod 4$.

Since $a^2 + b^2 = c^2 \not\equiv 2 \pmod 3$, one of a, b is divisible by 3, so that $ab \equiv 0 \pmod 3$.

Since, modulo 7, any square is congruent to 0, 1, 2 or 4, one can check that $a^2 + b^2$ is congruent to a square only if either $a \equiv b \pmod 7$, $a \equiv -b \pmod 7$ or $ab \equiv 0 \pmod 7$. The result follows from these assertions.

Problem 243. *First solution.* Fix a value of $A \leq \frac{\pi}{2}$. Then $\cos A > 0$ and $\cos(B+C)$ is fixed. Since $\sin B \sin C = \frac{1}{2}(\cos(B-C) - \cos(B+C))$, the maximum value of $\sin^2 A + \sin B \sin C \cos A$ is assumed when $B = C$. In this case $A = \pi - 2B$, $\sin A = \sin 2B$, $\cos A = -\cos 2B$ and

$$\sin^2 A + \sin B \sin C \cos A$$

$$= \sin^2 2B - \sin^2 B \cos 2B$$

$$= \sin^2 B(4\cos^2 B - 2\cos^2 B + 1)$$

$$= \sin^2 B(2\cos^2 B + 1)$$

$$= \sin^2 B(3 - 2\sin^2 B).$$

The function $u(3-2u) = \frac{9}{8} - 2(u-\frac{3}{4})^2$ attains its maximum value when $u = \frac{3}{4}$. Hence $\sin^2 B(3 - 2\sin^2 B)$ assumes its maximum when $B = \frac{\pi}{3}$.

Thus when $A \leq \frac{\pi}{2}$,

$$\sin^2 A + \sin B \sin C \cos A$$

assumes its maximum of $\frac{9}{8}$ when $A = B = C = \frac{\pi}{3}$.

If $A \geq \frac{\pi}{2}$, then $\cos A$ is negative while $\sin B$ and $\sin C$ are positive. The expression $\sin^2 A + \sin B \sin C \cos A$ then cannot exceed 1.

Second solution. Let R be the circumference and a, b, c the lengths of the sides opposite A, B, C of the triangle. Using the formulae $2bc \cos A = b^2 + c^2 - a^2$ and $a = 2R \sin A$, etc., we convert the expression to be maximized to $\dfrac{a^2 + b^2 + c^2}{8R^2}$.

An equivalent problem is to maximize the sum of squares of the sides for any inscribed triangle in a circle of radius R. Accordingly, let (x_1, y_1), (x_2, y_2), (x_3, y_3) be the coordinates of the vertices of a triangle inscribed in the circle of equation $x^2 + y^2 = R^2$. Then

$$a^2 + b^2 + c^2$$

$$= (x_1 - x_2)^2 + (x_2 - x_3)^2 + (x_3 - x_1)^2$$

$$+ (y_1 - y_2)^2 + (y_2 - y_3)^2 + (y_3 - y_1)^2$$

$$= 3(x_1^2 + x_2^2 + x_3^2) - (x_1 + x_2 + x_3)^2$$

$$+ 3(y_1^2 + y_2^2 + y_3^2) - (y_1 + y_2 + y_3)^2$$

$$= 9R^2 - (x_1 + x_2 + x_3)^2$$

$$- (y_1 + y_2 + y_3)^2,$$

since $x_i^2 + y_i^2 = R^2$ ($i = 1, 2, 3$).

Thus $a^2 + b^2 + c^2 \leq 9R^2$, with equality if and only if $x_1 + x_2 + x_3 = 0 = y_1 + y_2 + y_3$, or if the centroid of the triangle coincides with the circumcenter (i.e., the triangle is equilateral).

Therefore

$$\max \frac{a^2 + b^2 + c^2}{8R^2} = \frac{9}{8}.$$

Rider 1. Prove more generally that

$$yza^2 + zxb^2 + xyc^2 \leq (x + y + z)^2 R^2,$$

where x, y, z are arbitrary real numbers.

Rider 2. Prove even more generally that

$$(x + y + z)\left(xR_1^2 + yR_2^2 + zR_3^2\right)$$
$$\geq yza^2 + zxb^2 + xyc^2,$$

where x, y, z are arbitrary real numbers and R_1, R_2, R_3 are the distances from an arbitrary point P to the vertices A, B, C, respectively, of the triangle.

Problem 244. *First solution.* Without loss of generality, we can take one of the three points, A, B, C, say A, to be fixed on the x-axis, as shown in Figure 155. Let $\angle AOB = \theta$ and $\angle AOC = \phi$, measured counterclockwise from AO. Then both θ and ϕ fall in the interval $[0, 2\pi]$. $\theta = \phi$ with probability zero, and the cases $\theta > \phi$ and $\phi > \theta$ are equiprobable. Consequently, we shall assume that $\phi > \theta$. This gives rise to the uniform probability space $\{(\phi, \theta) | 0 < \phi < 2\pi, \theta < \phi\}$.

For $\triangle ABC$ to be acute, $\theta < \pi$, and C must lie on arc DB', where B' is the centrosymmetric image of B, so that $\pi < \phi < \theta + \pi$. (If C is between B and D, then $\angle CBA$ would be obtuse; if C is between B' and A, $\angle BAC$ would be obtuse.) These restrictions require the point (ϕ, θ) to

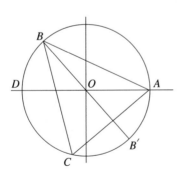

FIGURE 155

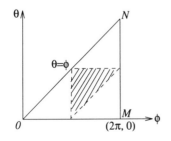

FIGURE 156

lie on the triangle whose vertices are the midpoints of ON, NM, and MO, a triangle with area one-quarter that of $\triangle OMN$. See Figure 156. Hence the required probability is $\frac{1}{4}$.

Second solution (by calculus). Let the circumference of the circle be 1. Let t be the shorter arc length between points A and B; t is chosen uniformly in the interval $[0, \frac{1}{2}]$. If $\triangle ABC$ is to be acute, C must fall in an arc of length t as indicated in Figure 157. This occurs with probability t. Thus the probability that $\angle ABC$ is acute is

$$\left(\frac{1}{2}\right)^{-1} \int_0^{\frac{1}{2}} t\, dt = \frac{1}{4}.$$

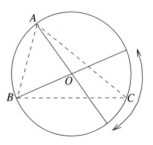

FIGURE 157

Problem 245. Such a coloring is possible. Paint (x, y) red if $x + y$ is even, white if x is odd and y is even, and blue if x is even and y is odd. Clearly, condition (a) is satisfied. Now suppose (x_1, y_1) is red, (x_2, y_2) is white and (x_3, y_3) is blue. Then $x_2 - x_1$ and $y_2 - y_1$ have opposite parity; $x_3 - x_2$ and $y_3 - y_2$ are both odd. Hence

$$(y_2 - y_1)(x_3 - x_2) \neq (y_3 - y_2)(x_2 - x_1),$$

so that $\frac{y_3 - y_2}{x_3 - x_2} \neq \frac{y_2 - y_1}{x_2 - x_1}$ (i.e., the three points are not collinear).

Problem 246. Look at the situation from the point of view of the walker. Suppose the wind is blowing at a velocity v in a direction θ measured from the easterly direction. The relative velocity of the air has two components, a component $\overrightarrow{OW}$ due to the wind and a component $\overrightarrow{OP}$ initially and $\overrightarrow{OQ}$ finally (acting southwards) due to his walking. Referring to Figure 158 and using the fact that

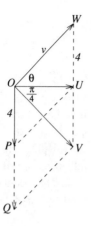

FIGURE 158

$\overline{WV} = \overline{OQ} = 8$, $\overline{OU} = \overline{UV} = \overline{WV} - \overline{WU} = 4$, we see that $\theta = \frac{\pi}{4}$ and $v = 4\sqrt{2}$. Thus the wind is blowing to the northeast at $4\sqrt{2}$ kph.

Problem 247. The configuration consists of three spheres, those of radius R, each touching the other two, along with a fourth smaller sphere, of radius r, nestling among the first three and the table. Since the point of contact of the fourth sphere with the table is the circumcenter of the equilateral triangle of sidelength $2R$ formed by the points of contact of the larger spheres and the table, it is at a distance $\frac{2R}{\sqrt{3}}$ from each of the latter contact points.

First solution. $R+r$ is the hypotenuse of a right triangle of base $\frac{2R}{\sqrt{3}}$ and height $R - r$ (see Figure

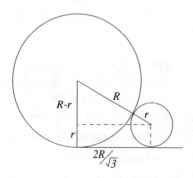

FIGURE 159

159). Hence

$$(R + r)^2 = (R - r)^2 + \frac{4}{3}R^2,$$

whereupon $r = \frac{R}{3}$.

Second solution. In Figure 160 P and Q are centers of a large and the small spheres, and the points of contact are T and U. $VW \perp PQ$, $PV \perp VQ$, and $\overline{VW} = \overline{TV} = \overline{VU} = \frac{R}{\sqrt{3}}$. Since $\overline{VW}^2 = \overline{PW} \cdot \overline{QW}$, $\frac{R^2}{3} = Rr$, so that $r = \frac{R}{3}$.

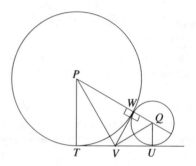

FIGURE 160

Problem 248. Suppose (a, b, c, d) is a solution to the equation, with $1 \le a \le b \le c \le d$. Then a is a root of the quadratic $x^2 - bcdx + (b^2 + c^2 + d^2)$. By the Factor Theorem, it is divisible by $x - a$. Indeed

$$x^2 - bcdx + (b^2 + c^2 + d^2)$$
$$= (x - a)(x - (bcd - a)).$$

Hence $(bcd - a, b, c, d)$ is a second solution.

If $2 \le a \le b \le c \le d$, then $bcd - a \ge 2a^2 - a > a$, so that $(bcd - a) + b + c + d > a + b + c + d$. Applying this procedure repeatedly, taking a as the minimum of a, b, c, d, gives a chain of distinct solutions. Starting with $(2, 2, 2, 2)$ gives $(2, 2, 2, 6)$, $(2, 2, 6, 22)$, $(2, 6, 22, 262)$, $(6, 22, 262, 34582)$, and so on.

Problem 249. Let $\angle BAC = 60°$ and BA be ten feet longer than CA. Let D be the second intersection point of the two circles. See Figure

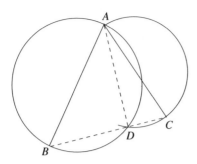

FIGURE 161

161. Since AC is a diameter of circle ADC, $CD \perp AD$. Since AB is a diameter of circle ADB, $BD \perp AD$. Hence B and C lie on the same perpendicular to AD, so that the distance from D to BC produced is 0.

Problem 250. *First solution.* Picture one of the equal sides as the base and let the other side swing in a semi-circle. Since the base is fixed, the area is largest when the altitude is longest, i.e., when the equal sides are perpendicular. Hence the length of the third side will be $\sqrt{2}$ times that of each equal side for the maximum area.

Second solution. Let θ be the angle between the equal sides, both of length a. The area of the triangle is $\frac{1}{2}a^2 \sin \theta$. This is maximum when $\theta = \frac{\pi}{2}$ and the third side has length $a\sqrt{2}$.

Problem 251. *First solution.* Our first solution is motivated by the theorem that the tangents drawn from an external point to a circle are of equal length. In Figure 162, $\overline{UP} = \overline{UR}$ and

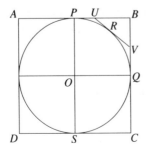

FIGURE 162

$\overline{VR} = \overline{VQ}$. Thus, $\overline{UV} = \overline{PU} + \overline{VQ}$. We now establish the converse theorem.

Proposition. *Let $ABCD$ be a square and let P and Q be the midpoints of AB and BC respectively. Suppose U and V are points in the segments PB and BQ such that $\overline{UV} = \overline{PU} + \overline{VQ}$. Then UV is tangent to the inscribed circle of the square.*

Proof. The three possibilities for the chord UV are:

    (1) not to intersect the circle;

    (2) to be tangent to the circle;

    (3) to intersect the circle in two points.

We rule out cases (1) and (3). If either of these two cases be valid, consider a parallel chord $U'V'$ which is tangent to the circle. Then, as pointed out above, $\overline{U'V'} = \overline{PU'} + \overline{V'Q}$. In the case of (1), $\overline{UV} < \overline{U'V'}$, $\overline{PU} > \overline{PU'}$, $\overline{QV} > \overline{QV'}$, which contradicts the hypothesis. Similarly, in the case of (3), $\overline{UV} > \overline{U'V'}$, $\overline{PU} < \overline{PU'}$, $\overline{QV} < \overline{QV'}$, which again contradicts the hypothesis. Hence case (2) must be correct.

(A direct proof can be obtained by choosing R on UV so that $\overline{PU} = \overline{UR}, \overline{RV} = \overline{VQ}$, by then showing that $\angle BUV = 2\angle URP$, $\angle BVU = 2\angle VRQ$, and hence that $\angle PRQ = 135°$ and $PQRS$ is a concyclic quadrilateral, S being the midpoint of CD.)

Returning to the problem, we let $\overline{DF} = a$, $\overline{AE} = a + p$, $\overline{CH} = a + q$ and $\overline{EH} = r$ (see Figure 163). Since $\triangle EAD$ is similar to $\triangle FCH$, $\dfrac{\overline{AD}}{\overline{AE}} = \dfrac{\overline{CH}}{\overline{CF}}$ so that

$$\frac{2a}{a+p} = \frac{q+a}{a},$$

whence

$$q = \frac{a^2 - ap}{a + p}.$$

By Pythagoras' Theorem applied to $\triangle BEH$,

$$r^2 = (a - p)^2 + (a - q)^2$$
$$= (a - p)^2 + \left(\frac{2ap}{a+p}\right)^2$$
$$= \frac{(a^2 - p^2)^2 + 4a^2p^2}{(a+p)^2} = \frac{(a^2 + p^2)^2}{(a+p)^2},$$

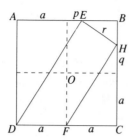

FIGURE 163

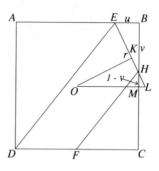

FIGURE 164

whence

$$r = \frac{a^2 + p^2}{a + p} = p + \frac{a^2 - ap}{a + p} = p + q.$$

Now apply the proposition.

Second solution. The inscribed circle will be tangent to EH if and only if the distance from O (the center of the square) to EH is equal to a, the radius of the circle. This can be found by dividing twice the area of triangle OEH by the length of EH to get the altitude of the triangle from O. Introduce Cartesian coordinates with the origin at O and the axes parallel to the sides of the square. Then $E = (p, a)$, $H = (a, q)$ and the area of $\triangle OEH$ is $\frac{a^2 - pq}{2}$.

As in the first solution, the length of EH is $p + q$, and $pq = a^2 - a(p + q)$. Thus $a^2 - pq = a(p + q)$. Hence the distance from O to EH is $\frac{a^2 - pq}{p + q} = a$, as required.

Third solution. Introduce Cartesian coordinates with the origin at O as in the second solution. The line through E and H has equation

$$(a - q)x + (a - p)y = a^2 - pq,$$

and the distance from the origin to this line is

$$\frac{a^2 - pq}{\sqrt{(a - q)^2 + (a - p)^2}} = \frac{a(p + q)}{p + q} = a,$$

where the computations are as in the first solution. Thus the distance from O to EH is equal to the radius of the inscribed circle.

Fourth solution. (E. Michael Thirian) Let the length of a side of the square be 2. Let $\overline{EB} = u$

and $\overline{BH} = v$ so that $\overline{AE} = 2 - u$ and $\overline{CH} = 2 - v$. Let OK and OM be the perpendiculars from the center O of the square to EH and BC respectively. EH and OM produced meet in L. See Figure 164. Since $\triangle AED \sim \triangle CFH$, $\overline{AE} \cdot \overline{CH} = \overline{AD} \cdot \overline{FC}$, $(2 - u)(2 - v) = 2$, or $uv - 2u - 2v + 2 = 0$. From the similar triangles EBH and LKO, $\overline{EH} \cdot \overline{KO} = \overline{BH} \cdot \overline{LO}$. Since

$$\overline{LO} = 1 + \frac{u(1 - v)}{v} = \frac{u + v - uv}{v},$$

$$\overline{KO}^2 = \frac{v^2 \left\{ \frac{u + v - uv}{v} \right\}^2}{u^2 + v^2} = \frac{(u + v - uv)^2}{u^2 + v^2}$$

$$= \frac{u^2 + v^2 + uv(uv - 2u - 2v + 2)}{u^2 + v^2}$$

$$= 1.$$

Thus, the distance from O to EH is equal to the distance from O to a side of the square and the result follows.

Remark. By considering an orthogonal projection of the entire figure, we also have the following equivalent result. Suppose $ABCD$ is a parallelogram circumscribing an ellipse which touches the parallelogram at the midpoints of the sides. If F is the midpoint of DC, H is on BC and E is on AB with $FH \parallel DE$, then EH is tangent to the conic.

Problem 252. *First solution.* Since $a \log a = \log b$ and $b \log b = \log a$,

$$ab \cdot \log b = a \log a = \log b.$$

Hence

$$(\log b)(ab - 1) = 0.$$

Either $\log b = 0$, in which case $b = 1$ and $a = b^b = 1^1 = 1$, or else $a^{a+1} = 1$. In the latter case, a must be equal to 1. (For, if a is positive and unequal to 1, taking logarithms to base a leads to $a + 1 = 0$ which is false.) Thus in both cases $a = b = 1$.

Second solution. Since $a^{ab} = b^b = a$, either $a = 1$, $b = 1$ or $ab = 1$. If $a \neq 1$, $b \neq 1$, then, say $a < 1$, $b > 1$; now $b = a^a < 1$ while $b > 1$, a contradiction.

Problem 253. Let x be the required number, so that, for some integer y, $x^2 = 10000y + 9009$. Then x must have the form $10a \pm 3$. Therefore $100a^2 \pm 60a + 9 = 10000y + 9009$, whence $2a(5a \pm 3) = 100(10y + 9) \equiv 0 \pmod{25}$. Since $5a \pm 3$ is not a multiple of 5, a must be a multiple of 25, say $a = 25b$; then

$$b(125b \pm 3) = 20y + 18.$$

In the case $b(125b + 3) = 20y + 18$, b is a solution only if the left side has 8 as its last digit. The smallest possible value of b for which y is an integer is 6:

$$6 \cdot 753 = 20 \cdot 225 + 18.$$

In the case $b(125b - 3) = 20y + 18$, we see that b must have either 4 or 9 as its last digit. But $b = 4$ does not lead to an integer value of y, while $b = 9$ leads to $y = 504$.

Hence $b = 6$, $x = 250b + 3 = 1503$ and $x^2 = 2259009$.

Remarks. A more systematic approach uses congruences. With x as above,

$$x^2 \equiv 259 \pmod{625}.$$

In particular,

$$x^2 \equiv 259 \equiv 4 \pmod{5}$$

so that $x \equiv 2$ or $x \equiv 3 \pmod 5$. In the first case, set $x = 5a + 2$ to get $25a^2 + 20a \equiv 255 \pmod{625}$ or $5a^2 + 4a \equiv 51 \pmod{125}$. Solving $5a^2 + 4a \equiv 51 \pmod 5$ tells us that $a = 5b - 1$ for some b. Continuing on in this way, we find that

$x \equiv \pm 253 \pmod{625}$. Since $x^2 \equiv 1 \pmod{16}$, we must have $x \equiv \pm 1 \pmod 8$. Thus

$$x = 625u + p = 8v + r$$
$$= 5000w - 624p + 625r$$

where $p = \pm 253$, $r = \pm 1$. Since this implies that $x \equiv \pm 2247$ or $\pm 1503 \pmod{5000}$, we can pick out the proper answer.

(Show that the smallest square ending in 009009 is $126503^2 = 16003009009$.)

Problem 254. Let $\overline{AB} = a$, $\overline{FD} = b$. It is easy to see that

$$\overline{AE} = \frac{x}{\sin \alpha}, \qquad \overline{EF} = \frac{x}{\sin \beta}$$

and hence

$$\overline{EB} = \frac{a \sin \alpha - x}{\sin \alpha}, \qquad \overline{ED} = \frac{b \sin \beta - x}{\sin \beta}.$$

From $\overline{EB} \cos \alpha = \overline{ED} \cos \beta$, it follows that

$$\frac{a \sin \alpha - x}{\sin \alpha \cos \beta} = \frac{b \sin \beta - x}{\sin \beta \cos \alpha}.$$

Solving for x yields

$$x = \frac{a \cos \alpha - b \cos \beta}{\cot \alpha - \cot \beta}.$$

Problem 255. One general law would be: for $n = 2, 3, 4, \ldots$

$$\prod_{k=2}^{n} \frac{1 - \frac{1}{k^3}}{1 + \frac{1}{k^3}} = \frac{2}{3} \left(1 + \frac{1}{n(n+1)} \right).$$

First proof (by induction). This evidently holds when $n = 2$. Assume it holds for $n = m$. Then, when $n = m + 1$,

$$\prod_{k=2}^{m+1}\left(\frac{1-\frac{1}{k^3}}{1+\frac{1}{k^3}}\right) = \frac{2}{3}\left(1+\frac{1}{m(m+1)}\right) \times$$

$$\left(\frac{1-\frac{1}{(m+1)^3}}{1+\frac{1}{(m+1)^3}}\right)$$

$$= \frac{2}{3}\left(\frac{m^2+m+1}{m^2+m}\right) \times$$

$$\left(\frac{m^3+3m^2+3m}{(m+2)(m^2+m+1)}\right)$$

$$= \frac{2}{3}\left(\frac{m^2+3m+3}{(m+1)(m+2)}\right)$$

$$= \frac{2}{3}\left(1+\frac{1}{(m+1)(m+2)}\right).$$

The result then follows by induction.

Second proof (direct).   We see that

$$\prod_{k=2}^{n}\left(\frac{1-\frac{1}{k^3}}{1+\frac{1}{k^3}}\right)$$

$$= \prod_{k=2}^{n}\frac{k^3-1}{k^3+1}$$

$$= \left(\prod_{k=2}^{n}\frac{k-1}{k+1}\right)\left(\prod_{k=2}^{n}\frac{k^2+k+1}{k^2-k+1}\right)$$

$$= \frac{2}{n(n+1)}\left(\prod_{k=2}^{n}(k^2+k+1)\right) \times$$

$$\left(\prod_{k=2}^{n}\frac{1}{(k^2-k+1)}\right).$$

Since

$$\prod_{k=2}^{n}\frac{1}{k^2-k+1}$$

$$= \prod_{j=1}^{n-1}\frac{1}{(j+1)^2-(j+1)+1}$$

$$= \prod_{j=1}^{n-1}\frac{1}{j^2+j+1},$$

the given product is equal to

$$\frac{2}{n(n+1)}\left(\prod_{k=2}^{n}(k^2+k+1)\right) \times$$

$$\left(\prod_{k=1}^{n-1}\left(\frac{1}{k^2+k+1}\right)\right)$$

$$= \frac{2(n^2+n+1)}{n(n+1)3} = \frac{2}{3}\left(1+\frac{1}{n(n+1)}\right).$$

(If the products above offer difficulty, write them out in full for some small n, say 4 or 5, in order to understand what the notation is conveying.)

Problem 256.   Clearly

$$x^n - n(x-1) - 1$$

$$= x^n - 1 - n(x-1)$$

$$= (x-1)(x^{n-1}+x^{n-2}+\cdots+1)$$

$$\quad - n(x-1)$$

$$= (x-1)(x^{n-1}+x^{n-2}+\cdots+1-n).$$

The second factor is 0 when $x = 1$, so it has $x-1$ as a factor, and we are done.

Indeed, carrying out a "long division" of the second factor by $x - 1$ leads to

$$x^n - n(x-1) - 1 = (x-1)^2 \times$$

$$(x^{n-2}+2x^{n-3}+\cdots+(n-2)x+n-1).$$

Problem 257.   Since every entry in the body of the table is the sum of the integers heading the row and column containing it, the sum of the six entries (no two in the same row or column) is just the sum of the top "headings" and the side "headings," namely

$$(1+9+3+2+4+8)$$

$$+(2+7+3+5+8+7) = 59.$$

Problem 258.   The pattern of circles and hexagons is uniform throughout the plane (see Figure 165). Hence the percentage desired is simply

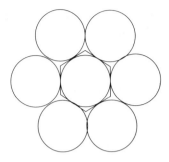

FIGURE 165

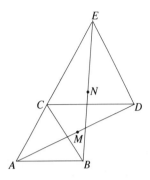

FIGURE 166

the ratio of the area of a circle to its regular circumscribed hexagon: $\dfrac{\pi}{6/\sqrt{3}} = \dfrac{\pi\sqrt{3}}{6}$ or about 91%.

Problem 259. $a + b\sqrt{2} = -c\sqrt{3}$ implies that $a^2 + 2ab\sqrt{2} + 2b^2 = 3c^2$, so that $2\sqrt{2}ab = 3c^2 - a^2 - 2b^2$. Since $\sqrt{2}$ is irrational, we must have $ab = 0$ and $3c^2 - a^2 - 2b^2 = 0$. If $a = 0$, then $3c^2 = 2b^2$, from which, by the irrationality of $\sqrt{\frac{3}{2}}$, follows that $b = c = 0$. If $b = 0$, then $3c^2 = a^2$, from which follows that $a = c = 0$.

Riders. (a) Show that the only integers a, b, c, d for which $a + b\sqrt{2} + c\sqrt{3} + d\sqrt{5} = 0$ are $a = b = c = d = 0$.

(b) Does a similar result follow from $a + b\sqrt{2} + c\sqrt{3} + d\sqrt{6} = 0$?

Problem 260. Observe that when $xyz \neq 0$ then $x^2 + y^2 + z^2 = (x + y + z)^2$ is equivalent to $yz + zx + xy = 0$, which is equivalent to $\frac{1}{x} + \frac{1}{y} + \frac{1}{z} = 0$. Since the last condition is satisfied when

$$x = \frac{1}{b-c}, \quad y = \frac{1}{c-a}, \quad z = \frac{1}{a-b},$$

it follows that

$$\frac{1}{(b-c)^2} + \frac{1}{(c-a)^2} + \frac{1}{(a-b)^2}$$
$$= \left(\frac{1}{b-c} + \frac{1}{c-a} + \frac{1}{a-b}\right)^2,$$

and the number within the brackets is of course rational.

Problem 261. A 60° counterclockwise rotation of the plane about C carries A to B and D to

E, hence AD to BE. Thus, M is carried to N. Hence, $\angle MNC = 60°$ and $CM = CN$, from which the result follows. See Figure 166.

Problem 262. Denoting the given polynomial by $P(x)$, we have

$$P(x) = (x - a_1)(x - a_3)(x - a_5)$$
$$+ (x - a_2)(x - a_4)(x - a_6).$$

Hence,

$$P(x) > 0 \quad \text{for} \quad x > a_1,$$
$$P(x) < 0 \quad \text{for} \quad a_2 > x > a_3,$$
$$P(x) > 0 \quad \text{for} \quad a_4 > x > a_5,$$
$$P(x) < 0 \quad \text{for} \quad a_6 > x.$$

By the continuity of $P(x)$, there are three real distinct roots.

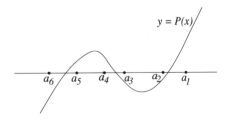

FIGURE 167

Problem 263.

$$4n^3 + 6n^2 + 4n + 1$$
$$= (n+1)^4 - n^4$$
$$= \{(n+1)^2 - n^2\}\{(n+1)^2 + n^2\}$$
$$= (2n+1)\{(n+1)^2 + n^2\}.$$

Problem 264. The product of the elements of any row, column, or diagonal equals k; in particular

$$aei = beh = ceg = k,$$

so that $a = \frac{k}{ei}$, $b = \frac{k}{eh}$, $c = \frac{k}{eg}$. Therefore

$$k = abc = \frac{k}{ei} \cdot \frac{k}{eh} \cdot \frac{k}{eg} = \frac{k^3}{e^3 ghi} = \frac{k^3}{e^3 k},$$

i.e., $k = e^3$, a perfect cube.

Problem 265.

If the criminal is	Then the statements of Andy are	Bob are	Carl are	Dave are
Andy	False	True	False	True
Bob	False	False	False	True
Carl	True	True	False	True
Dave	False	True	True	False

Thus, in case (a) Bob did it; in case (b) Carl did it.

Problem 266. Suppose there were such a loading of dice. Each of the 11 outcomes would occur with probability $p = \frac{1}{11}$. A 2 can be obtained only if both dice show 1; similarly, for 12 both dice must show 6.

Suppose the probabilities of rolling 1 on the two dice are u and v, and of rolling 6 are x and y respectively. Then $p = uv = xy$. A 7 can be obtained rolling a 1 and 6, as well as in other ways. Therefore the probability of rolling a 7 would be

$$p \geq uy + vx = u\left(\frac{p}{x}\right) + \left(\frac{p}{u}\right)x$$
$$= p\left(\frac{u}{x} + \frac{x}{u}\right) \geq 2p.$$

(The last inequality because $a + \frac{1}{a} \geq 2$ when $a > 0$.)

This contradiction shows that it is impossible to load the dice so that all outcomes are equally likely.

Rider. Can you load the dice so that all outcomes but one are equally likely?

Problem 267. (a) Recall that $|u| = u$ if $u \geq 0$ and $|u| = -u$ if $u \leq 0$, and hence $|-u| = |u|$. It follows that the region R determined by

$$|x| + |y| + |x + y| \leq 2 \qquad (1)$$

is symmetrical with respect to the origin, i.e., if (a, b) is in R then so is $(-a, -b)$; it suffices to investigate (1) when $y \geq 0$.

In the quadrant $x \geq 0$, $y \geq 0$ we have $x + y \geq 0$ and (1) becomes $x + y + (x + y) \leq 2$; the part of R in this quadrant is given by

$$0 \leq x, \qquad 0 \leq y, \qquad x + y \leq 1,$$

the triangle AOB in Figure 168.

In the quadrant $x \leq 0$, $y \geq 0$ either (i) $x + y \geq 0$ or (ii) $x + y \leq 0$. If (i) we have $-x + y + x + y \leq 2$ i.e., $y \leq 1$, so

$$x \leq 0, \qquad 0 \leq y \leq 1, \qquad x + y \geq 0,$$

(and parenthetically, $-1 \leq x$) determining triangle BOC. If (ii), we have $-x + y - (x + y) \leq 2$, i.e., $-1 \leq x$, so

$$-1 \leq x \leq 0, \qquad 0 \leq y, \qquad x + y \leq 0,$$

(and parenthetically, $y \leq 1$) determining triangle COD.

(Indeed, taking the parenthetical inequalities into account shows that

$$x \leq 0, \qquad 0 \leq y, \qquad |x| + |y| + |x + y| \leq 2$$

is equivalent to

$$-1 \leq x \leq 0, \qquad 0 \leq y \leq 1,$$

determining the square $OBCD$.)

The "lower" half of R is picked up by reflecting the upper half (Figure 168) in the origin

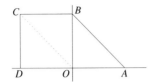

FIGURE 168

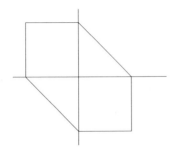

FIGURE 169

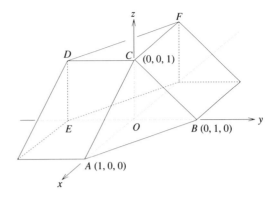

FIGURE 170

(i.e., rotating the region in Figure 168 about O). R is the hexagon in Figure 169, of area 3.

(b) The analysis is similar, but there are more cases to consider. As before, the region R determined by

$$|x| + |y| + |z| + |x + y + z| \leq 2 \qquad (2)$$

is symmetrical with respect to O i.e., if (a, b, c) is in R so is $(-a, -b, -c)$, and we can focus on the four upper octants, where $z \geq 0$.

A brief look at the four upper octants follows; the details are left to the reader.

$x \geq 0$, $y \geq 0$, $z \geq 0$. These and (2) are equivalent to

$$x \geq 0, \quad y \geq 0, \quad z \geq 0, \quad x + y + z \leq 1,$$

the tetrahedron $OABC$ (Figure 170), with volume $\frac{1}{6}$.

$x \geq 0$, $y \leq 0$, $z \geq 0$. These and (2) are equivalent to

$$0 \leq x \leq 1, \quad -1 \leq y \leq 0,$$
$$0 \leq z, \quad x + z \leq 1,$$

the "wedge" (half of the cube which has AD as one long diagonal) with volume $\frac{1}{2}$.

$x \leq 0$, $y \geq 0$, $z \geq 0$. These and (2) are equivalent to

$$-1 \leq x \leq 0, \quad 0 \leq y \leq 1,$$
$$0 \leq z, \quad y + z \leq 1,$$

the wedge (half of the cube which has BF as one long diagonal) with volume $\frac{1}{2}$.

$x \leq 0$, $y \leq 0$, $z \geq 0$. These and (2) are equivalent to

$$-1 \leq x \leq 0, \quad -1 \leq y \leq 0,$$
$$0 \leq z \leq 1, \quad -1 \leq x + y,$$

the wedge (half the cube which has EF as one diagonal) with volume $\frac{1}{2}$.

The lower half of R is obtained by reflecting the upper half in O. The solid region R is seen in Figure 171; it has volume

$$2\left(\frac{1}{6} + \frac{1}{2} + \frac{1}{2} + \frac{1}{2}\right) = \frac{10}{3}.$$

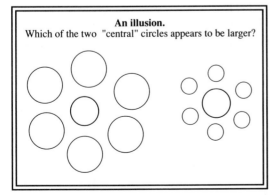

An illusion.
Which of the two "central" circles appears to be larger?

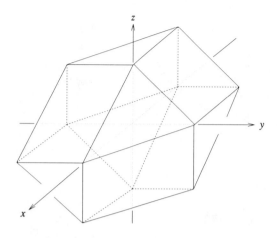

FIGURE 171

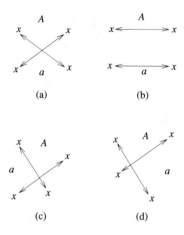

FIGURE 172

Problem 268. *First solution.* (Using the Principle of Inclusion and Exclusion.) Assuming that all seats at the table are identical, and using one person as a "marker," there are $5! = 120$ ways of seating the other 5 people without restriction. If a particular couple is to be kept together, the remaining 4 people can be seated in $4!$ ways. However, for each of these ways, we can switch the positions of the members of the couple. Hence, the total number of ways of seating the people with a particular couple kept together is $2 \cdot 4! = 48$.

There are $2^2 \cdot 3! = 24$ ways of keeping 2 particular couples together, for, using one couple as marker, there are $3!$ ways of arranging the three units consisting of the remaining couple and the other two individuals.(The factor 2^2 results from switching positions of husband and wife of the two couples.)

Finally , there are $2^3 \cdot 2! = 16$ ways of seating the 6 people so that all three couples are kept together.

By the Principle of Inclusion and Exclusion, the number of seating arrangements for which no couple is kept together is

$$120 - 3 \cdot 48 + 3 \cdot 24 - 16 = 32.$$

Second solution. Let the couples be (A, a), (B, b), (C, c). Use A as a marker. The following diagrams show the different types of arrange-

ments; the double arrows join the two members of the same couple.

There are 4 ways for which either B or b sits on the left of A, and 4 ways for which either B or b sits on the right of A. This arrangement type can be realized in 8 ways. See Figure 172(a).

There are 4 ways for which A is flanked by (B, b) and 4 ways for which A is flanked by (C, c). Again, this arrangement is realizable in 8 ways. See Figure 172(b).

Similarly, each of Figure 172(c) and Figure 172(d) can be realized in 8 ways. Hence, all told, there are 32 arrangements for which the couples are kept apart.

Rider. Solve the same problem for 3 replaced by n.

Problem 269. For any route, there is a second route which is symmetrical about the bisector of $\angle BAC$ and can be travelled in at least as short a time. (This is because any path μ must cross this bisector at least once, say at a point G. Suppose the part of μ from B to G can be traversed in at least as short a time as the part of μ from G to C. Then we can get a better path ν by replacing that part of μ from G to C by the mirror image of that part of μ from B to G. A similar procedure can be used if the latter part of μ takes the shorter time.)

Hence we need only consider paths symmetrical about the bisector of $\angle BAC$. Since the

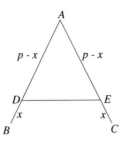

FIGURE 173

shortest distance between two points lies along a straight line, an optimum path is of the type pictured in Figure 173, where $AD = AE$ and $BD = CE$. Let $\overline{BD} = x$. Since $\overline{AB} = p$ and $\overline{BC} = q$, consideration of the similar triangles ADE and ABC leads to

$$\overline{DE} = \frac{(p-x)q}{p}.$$

The time $t(x)$ taken to traverse the path $BDEC$ is given by

$$t(x) = \frac{2x}{v} + \frac{(p-x)q}{pw} = \left(\frac{2}{v} - \frac{q}{pw}\right)x + \frac{q}{w}.$$

When $\frac{2}{v} - \frac{q}{pw} \geq 0$, t is an increasing function of x, so that the time is smallest when $x = 0$.

When $\frac{2}{v} - \frac{q}{pw} \leq 0$, t is a decreasing function of x, so that the time is smallest when $x = p$.

Problem 270. *First solution.* Suppose there were such a polynomial $p(x)$. Noting that $\log x$ is an increasing function (i.e., $\log u < \log v$ when $0 < u < v$) and that $\log x < x$ (from the definition of the log function) we have

$$p(n) < n \log n < n^2.$$

It follows that the degree of the polynomial $p(x)$ is at most 2. (A justification of this last assertion is given at the end, within the brackets [ ].)

Evidently $p(n)$ is neither constant nor linear. There remains the case to consider that $p(x)$ is a quadratic function. In this case we deduce from $p(1) = \log 1 = 0$ that

$$p(x) = (x - 1)(ax + b).$$

Evaluating at $x = 2$ and $x = 4$ we have

$$\log 2 = p(2) = 2a + b,$$

$$\log 24 = p(4) = 3(4a + b),$$

and solving yields

$$a = \frac{1}{6}\log 3, \qquad b = \log 2 - \frac{1}{3}\log 3.$$

Thus

$$p(x) = (x - 1)\left(\frac{x\log 3}{6} + \log 2 - \frac{1}{3}\log 3\right)$$

and hence

$$\log 2 + \log 3 = p(3)$$
$$= 2\left(\frac{3\log 3}{6} + \log 2 - \frac{1}{3}\log 3\right)$$

which reduces to $3\log 2 = 2\log 3$ i.e., $2^3 = 3^2$, a contradiction.

[We indicate here why a polynomial

$$p(x) = a_0 x^m + a_1 x^{m-1} + \cdots + a_m$$

of degree $m \geq 1$ has the property

$$|p(x)| > x^{m-1} \quad \text{for sufficiently large } x.$$

First write $p(x)$ in the form

$$p(x) = a_0 x^m \left(1 + \frac{b_1}{x} + \frac{b_2}{x^2} + \cdots + \frac{b_m}{x^m}\right).$$

There once was a hairy baboon
Who always breathed down a bassoon
For he said "It appears
That in millions of years
I shall certainly hit on a tune."

A. Eddington

By taking x sufficiently large we can insure that the expression within the brackets is greater than $\frac{1}{2}$, so that then

$$|p(x)| > \frac{1}{2}|a_0|x^m.$$

Now choose x so large that $\frac{x|a_0|}{2} > 1$, and then

$$|p(x)| > x^{m-1}.$$

It follows that if $p(x) < x^2$ for large x then $m \leq 2$.]

Second solution. Suppose

$$\log n! = p(n) = \sum_{r=0}^{m} a_r n^r$$

(where $a_m \neq 0$) for $n = 1, 2, 3, \ldots$. Then

$$\log(n+1)! = \sum_{r=0}^{m} a_r(n+1)^r.$$

It follows that

$$\log(n+1) = \sum_{r=0}^{m} a_r(n+1)^r - \sum_{r=0}^{m} a_r n^r,$$

which is a polynomial of degree $\leq (m-1)$. By Problem 123, $\log x$ cannot be expressed as $f(x)$ for some polynomial $f(x)$. Thus we are led to a contradiction, and the result follows.

Remark. An alternative argument for showing that $\log n$ is not a polynomial in n is sketched. From $\log n = b_r n^r + b_{r-1}n^{r-1} + \cdots + b_1 n + b_0$ and $n = e^p$ we have

$$p = b_r e^{pr} + b_{r-1}e^{p(r-1)} + \cdots + b_0,$$

whence

$$\frac{p}{e^{pr}} = b_r + b_{r-1}e^{-p} + \cdots.$$

As $p \to \infty$, the left side tends to 0, the right to b_r, yielding a contradiction.

Problem 271. Let us see if the values of $f(n)$ for small n will reveal something useful (Table 3).

It seems that $\dfrac{f(3)}{f(2)} = \dfrac{8}{3}$ is the minimum and one way of proving this is to show that $\dfrac{f(n+1)}{f(n)} > 3$ for $n = 5, 6, 7, \ldots$. Indeed, for $n \geq 6$,

$$f(n+1) = 1^{n+1} + 2^n + 3^{n-1}$$
$$+ \cdots + n^2 + (n+1)$$
$$> 1^{n+1} + 2^n + 3^{n-1} + 4^{n-2} + 5^{n-3}$$
$$+ 6^{n-4} + \cdots + (n-1)^3 + n^2$$
$$> 1^{n+1} + 2^n + 3^{n-1} + 4^{n-2} + 5^{n-3}$$
$$+ 3(6^{n-5} + 7^{n-6} + \cdots + (n-1)^2 + n)$$
$$= 1^{n+1} + 2^n + 3^{n-1} + 4^{n-2} + 5^{n-3}$$
$$+ 3(f(n) - 1^n - 2^{n-1} - 3^{n-2} - 4^{n-3} - 5^{n-4})$$
$$= 3f(n) + 2(5^{n-4} - 1) + 2^{n-1}(2^{n-5} - 1)$$
$$> 3f(n).$$

Problem 272. There are five ways of interpreting $a^{b^{c^d}}$, each of which expresses a raised to some power:

$$a^{\left(b^{(c^d)}\right)}; \qquad a^{\left((b^c)^d\right)} = a^{\left(b^{cd}\right)};$$
$$(a^b)^{(c^d)} = a^{(bc^d)}; \qquad ((a^b)^c)^d = a^{bcd};$$
$$\left(a^{(b^c)}\right)^d = a^{(b^c d)}.$$

Since $a^x \leq a^y$ if and only if $x \leq y$ when a, x, y all exceed 1, we have to compare the sizes of bcd, $b^c d$, bc^d, b^{cd}, $b^{(c^d)}$.

n	1	2	3	4	5	6	7
$f(n)$	1	3	8	22	65	209	732
$\dfrac{f(n+1)}{f(n)}$	3	$\dfrac{8}{3} \doteq 2.66$	$\dfrac{22}{8} = 2.75$	$\dfrac{65}{22} \doteq 2.95$	$\dfrac{209}{65} \doteq 3.22$	$\dfrac{732}{209} \doteq 3.50$	

TABLE 3

Observe that, for every natural number $n \geq 2$, $n \leq 2^{n-1} \leq b^{n-1}$. This can be applied to show that

$$bcd \leq b^c d \leq b^{cd} \leq b^{(c^d)}, \qquad bcd \leq bc^d.$$

Only the pairs $(b^c d, bc^d)$, (bc^d, b^{cd}) and $(bc^d, b^{(c^d)})$ remain to be considered. $b^c d$ and bc^d satisfy no general inequality, as exemplified by $3^2 \cdot 2 > 3 \cdot 2^2$ and $2^2 \cdot 3 < 2 \cdot 2^3$. As before, we have $bc^d \leq b^{c^d}$.

It remains to show that $bc^d \leq b^{cd}$. Use induction on d. For $d = 1$, the inequality is clear. If $bc^d \leq b^{cd}$, then

$$bc^{d+1} = cbc^d \leq cb^{cd} \leq b^c b^{cd} = b^{c(d+1)}.$$

Rider. When can there be equality for the various pairs?

Problem 273. The information is nicely displayed in a Venn diagram. Assume a total population of 1000 inside the rectangle (Figure 174).

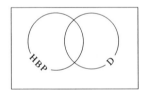

FIGURE 174

The left circle HBP contains those with high blood pressure and the right circle contains those who drink. Of the 35 people with high blood pressure, $0.8 \times 35 = 28$ drink and $35 - 28 = 7$ do not. We therefore place a 28 and 7 indicated in Figure 175.

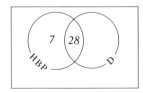

FIGURE 175

The remaining $1000 - 35 = 965$ with normal blood pressure fall into two classes: those who drink—$0.6 \times 965 = 579$ of these, and $965 - 579 = 386$ who do not. The complete picture is in Figure 176. Now we compute the percentage of drinkers who have high blood pressure:

$$100 \times \frac{28}{28 + 579} = 4.6\ldots\%.$$

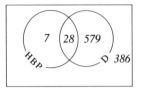

FIGURE 176

Problem 274. Choose the last number s_n first. There are three possibilities: n, $n - 1$, $n - 2$. Having chosen s_n, there are three ways of choosing s_{n-1} from those of n, $n-1$, $n-2$, $n-3$ not taken for s_n. Likewise, with s_n and s_{n-1} chosen, there are three ways of choosing s_{n-2}. We can continue on, choosing each s_i from three available possibilities until we have chosen s_3; now only s_2 and s_1 remain to be determined. Only 2 possibilities are left for s_2 and, with s_2 chosen, s_1 is determined. Hence the number of arrangements satisfying the conditions are

$$\underbrace{3 \times 3 \times 3 \times \cdots \times 3}_{n-2 \text{ 3's}} \times 2 \times 1 = 3^{n-2} \cdot 2!$$

Rider. What is the solution if the condition is: $s_k \geq k - m$, where m is some fixed number between 1 and n, for all $k = 1, 2, \ldots, n$?

Problem 275. *First solution.* Since $(b - c)^2 + (c - a)^2 + (a - b)^2 \geq 0$ and $a + b + c > d$,

$$3(a^2 + b^2 + c^2)$$
$$= (a^2 + b^2 + c^2) + 2(a^2 + b^2 + c^2)$$
$$\geq (a^2 + b^2 + c^2) + 2(bc + ca + ab)$$
$$= (a + b + c)^2 > d^2.$$

Second solution. Consider any $(n + 1)$-gon, not necessarily coplanar nor non–self-intersecting, with sides $a_1, a_2, \ldots, a_{n+1}$. By the Hölder inequality, given any $m > 1$,

$$(a_1^m + a_2^m + \cdots + a_n^m)^{\frac{1}{m}} (1 + 1 + \cdots + 1)^{1 - \frac{1}{m}}$$
$$\geq a_1 + a_2 + \cdots + a_n,$$

with equality if and only if $a_1 = a_2 = \cdots = a_n$. Since, by an extended triangle inequality,

$$a_1 + a_2 + \cdots + a_n \geq a_{n+1},$$
$$a_1^m + a_2^m + \cdots + a_n^m \geq \frac{a_{n+1}^m}{n^{m-1}},$$

with equality only when $a_1 = a_2 = \cdots = a_n = \dfrac{a_{n+1}}{n}$, in which case the polygon is degenerate.

For $m = 2$, $n = 3$, we have the situation of the problem. (See also Problem 466.)

Problem 276. Let the chosen numbers be named in ascending order of magnitude: $a_1 < a_2 < a_3 < \cdots < a_8$. If no three of the seven integers

$$a_2 - a_1, \; a_3 - a_2, \; a_4 - a_3, \ldots, \; a_8 - a_7$$

are equal, then at most two are equal to 1, at most two are equal to 2 and so on. Hence

$$a_8 - a_1 = (a_2 - a_1) + (a_3 - a_2)$$
$$+ (a_4 - a_3) + \cdots + (a_8 - a_7)$$

must be not less than $1+1+2+2+3+3+4 = 16$. On the other hand, since $1 \leq a_1 < a_8 \leq 16$,

$$a_8 - a_1 \leq 15,$$

which contradicts the earlier statement.

Thus, in fact, the following stronger result holds: there is a k for which $a_i - a_{i-1} = k$ has three solutions when the a_i are chosen in ascending order of magnitude.

Problem 277. The answer to both questions (i) and (ii) is NO. It is enough to prove this in case (ii). The explanation depends in part on the fact that perfect squares leave remainders 0 or 1 upon division by 4. Thus, we examine the possibilities modulo 4. This has the advantage that there are only 3 cases to consider:

(a) $k \equiv 0 \pmod 4$;

(b) $k \equiv 2 \pmod 4$;

(c) k is odd.

In the case (a), for any integers x and y,

$$x^2 + kxy + y^2 \equiv x^2 + y^2 \pmod 4.$$

This means that $x^2 + kxy + y^2$ differs from the sum of two squares by a multiple of 4. But the sum of two squares can never leave a remainder 3 when divided by 4. Hence $x^2 + kxy + y^2$ can never assume any of the values $\{3, 7, 11 \ldots\}$.

In the case (b), for any integers x and y,

$$x^2 + kxy + y^2 \equiv x^2 + 2xy + y^2$$
$$= (x + y)^2 \pmod 4.$$

Hence $x^2 + kxy + y^2$ must leave a remainder 0 or 1 upon division by 4, and so it never assumes one of the values $\{2, 3, 6, 7, 10, 11, \ldots\}$.

Now suppose case (c) holds and k is odd. Then $k \equiv \pm 1 \pmod 4$, so that for any integers x and y

$$x^2 + kxy + y^2 \equiv x^2 \pm xy + y^2$$
$$\equiv x(x \pm y) + y^2 \pmod 4.$$

When y is odd, then y^2 is odd and $x(x \pm y)$ must be even, so that $x^2 + kxy + y^2$ is odd. On the other hand, when y is even, either $x(x \pm y)$ is odd or is divisible by 4, while y^2 is divisible by 4, so that $x^2 + kxy + y^2$ is odd or divisible by 4.

Thus, whether y is odd or even, $x^2 + kxy + y^2$ can never leave remainder 2 when divided by 4, and so can never be one of the values $\{2, 6, 10, 14, \ldots\}$.

Therefore, whatever particular value of k is chosen, there is some positive integer n for which

$$x^2 + kxy + y^2 \neq n.$$

Problem 278. The rational root can be written in the form $\frac{p}{q}$, where p and q are integers whose greatest common divisor is 1. Thus, p and q cannot both be even, and

$$ap^2 + bpq + cq^2 = 0.$$

Suppose, if possible, that a, b, c are all odd integers. Since the left-hand side of the equation is an even integer, either none or else exactly two of its terms must be odd.

If p and q were both odd, then every term on the left-hand side would be odd, which is impossible.

If one of p and q, say p, were even, then ap^2 and bpq would both be even while cq^2 would be odd, which is impossible.

Hence a, b, c cannot all be odd.

Problem 279. Suppose all three men have different weights, B, K, M (given by their respective initials). Then their statements can be written

   Barbeau: $B > K$, $K > M$.
   Klamkin: $M > K$, $M > B$.
   Moser: $K > M$, $M = B$.

Since each makes two statements, the lightest must make two correct statements, the heaviest none and the other exactly one.

One of Moser's statements ($M = B$) is incorrect, hence he is not the lightest. Barbeau also cannot be the lightest, for then his statement that $B > K$ would be wrong. Nor can Barbeau be the heaviest, for then $B > K$ would be correct. Hence Barbeau is in the middle, weightwise. This gives the ordering $M > B > K$.

Now, it may happen that at least two have the same weight, as we are given no information about the relative veracity of two who weigh the same. In this case, of the seven possibilities $K = M = B$, $K > M = B$, $M = B > K$, $B > K = M$, $K = M > B$, $M > K = B$, $K = B > M$, the first two ($K = M = B$ and $K > M = B$) are possible orderings.

Problem 280. For each x and y, $f(x, y) = kf(y, x) = k \cdot k \cdot f(x, y) = k^2 f(x, y)$. Since there are values of x and y for which $f(x, y) \neq 0$, k^2 must equal 1. Hence $k = 1$ or $k = -1$.

An example with $k = 1$ is $f(x, y) = x + y$; an example with $k = -1$ is $f(x, y) = x - y$.

Rider. Solve the same problem for a polynomial $f(x, y, z)$ of three variables for which the condition is $f(x, y, z) = kf(y, z, x)$.

Problem 281. The key to the solution is to begin by minimizing the distances to two opposite

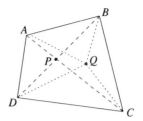

FIGURE 177

vertices. If A and C are two vertices, and P is any point on the segment AC and Q is any point not on the segment (Figure 177), we have that $\overline{AP} + \overline{PC} = \overline{AC} < \overline{AQ} + \overline{QC}$, by the Triangle Inequality. This suggests that the required point is the intersection of the two diagonals. Indeed, if the four vertices are A, B, C, D and AC and BD intersect in P, then for any point Q,

$$\overline{AP} + \overline{BP} + \overline{CP} + \overline{DP} = \overline{AC} + \overline{BD}$$
$$\leq \overline{AQ} + \overline{CQ} + \overline{BQ} + \overline{DQ}$$

with equality only if $P = Q$.

Rider. Solve the same problem when the quadrilateral (say $ABCD$ as in Figure 178) is not convex.

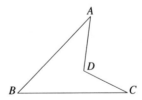

FIGURE 178

Problem 282. More generally, let the latitude of the place be λ, and the latitude of the Arctic Circle be α. Let $\lambda \leq \alpha$. (α is $66°$ $33'$.) In Figure 179, O is the center of the earth. If r is the radius of the earth, we have that

$$\overline{OA} = r \sin \lambda, \qquad \overline{AE} = r \cos \lambda$$
$$\overline{OC} = r \sin \alpha, \qquad \overline{CD} = r \cos \alpha.$$

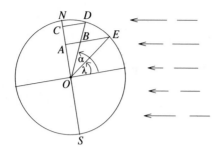

FIGURE 179

Hence

$$\overline{AB} = \frac{r\sin\lambda}{r\sin\alpha}\, r\cos\alpha = \frac{r\sin\lambda}{\tan\alpha}.$$

What proportion of the time is a point on the surface at latitude λ in the sunlight? It will be in the sunlight from the time it is at position F until it is at position G (in Figure 180). Assuming uniform rotation, this will be the ratio of the short arc $\overset{\frown}{FG}$ to the whole circumference of the cross-section, or the ratio of $\angle GAF$ to 2π.

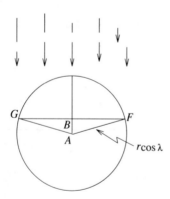

FIGURE 180

Since $\angle GAF = 2\arccos\dfrac{\overline{AB}}{\overline{AF}}$, the number of hours of sunlight at a point of north latitude λ is

$$\frac{24}{2\pi}2\arccos\frac{\overline{AB}}{\overline{AF}} = \frac{24}{\pi}\arccos\frac{\tan\lambda}{\tan\alpha}.$$

When $\alpha = 66°\ 33'$, $\lambda = 43°\ 45'$, this is about 8 hours and 45 minutes.

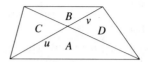

FIGURE 181

Problem 283. Denote the areas of the four triangles (into which the trapezoid is divided by the diagonals) by A, B, C, D as in Figure 181. If one diagonal divides the other in the ratio $\frac{u}{v}$ (as indicated), then

$$\frac{C}{B} = \frac{A}{D} = \frac{u}{v},$$

so $A \cdot B = C \cdot D$. Since triangles with the same base and between the same parallels have equal area, $A + C = A + D$, whence $C = D = \sqrt{AB}$. Thus, the area of the trapezoid is

$$A+B+C+D = A+B+2\sqrt{AB} = (\sqrt{A}+\sqrt{B})^2.$$

Rider. Given a quadrilateral, with triangular divisions lettered as in the diagram, for which the area is $(\sqrt{A} + \sqrt{B})^2$, must the figure be a trapezoid?

Problem 284. To get an idea how to proceed, look at the case $n = 2$. Then we have to find the sum of the digits in the numbers $1, 2, 3, \ldots, 96, 97, 98, 99$. Look at the numbers from the top down. 99 has the digital sum 18; 98 has the digital sum 17, which with 1 added gives 18; 97 has the digital sum 16, which is 2 short of 18. When we get into the 80's, we note that the digital sum of 87, for example, is $1 + 2$ short of 18, and $1 + 2$ is, of course, the digital sum of $12 = 99 - 87$. This suggests that we augment the set of integers by 0 and pair them off: $(0, 99), (1, 98), (2, 97), \ldots, (12, 87), \ldots, (32, 67),$ $\ldots, (48, 51), (49, 50)$, the sum of each pair being 99. There are 50 pairs and the digital sums of the two members of each pair add up to $9 + 9 = 18$. The answer, for $n = 2$, is therefore

$$9 \cdot 2 \cdot 50 = 900.$$

In general, pair off k and $10^n - 1 - k$ $(k = 0, 1, 2, \ldots, 5 \cdot 10^{n-1} - 1)$. Then the units digits of the two numbers of each pair add up to 9, as do the tens digits, and so on. Since there are up to n digits in the numbers of each pair, the sum of the digits in each pair is $9n$. Since there are $\frac{1}{2} \cdot 10^n$ pairs, the total digital sum of all integers is

$$\frac{9n \cdot 10^n}{2} = 5 \cdot 9 \cdot n \cdot 10^{n-1}.$$

Problem 285. *First solution.* The matter hinges on finding the altitude of the shaded isosceles triangle AGC. This can be found by bisecting the right angle $\angle ABC$; the bisector passes through G and is perpendicular to AC. Hence, we seek $\overline{GD}$.

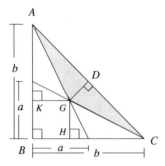

FIGURE 182

Let $GH \perp BC$, as in Figure 182, and denote $\overline{GH}$ by x. Since $\dfrac{\overline{GH}}{\overline{HC}} = \dfrac{\overline{EB}}{\overline{BC}}$,

$$\frac{x}{b-x} = \frac{a}{b},$$

and hence

$$x = \frac{ab}{a+b}.$$

Then

$$\overline{GD} = \overline{BD} - \overline{BG} = \frac{b}{\sqrt{2}} - \sqrt{2}x,$$

so that the area of triangle AGC is

$$\frac{1}{2}\overline{AC} \cdot \overline{GD} = \frac{1}{2}(\sqrt{2}b)\left(\frac{b}{\sqrt{2}} - \sqrt{2}x\right)$$
$$= \frac{b}{2}(b - 2x) = \frac{b^2}{2}\left(1 - \frac{2a}{a+b}\right)$$
$$= \frac{b^2}{2}\left(\frac{b-a}{b+a}\right).$$

Second solution. Using the notation and diagram of the first solution, we have that

$$\text{Area } AGC = \text{Area } ABC - 2 \cdot \text{Area } GBC$$
$$= \frac{b^2}{2} - xb = \frac{b^2}{2}\left(1 - \frac{2a}{a+b}\right).$$

Problem 286. Any 9 points inside the pentagon will do! (Make a diagram.)

Problem 287. *First solution.* The expression vanishes if we set $x = 0$, $y = 0$ or $x = -y$. By the Factor Theorem, x, y and $x + y$ are all factors. This can also be seen directly, since $x + y$ divides $x^7 + y^7$:

$$(x + y)^7 - (x^7 + y^7)$$
$$= (x + y)\{(x + y)^6$$
$$\quad - (x^6 - x^5y + x^4y^2 - x^3y^3 + x^2y^4 - xy^5 + y^6)\}$$
$$= (x + y)\times$$
$$\quad \{7x^5y + 14x^4y^2 + 21x^3y^3 + 14x^2y^4 + 7xy^5\}$$
$$= 7xy(x + y)(x^4 + 2x^3y + 3x^2y^2 + 2xy^3 + y^4)$$
$$= 7xy(x + y)(x^2 + xy + y^2)^2.$$

Second solution. As above x, y and $x + y$ are factors. Let $x = \omega y$ where ω is "the imaginary cube root of unity" ($\omega = \frac{1}{2}(-1 + i\sqrt{3})$, $\omega^3 = 1$, $\omega^2 + \omega + 1 = 0$). Then

$$(x + y)^7 - (x^7 + y^7)$$
$$= y^7(\omega + 1)^7 - y^7(\omega^7 + 1)$$
$$= y^7(-\omega^2)^7 - y^7(\omega + 1)$$
$$= -y^7(\omega^{14} + \omega + 1)$$
$$= y^7(\omega^2 + \omega + 1) = 0.$$

Hence $x - \omega y$ is a factor. In fact, we can say more. Let

$$g(x) = (x + y)^7 - x^7 - y^7.$$

Then

$$g'(x) = 7\{(x + y)^6 - x^6\}$$

and

$$g'(\omega y) = 7y^6\{(\omega + 1)^6 - \omega^6\}$$
$$= 7y^6\{(-\omega^2)^6 - \omega^6\} = 0,$$

so $x - \omega y$ is a factor of $g(x)$. Therefore $(x - \omega y)^2$ divides $(x+y)^7 - (x^7 + y^7)$. Since the coefficients are real, $(x+y)^7 - (x^7 + y^7)$ is also divided by the complex conjugate $(x - \omega^2 y)^2$. Therefore

$$(x - \omega y)^2 (x - \omega^2 y)^2 = (x^2 + xy + y^2)^2$$

is a factor. The result follows.

Rider. Investigate the factorization of $(x + y)^n - (x^n + y^n)$ for other integer values of n.

Problem 288. *First solution.* Since

$$BC \parallel FH, \quad CD \parallel HK, \quad DE \parallel KG,$$

it follows that

$$\frac{\overline{AB}}{\overline{AF}} = \frac{\overline{AC}}{\overline{AH}} = \frac{\overline{AD}}{\overline{AK}} = \frac{\overline{AE}}{\overline{AG}}$$

so that $AFHKG$ is a "reduction" of $ABCDE$ under a homothetic transformation with fixed point A. Hence $AFHKG$ and $ABCDE$ are similar, so that the pentagon $AFHKG$ is regular.

Second solution. Since $\angle BAE = 108°$ and $\overline{AB} = \overline{AE}$, we have $\angle ABE = \angle AEB = 36°$. Similarly, $\angle BAC = \angle EAD = 36°$ so that $\angle CAD = 36°$.

Since $FH \parallel AD, \angle FHA = \angle CAD = 36°$. Since $\angle ABE = \angle BAC = 36°, \angle BHA = 108°$. Thus, $\angle FHB = \angle HFB = 72°$, so $\angle AHK = 72°$.

Similarly $\angle AKH = 72°$. By isosceles triangles, $\overline{BF} = \overline{BH} = \overline{AH} = \overline{AK}$, so $\triangle BFH \cong$

$\triangle AHK$. Hence $\overline{FH} = \overline{HK}$. Clearly $\overline{AF} = \overline{FH}$. Similarly $\overline{AG} = \overline{GK} = \overline{HK}$.

Thus, the pentagon $AFHKG$ has all its sides and all its angles equal: it is regular.

Problem 289. Suppose that there were a solution to

$$\begin{array}{ccccc}
T & H & R & E & E \\
+ & & F & I & V & E \\
\hline
E & I & G & H & T \\
\end{array}$$

Since T and E are distinct, there must be a carry of 1 into the leftmost column. Thus, $E = T + 1$. Now look at the right column. Since $E > T$, $E + E$ must exceed 10, so that

$$E + E = 10 + T = 9 + E.$$

Thus $E = 9$ and $T = 8$. Now using the 1 carried from the right column we find, from the next column,

$$1 + 9 + V = 10 + H.$$

This yields $V = H$, contradicting the given $V \neq H$.

Problem 290. Let d be the greatest common divisor of a, b, c; let g be the greatest common divisor of x, y, z. Since d divides both b and c, d must divide x. Similarly d must divide y and also z. Hence d, being a common divisor of x, y and z, does not exceed g. On the other hand, since g divides x, g must divide each of b and c. Since g divides y, g must divide a and c. Hence g must divide a, b and c. Therefore $g \leq d$. Thus $g = d$.

Problem 291. $13\frac{1}{2}$ days.

Rider. In such a situation, it is more natural that there be some limiting factor on the growth which takes drastic effect for larger populations. We might suppose, for example, that if N is the population at a given time, then 12 hours later, the population is $N(2 - \frac{N}{K})$, where K is some constant.

Are there values of K which would prevent a population of 1,000,000 from ever increasing to 2,000,000? If so, what is the largest such value? For what K will the growth law permit a culture

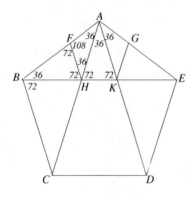

FIGURE 183

of $1,000,000$ bacteria to increase to $2,000,000$ bacteria in 24 hours?

Problem 292. The condition on f can be formulated as follows: if $a \neq b$, then

$$\frac{f(b) - f(a)}{b - a} \geq 0.$$

First, let us try to find a solution. Looking at the form of the equation, we might try $x = c - f(u)$, where u is to be found. This leads to $f(u) = f(c - f(u) + f(c))$, from which it is clear that $u = c$ will work. Thus

$$x = c - f(c)$$

is a solution.

Is there another solution? Suppose that $x = v \neq c - f(c)$ is a second solution. Taking account of the condition on f and the fact that $v + f(c) \neq c$, we have

$$\frac{f(v + f(c)) - f(c)}{v + f(c) - c} \geq 0. \qquad (1)$$

However, since v is a solution, $f(v + f(c)) = c - v$, so $f(v + f(c)) - f(c) = -(v + f(c) - c)$. Thus (1) implies that $-1 \geq 0$. Therefore there is no second solution!

Problem 293. *First solution.* Join AE. Since $\overline{AF} = 2\overline{CF}$, $[AEF] = 2[FEC]$. ($[ABC]\ldots$ denotes the area of polygon $ABC\ldots$.) Also

$$[ABE] = [AEC] = [AEF] + [FEC] = 3[FEC].$$

Hence

$$[ABEF] = [ABE] + [AEF] = 5[FEC].$$

Second solution. Let X and Y be the respective feet of the perpendiculars from F and A to BC. See Figure 184. Since $\triangle AYC$ is similar to $\triangle FXC$, $\overline{AY} = 3\overline{FX}$. Hence

$$[ABC] = 2[AEC] = 2 \cdot 3[FEC] = 6[FEC],$$

so that $[ABEF] = 5[FEC]$.

Problem 294.

$$(\log_3 169) \times (\log_{13} 243)$$
$$= (2\log_3 13) \times (5\log_{13} 3)$$
$$= 10 \log_3 13 \log_{13} 3 = 10.$$

Rider. Analyze the following "proof": if $a = \log_3 169$ and $b = \log_{13} 243$, then $3^a = 169$ and $13^b = 243$, whence $169 - 13^b + 243 - 3^a = 0$. Thus, $13^2 - 13^b + 3^5 - 3^a = 0$, so $b = 2$, $a = 5$ and $ab = 10$.

Problem 295. *First solution.* Extend BA to D so that $\overline{AD} = c$. Clearly $\triangle AA_1A_2 \cong \triangle ACD$ and we have

$$[AA_1A_2] = [ACD] = [ACB] = \frac{1}{2}ab.$$

Similarly, $[BB_2B_1] = \frac{1}{2}ab$. Of course $[CC_1C_2] = \frac{1}{2}ab$. The area of the hexagon is

$$a^2 + b^2 + c^2 + 4\left(\frac{1}{2}ab\right) = 2(a^2 + ab + b^2).$$

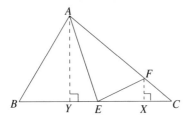

FIGURE 184

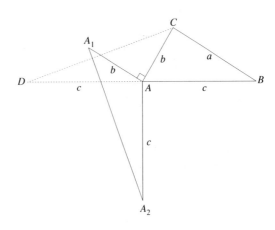

FIGURE 185

Second solution. The area of $\triangle AA_1A_2$ can also be obtained by trigonometry. Since

$$\angle A_1AA_2 + \angle CAB = 180°,$$

the area is

$$\frac{1}{2}\overline{AA_1} \cdot \overline{AA_2} \sin \angle A_1AA_2 = \frac{1}{2}bc \sin \angle CAB$$

$$= \frac{bc}{2} \cdot \frac{a}{c} = \frac{1}{2}ab.$$

Similarly, the area of triangle BB_1B_2 is $\frac{1}{2}ab$. We complete the result as in the first solution.

Rider. Would the area of the three triangles AA_1A_2, BB_1B_2 and CC_1C_2 necessarily be equal when triangle ABC does not have a right angle?

Problem 296. If the three numbers are a, b, c with $a < b < c$, it suffices to show that $c^2 = a^2 + b^2$. Taking $a = 6 \cdot 10^{n+2}$, $b = 1125 \cdot 10^{2n+1} - 8$, and $c = 1125 \cdot 10^{2n+1} + 8$, we have

$$a^2 + b^2 = 36 \cdot 10^{2n+4} + (1125 \cdot 10^{2n+1})^2$$

$$- 18000 \cdot 10^{2n+1} + 64$$

$$= (36000 - 18000)10^{2n+1}$$

$$+ (1125 \cdot 10^{2n+1})^2 + 64$$

$$= (1125 \cdot 10^{2n+1})^2 + 18000 \cdot 10^{2n+1} + 64$$

$$= c^2,$$

as desired.

Problem 297. How can two players, call them A and B, get to play together? Either (a): they can be chosen as a competing pair in the first round, or (b): both can play and win over separate opponents in the first round and be among the half of the players who continue in the tournament. These cases are mutually exclusive.

Consider case (a) from A's point of view. If the choice of competing pairs is unbiased, A has as much chance of playing with B as with anyone else in the tournament. Thus, if there are n players in all, A has probability $\frac{1}{n-1}$ of playing against B in the first round.

If case (b) occurs, the problem is reduced to consideration of the same situation with half as many players. Thus, we can reduce the problem with 32 players to the same problem with 16, then 8, then 4, then 2 players.

With this in mind, for $k = 1, 2, 3, 4, 5$, let p_k be the probability that, if A and B are among 2^k players in the tournament, then they will play together.

If $k = 1$, there are only two players, A and B, who must therefore compete. Hence $p_1 = 1$.

Suppose $k = 2$. Then there are 4 players and there is one chance in three that A and B play together in the first round. Otherwise, they play against separate opponents (of equal skill to themselves) with an even chance of winning. They both survive the first round with probability $(\frac{1}{2})^2 = \frac{1}{4}$. In the second round, being the only players remaining, they compete. Hence

$$p_2 = \frac{1}{3} + (\frac{2}{3})(\frac{1}{4})p_1 = \frac{1}{2}$$

(where $\frac{2}{3}$ is the probability they do not play each other in the first round).

Now let $k \geq 2$. There are 2^k players and A plays B in the first round with probability $\frac{1}{2^k-1}$. Hence, with probability $\frac{2^k-2}{2^k-1} = \frac{2(2^{k-1}-1)}{2^k-1}$, A and B have separate opponents, vanquishing them with probability $(\frac{1}{2})^2 = \frac{1}{4}$. Given that both are among the 2^{k-1} to survive the first round, they will later play together with probability p_{k-1}. Thus

$$p_k = \frac{1}{2^k - 1} + \frac{2(2^{k-1} - 1)}{2^k - 1} \cdot \frac{1}{4} \cdot p_{k-1},$$

so $2(2^k - 1)p_k = 2 + (2^{k-1} - 1)p_{k-1}$. Hence

$$2^k(2^k - 1)p_k$$

$$= 2^k + 2^{k-1}(2^{k-1} - 1)p_{k-1}$$

$$= 2^k + 2^{k-1} + 2^{k-2}(2^{k-2} - 1)p_{k-2}$$

(and by recursion on k)

$$= 2^k + 2^{k-1} + 2^{k-2} + \cdots + 2 \cdot 1 \cdot p_1$$

$$= 2(2^{k-1} + 2^{k-2} + \cdots + 1)$$

$$= 2(2^k - 1),$$

so that

$$p_k = \frac{1}{2^{k-1}}.$$

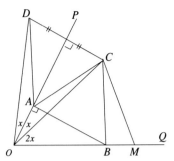

FIGURE 186

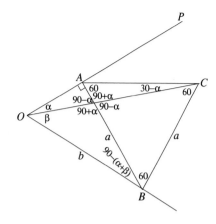

FIGURE 187

In particular, the required chance that A and B among 32 players get to play together is one in sixteen.

Problem 298. *First solution.* (John Im) Suppose we have a configuration in which the angle is trisected. Let $\angle POC = x°$ so that $\angle COQ = 2x°$. Let D be the image of C under the reflection in line OP, and let M be the point on OQ for which $\overline{OM} = \overline{OC}$. It is straightforward to establish

$$\triangle ACD \cong \triangle ABC \quad \text{and} \quad \triangle OCD \cong \triangle OMC,$$

so that

$$\overline{CB} = \overline{AC} = \overline{CD} = \overline{CM}$$

and

$$\angle CBM = \angle CMB.$$

Also,

$$\angle CBM = 2x + (30° + x) = 30° + 3x,$$

$$\angle CMB = \angle OCM = \frac{1}{2}(180° - 2x) = 90° - x.$$

Hence $30° + 3x = 90° - x$ so that $x = 15$. Hence we must have $\angle POQ = 45°$. (See *Crux Mathematicorum* 7 (1981) 100–101.)

Second solution. Let $\alpha = \angle AOC$ and $\beta = \angle COB$. Then $0 < \alpha, \beta \leq 90°$ and $\alpha + \beta \leq 90°$. We can label other angles as in Figure 187. Let $\overline{AC} = \overline{AB} = \overline{BC} = a$, $\overline{OB} = b$ and $\overline{OC} = c$. Applying the Law of Sines respectively to triangles OBC and AOB, we have

$$\frac{\sin \beta}{\sin(30° + \alpha)} = \frac{a}{b}$$

and

$$\sin(\alpha + \beta) = \frac{a}{b},$$

so that

$$\sin \beta = \sin(\alpha + \beta) \sin(30° + \alpha)$$
$$= \frac{1}{2}\{\cos(30° - \beta) - \cos(30° + 2\alpha + \beta)\}. \quad (1)$$

Applying the Law of Sines respectively to triangles OBC and OAC, we have

$$\frac{\sin \beta}{\sin(150° - \alpha - \beta)} = \frac{a}{c} \quad \text{and} \quad \frac{\sin \alpha}{\sin 150°} = \frac{a}{c}.$$

Hence

$$\frac{\sin \alpha}{\sin 30°} = \frac{\sin \alpha}{\sin 150°} = \frac{\sin \beta}{\sin(150° - \alpha - \beta)}$$
$$= \frac{\sin \beta}{\sin(30° + \alpha + \beta)},$$

so that, using $\sin 30° = \frac{1}{2}$,

$$\sin \beta = 2 \sin \alpha \sin(30° + \alpha + \beta)$$
$$= \cos(30° + \beta) - \cos(30° + 2\alpha + \beta). \quad (2)$$

From (1) and (2), we get

$$2\cos(30° + \beta) - 2\cos(30° + 2\alpha + \beta)$$
$$= \cos(30° - \beta) - \cos(30° + 2\alpha + \beta)$$

or

$$\cos(30° + 2\alpha + \beta)$$
$$= 2\cos(30° + \beta) - \cos(30° - \beta). \quad (3)$$

Now suppose the method actually gives a trisection of the angle $\angle POQ$. Then $2\alpha = \beta$ and equation (3) becomes

$$\cos(30° + 2\beta) = 2\cos(30° + \beta) - \cos(30° - \beta),$$

or

$$\cos(30° + 2\beta) - \cos(30° + \beta)$$
$$= \cos(30° + \beta) - \cos(30° - \beta),$$

or

$$-2\sin\left(30° + \frac{3\beta}{2}\right)\sin\frac{\beta}{2} = -2\sin 30° \sin\beta$$

$$= -2\sin\frac{\beta}{2}\cos\frac{\beta}{2}.$$

Either $\beta = 0$ or $\sin(30° + \frac{3\beta}{2}) = \cos\frac{\beta}{2} = \sin(90° \pm \frac{\beta}{2})$. Hence $\beta = 0$, $30° + \frac{3\beta}{2} = 90° - \frac{\beta}{2}$ or $30° + \frac{3\beta}{2} = 90° + \frac{\beta}{2}$, yielding three possibilities:

$$\alpha = \beta = 0°,$$
$$\alpha = 15°, \quad \beta = 30°,$$
$$\alpha = 30°, \quad \beta = 60°.$$

Conversely, it can be shown that equation (3) is satisfied in each of these cases. Thus the only acute angle $\angle POQ$ ($= \alpha + \beta$) trisected by the method is $45°$.

Third solution. In Figure 188, let D be the foot of the perpendicular from C to OA produced, and M the foot of the perpendicular from C to ABc. Let $\overline{OA} = 2c$, $\angle AOC = \alpha$, $\angle BOA = \gamma$. Then $\overline{AB} = 2c\tan\gamma$, so

$$\overline{CD} = \overline{AM} = \frac{1}{2}\overline{AB} = c\tan\gamma.$$

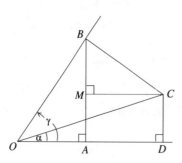

FIGURE 188

Thus,

$$\tan\alpha = \frac{\overline{CD}}{\overline{OD}} = \frac{c\tan\gamma}{2c + c\sqrt{3}\tan\gamma}$$
$$= \frac{\tan\gamma}{2 + \sqrt{3}\tan\gamma}.$$

Suppose the configuration is such that OC actually trisects the angle $\angle BOA$, i.e., $\gamma = 3\alpha$. Then, since

$$\tan 3\alpha = \frac{3\tan\alpha - \tan^3\alpha}{1 - 3\tan^2\alpha},$$

we would have

$$\tan\alpha = \frac{\tan 3\alpha}{2 + \sqrt{3}\tan 3\alpha}$$
$$= \frac{3\tan\alpha - \tan^3\alpha}{2 - 6\tan^2\alpha + 3\sqrt{3}\tan\alpha - \sqrt{3}\tan^3\alpha}.$$

Setting $x = \tan\alpha$ and cancelling out x, we obtain the equation

$$1 + 5x^2 = \sqrt{3}(3x - x^3).$$

We can clear out the surd by substituting $x = \sqrt{3}y$. This gives

$$1 + 15y^2 = 9y - 9y^3$$

or

$$3y(3y^2 + 5y - 3) + 1 = 0.$$

If there is a rational solution, it is certainly the case in lowest terms that 3 must divide the denominator. Accordingly, if $y = \frac{z}{3}$, then

$$z^3 + 5z^2 - 9z + 3 = 0.$$

Since $z^3 + 5z^2 - 9z + 3 = (z-1)(z^2 + 6z - 3)$, we obtain that $z = 1$, $-3 + 2\sqrt{3}$, $-3 - 2\sqrt{3}$, whence $x = \frac{\sqrt{3}}{3}$, $2 - \sqrt{3}$, $-2 - \sqrt{3}$. The third value for x, being negative, is extraneous. Hence we find the method works only for

$$\tan\alpha = 0°, \quad \frac{1}{\sqrt{3}}, \quad 2 - \sqrt{3},$$

which corresponds to $\alpha = 0°$, $30°$, $15°$ or $\gamma = 0°$, $90°$, $45°$.

Problem 299. Suppose there are k 1's. Then the number represented is

$$1 + 2 + 2^2 + 2^3 + \ldots + 2^{k-1} = 2^k - 1.$$

Its square is

$$(2^k - 1)^2 = 2^{2k} - 2^{k+1} + 1$$
$$= 2^{k+1}(2^{k-1} - 1) + 1$$
$$= 2^{k+1}(2^{k-2} + 2^{k-3} + \ldots + 1) + 1$$
$$= 2^{2k-1} + 2^{2k-2} + \ldots + 2^{k+1} + 1.$$

In base 2, this square is expressed

$$\underbrace{111 \ldots 11}_{k-1} \underbrace{000 \ldots 00}_{k} 1.$$

Problem 300. Before tackling the problem in general, get a feel for it by looking at the first few cases. For $k = 1$, 2 and 3, the respective equations are

$$\sin \frac{\pi}{2} = 1, \quad \sin \frac{\pi}{4} = \frac{1}{\sqrt{2}}, \quad \sin \frac{\pi}{6} \sin \frac{3\pi}{6} = \frac{1}{2},$$

as can be readily verified.

Now try to prove the $k = 4$ case. The left side is $\sin \frac{\pi}{8} \sin \frac{3\pi}{8}$. We note that

$$\sin \frac{3\pi}{8} = \cos(\frac{\pi}{2} - \frac{3\pi}{8}) = \cos \frac{\pi}{8},$$

so the expression becomes

$$\sin \frac{\pi}{8} \cos \frac{\pi}{8} = \frac{1}{2} \sin \frac{\pi}{4} = \frac{1}{2\sqrt{2}},$$

and the equation holds for $k = 4$.

When $k = 5$, the left side is

$$\sin \frac{\pi}{10} \sin \frac{3\pi}{10} \sin \frac{5\pi}{10}.$$

In this case, trying to substitute $\cos \frac{4\pi}{10}$ for $\sin \frac{\pi}{10}$ and $\cos \frac{2\pi}{10}$ for $\sin \frac{3\pi}{10}$ does not seem particularly helpful. However, we might try to get in a position to use the $\sin 2\theta = 2 \sin \theta \cos \theta$ identity by looking at

$$\sin \frac{\pi}{10} \sin \frac{2\pi}{10} \sin \frac{3\pi}{10} \sin \frac{4\pi}{10} \sin \frac{5\pi}{10}$$
$$= \cos \frac{4\pi}{10} \cos \frac{3\pi}{10} \cos \frac{2\pi}{10} \cos \frac{\pi}{10}.$$

Then (noting that $\sin \frac{5\pi}{10} = 1$),

$$\sin^2 \frac{\pi}{10} \sin^2 \frac{2\pi}{10} \sin^2 \frac{3\pi}{10} \sin^2 \frac{4\pi}{10}$$
$$= \left(\sin \frac{\pi}{10} \cos \frac{\pi}{10} \right) \left(\sin \frac{2\pi}{10} \cos \frac{2\pi}{10} \right) \times$$
$$\left(\sin \frac{3\pi}{10} \cos \frac{3\pi}{10} \right) \left(\sin \frac{4\pi}{10} \cos \frac{4\pi}{10} \right)$$
$$= \frac{1}{2^4} \sin \frac{\pi}{5} \sin \frac{2\pi}{5} \sin \frac{3\pi}{5} \sin \frac{4\pi}{5}$$
$$= \frac{1}{2^4} \sin \frac{\pi}{5} \sin \frac{2\pi}{5} \sin \frac{2\pi}{5} \sin \frac{\pi}{5}$$
$$= \frac{1}{2^4} \sin^2 \frac{2\pi}{10} \sin^2 \frac{4\pi}{10}.$$

Cancelling gives $\sin^2 \frac{\pi}{10} \sin^2 \frac{3\pi}{10} = \frac{1}{2^4}$, from which the $k = 5$ case follows.

Look at the general case. Let

$$A = \sin \frac{\pi}{2k} \sin \frac{2\pi}{2k} \sin \frac{3\pi}{2k} \cdots \sin \frac{(k-1)\pi}{2k}.$$

Since $\sin \frac{\pi}{2} = 1$ and

$$\sin \frac{r\pi}{2k} = \cos \left((k - r) \frac{\pi}{2k} \right)$$

for $r = 1, 2, \ldots, k - 1$,

$$A^2 = \left(\sin \frac{\pi}{2k} \cos \frac{\pi}{2k} \right) \left(\sin \frac{2\pi}{2k} \cos \frac{2\pi}{2k} \right) \times$$
$$\cdots \times \left(\sin \frac{(k-1)\pi}{2k} \cos \frac{(k-1)\pi}{2k} \right)$$
$$= \frac{1}{2^{k-1}} \sin \frac{\pi}{k} \sin \frac{2\pi}{k} \sin \frac{3\pi}{k} \cdots \sin \frac{(k-1)\pi}{k}$$
$$= \frac{1}{2^{k-1}} \sin^2 \frac{\pi}{k} \sin^2 \frac{2\pi}{k} \cdots \sin^2 \left[\frac{k}{2} \right] \frac{\pi}{k}$$
$$\left(\text{since } \sin \frac{(k-s)\pi}{k} = \sin \frac{s\pi}{k} \right.$$
$$\text{for } s = 1, 2, \ldots, \left[\frac{k}{2} \right] \right),$$
$$= \frac{1}{2^{k-1}} \sin^2 \frac{2\pi}{2k} \sin^2 \frac{4\pi}{2k} \cdots \sin^2 \left[\frac{k}{2} \right] \frac{2\pi}{2k}$$

Cancelling out the $\left[\frac{k}{2} \right]$ factors $\sin^2 \frac{2r\pi}{2k}$ ($r = 1, 2, \ldots, \left[\frac{k}{2} \right]$) from the two expressions for A^2 yields, for k even,

$$\sin^2 \frac{\pi}{2k} \sin^2 \frac{3\pi}{2k} \cdots \sin^2 \frac{(k-1)\pi}{2k} = \frac{1}{2^{k-1}}$$

$$\sin^2 \frac{\pi}{2k} \sin^2 \frac{3\pi}{2k} \ldots \sin^2 \frac{k\pi}{2k} = \frac{1}{2^{k-1}} \quad \text{for } k \text{ odd.}$$

Since

$$2\left[\frac{k+1}{2}\right] - 1 = \begin{cases} k-1 & \text{for } k \text{ even,} \\ k & \text{for } k \text{ odd,} \end{cases}$$

the required result follows.

Problem 301. (a) Since $\frac{1}{11} + \frac{1}{110} = \frac{11}{110} = \frac{1}{10}$ and $\frac{1}{41} + \frac{1}{1640} = \frac{41}{1640} = \frac{1}{40}$, the right-hand side is equal to

$$\left(\frac{1}{2} + \frac{1}{5} + \frac{1}{10}\right) + \left(\frac{1}{8} + \frac{1}{20} + \frac{1}{40}\right)$$

$$= \frac{4}{5} + \frac{1}{4}\left(\frac{1}{2} + \frac{1}{5} + \frac{1}{10}\right) = \frac{4}{5} + \frac{1}{5} = 1,$$

as required.

(b) Suppose that the representation uses the reciprocals of k distinct positive integers $x_1, x_2, \ldots, x_k$, where $x_i \equiv 2 \pmod{3}$ for $i = 1, 2, \ldots, k$ (i.e., x_i is one of $2, 5, 8, 11, \ldots$).

Since

$$1 = \frac{1}{x_1} + \frac{1}{x_2} + \cdots + \frac{1}{x_k},$$

multiplication by $x_1 x_2 x_3 \ldots x_k$ yields

$$x_1 x_2 x_3 \ldots x_k = \sum_{i=1}^{k} X_i, \qquad (*)$$

where $X_i = \frac{1}{x_i}(x_1 x_2 \ldots x_k)$ is the product of $k-1$ of the numbers x_j.

We obtain, from $(*)$, $2^k \equiv k 2^{k-1} \pmod{3}$, from which $k \equiv 2 \pmod{3}$.

Hence k must be one of $2, 5, 8, 11, \ldots$.

Since $\frac{1}{x_1} + \frac{1}{x_2} < \frac{1}{2} + \frac{1}{2} = 1$ for distinct integers x_1 and x_2 exceeding 1, there is no representation of the required type with only 2 terms.

Is it possible to find 5 numbers from $\{2, 5, 8, \ldots\}$ whose reciprocals add up to 1? Suppose yes, say

$$1 = \frac{1}{x_1} + \frac{1}{x_2} + \frac{1}{x_3} + \frac{1}{x_4} + \frac{1}{x_5}$$

with $2 \le x_1 < x_2 < x_3 < x_4 < x_5$. Then it must be that $x_1 \ge 2$, $x_2 \ge 5$, $x_3 \ge 8$, $x_4 \ge 11$, $x_5 \ge 14$, so that

$$\frac{1}{x_1} + \frac{1}{x_2} + \frac{1}{x_3} + \frac{1}{x_4} + \frac{1}{x_5}$$

$$\le \frac{1}{2} + \frac{1}{5} + \frac{1}{8} + \frac{1}{11} + \frac{1}{14}$$

$$= \left(\frac{1}{2} + \frac{1}{8}\right) + \left(\frac{1}{5} + \frac{1}{11}\right) + \frac{1}{14}$$

$$< \frac{5}{8} + \left(\frac{1}{5} + \frac{1}{10}\right) + \frac{3}{42}$$

$$< \frac{5}{8} + \frac{3}{10} + \frac{3}{40} = 1,$$

a contradiction.

Hence we need more than 5 numbers for the representation. The next possibility is 8, and (a) shows that there is indeed a representation with 8 terms.

Problem 302. Let the length of a side of the given square be 1. The pentagon $ACDFE$ (Figure 189) then has area $\frac{5}{8}$, so that the side length of the fitted square should be $\sqrt{\frac{5}{8}}$.

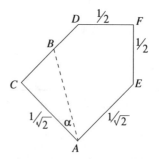

FIGURE 189

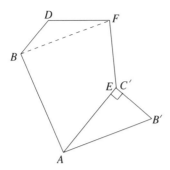

FIGURE 190

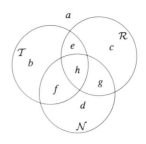

FIGURE 191

We try to determine B on CD so that AB is a side of the square. Since $\overline{AC} = \frac{1}{\sqrt{2}} < \sqrt{\frac{5}{8}} < 1 = \overline{AD}$ this is certainly possible.

Since $\overline{AB} = \sqrt{\frac{5}{8}}$, by Pythagoras' Theorem $\overline{CB} = \sqrt{\frac{1}{8}} = \frac{1}{2\sqrt{2}} = \frac{1}{2}\overline{CD}$. Hence $\alpha = \angle CAB = 30°$.

Cut along the line AB and move triangle ACB so that AC falls along AE, and the triangle occupies $AC'B'$ (Figure 190), where $C' = E$. We have that $\angle FC'B' = 360° - 90° - 135° = \angle BDF$, $\overline{BD} = \overline{B'C'}$, $\overline{DF} = \overline{EF} = \overline{C'F}$. Hence, make a second cut along BF and fit triangle BDF so that B, D and F fall on B', C' and F respectively. Then $ABFB'$ is the required fitted square.

We check that $ABFB'$ is a square. Clearly

$$\angle BAB' = \angle CAE = 90°$$

and

$$\angle BFB' = \angle DFE = 90°,$$

and by the Cosine Law

$$\overline{B'F}^2 = \overline{BF}^2 = \frac{1}{8} + \frac{1}{4} + \frac{1}{2\sqrt{2}}\left(\frac{1}{\sqrt{2}}\right)$$
$$= \frac{5}{8} = \overline{AB}^2 = \overline{AB'}^2.$$

Problem 303. In the Venn diagram (Figure 191), the three circles labelled $\mathcal{T}$, $\mathcal{R}$, $\mathcal{N}$ "contain" respectively those people who used Television, Radio, Newspapers, while a, b, c, d, e, f, g, h are

the number of people in the subsets: a is the number who used none of the news sources; b used Television but not Radio nor Newspapers; e used Television and Radio but not Newspapers; and so on.

From the Venn diagram we see that the information can be put in the following form:

$$b + e + f + h = 50 \qquad (1)$$
$$a + b + d + f = 61 \qquad (2)$$
$$a + b + c + e = 13 \qquad (3)$$
$$e + f + g + h = 74 \qquad (4)$$

and

$$N = a + b + c + d + e + f + g + h,$$

where a, b, c, d, e, f, g, h are all nonnegative integers.

Upper bound for N. By (2), $d \le 61$, and by (3) and (4),

$$a + b + c + e + f + g + h \le 13 + 74 = 87.$$

Hence $N \le 61 + 87 = 148$. This value of N is realized for $a = b = f = e = 0$, $d = 61$, $h = 50$, $c = 13$, $g = 24$.

Lower bound for N. Since

$$f \le b + e + f + h = 50,$$

we have

$$a + b + d = 61 - f \ge 11.$$

Hence

$$N = (a + b + d) + (e + f + g + h) + c$$
$$\ge 11 + 74 + 0 = 85.$$

Is $N = 85$ possible? If so, then all the inequalities in the above estimates must be equalities so that

$$a + b + d = 11, \quad c = 0, \quad f = 50.$$

But then $b = e = h = 0$ (from(1)), so that

$$a + b + c + e = a \leq 11,$$

which contradicts (3). Hence N cannot be 85.

Suppose $N = 86$. Then either $a + b + d = 11$ and $c = 1$, or $a + b + d = 12$ and $c = 0$.

In the former case, $a + b + d = 11$, $c = 1$ which, in turn, imply: $f = 50$; $b = e = h = 0$; $a + b + c + e \leq 12$, contradicting (3).

In the latter case, $a + b + d = 12$ and $c = 0$ which imply $f = 49$ and $b + e + h = 1$. Since $a + b + d = 12$ and $a + b + e = 13$ (by (3)), $e = d + 1$, so that $d = 0$, $e = 1$, $b = h = 0$. Hence $a = 12$, and, by (4), $g = 24$. Thus, $N = 86$ occurs when

$$a = 12, \quad e = 1, \quad f = 49,$$

$$g = 24, \quad b = c = d = h = 0.$$

Rider. Are there any impossible values for N between 86 and 148?

Problem 304. Let W be a point on the parabola such that WF is perpendicular to the axis. Let d be the directrix, a be the axis, and let $\mathrm{dist}(X, l)$ denote the perpendicular distance from the point X to the line l.

Case 1. P lies between V and W.

By the reflection property of parabolas, PQ bisects the angle formed by PF and the line

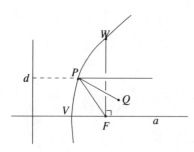

FIGURE 192

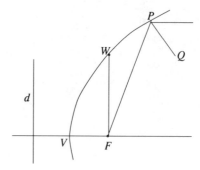

FIGURE 193

through P parallel to the axis. If PQ crosses the axis we would have

$$\overline{PQ} > \overline{PF} = \mathrm{dist}(P, d) > \mathrm{dist}(V, d) = \overline{VF},$$

a contradiction. Hence PQ cannot cross the axis.

Case 2. W lies between V and P. See Figure 193. If PQ crosses the axis, then

$$\overline{PQ} > \mathrm{dist}(P, a) \geq \overline{WF} = \mathrm{dist}(W, d)$$
$$= 2\,\mathrm{dist}(V, d) = 2\,\overline{VF},$$

again a contradiction.

Problem 305. *First solution.* Since $(x + y)^2 = (5 - z)^2$ and $xy = 3 - z(x + y) = 3 - z(5 - z)$, we have that

$$0 \leq (x - y)^2 = (x + y)^2 - 4xy$$
$$= 25 - 10z + z^2 - 12 + 20z - 4z^2$$
$$= -3z^2 + 10z + 13 = -(z + 1)(3z - 13).$$

Hence, $-1 \leq z \leq \frac{13}{3}$. Thus, the maximum possible value of z is $\frac{13}{3}$. This value can actually be achieved when $x - y = 0$, $x + y = 5 - \frac{13}{3} = \frac{2}{3}$, i.e., when $x = y = \frac{1}{3}$.

Second solution. Since $x + y = 5 - z$ and $xy = 3 + z^2 - 5z$, x and y are the two roots of the quadratic

$$t^2 + (z - 5)t + (z^2 - 5z + 3).$$

The condition that this quadratic have real roots is that

$$(z - 5)^2 - 4(z^2 - 5z + 3) \geq 0$$

or

$$3z^2 - 10z - 13 \leq 0.$$

This is the condition on z obtained in the first solution, so that, as above, $\frac{13}{3}$ is the maximum value of z.

Rider. If $x_1, x_2, \ldots, x_n$ are real numbers such that

$$x_1 + x_2 + \cdots + x_n = a,$$

$$\sum_{1 \leq i < j \leq n} x_i x_j = b^2,$$

determine the largest value that any one of the x_i's can be.

Problem 306. *First solution.* In these divisibility problems, it is generally useful to factor the divisor. Here we find that $2304 = 2^8 \cdot 3^2 = (2^4 \cdot 3)^2 = 48^2$. Consequently, we manipulate the expression in n to yield terms involving 48. Thus (using the Binomial Theorem)

$$7^{2n} - 2352n - 1 = (7^2)^n - 2304n - 48n - 1$$

$$= (1 + 48)^n - 48n - 1 - 2304n$$

$$= \binom{n}{2}48^2 + \binom{n}{3}48^3 + \cdots + 48^n - 2304n$$

$$= 2304 \left\{ \binom{n}{2} + \binom{n}{3}48 + \cdots + 48^{n-2} - n \right\}.$$

Second solution (by induction). When $n = 1$, the expression becomes $7^2 - 2352 - 1 = -2304$, and the result holds. Assume it holds for $n = k$. Then

$$7^{2(k+1)} - 2352(k+1) - 1$$

$$= \{7^{2k} - 2352k - 1\} + \{7^{2k+2} - 7^{2k}\}$$

$$\quad - \{7^2 - 1 + 2304\}$$

$$= \{7^{2k} - 2352k - 1\}$$

$$\quad + (7^2 - 1)(7^{2k} - 1) - 2304$$

$$= \{7^{2k} - 2352k - 1\} + (7^2 - 1)^2 \times$$

$$\quad (7^{2(k-1)} + 7^{2(k-2)} + \cdots + 1) - 2304.$$

Since $7^{2k} - 2352k - 1$ is divisible by 2304 and $(7^2 - 1)^2 = 2304$, the left side is divisible by 2304.

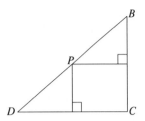

FIGURE 194

Problem 307. *First solution.* Let BCD be the base of the tetrahedron; this is a right-angled triangle with hypotenuse DB. Since triangles ABC, ABD and ACD are isosceles, point A must lie on the intersection of the right-bisecting planes of BC, BD and CD. Since these three planes meet BD at its midpoint P, we have $AP \perp \triangle BCD$. Hence the height of the tetrahedron is $\overline{AP} = \frac{1}{2}(5\sqrt{3})$. Thus the volume of the tetrahedron is

$$\frac{1}{3} \cdot \left(\frac{1}{2} \cdot 3 \cdot 4 \right) \cdot \left(\frac{1}{2} \cdot 5\sqrt{3} \right) = 5\sqrt{3}.$$

Second solution. Since A is equidistant from B, C and D, the orthogonal projection P of A onto the plane of B, C, D is also equidistant from B, C and D. Hence P is the circumcenter of $\triangle BCD$. Let $R = \overline{BP} = \overline{CP} = \overline{DP}$ and $h = \overline{AP}$. Then $R^2 + h^2 = 5^2$. The volume V of the tetrahedron is given by $V = \frac{1}{3}h \cdot \text{area}(\triangle BCD)$. But $4R \cdot \text{area}(\triangle BCD) = \overline{BC} \cdot \overline{BD} \cdot \overline{CD}$, i.e, $6 \cdot 4R = 3 \cdot 4 \cdot 5$. Therefore $R = \frac{5}{2}$, so $h = \sqrt{25 - \frac{25}{4}} = \frac{5\sqrt{3}}{2}$. Thus $V = 5\sqrt{3}$.

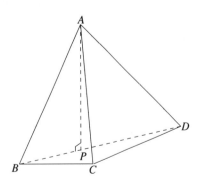

FIGURE 195

Rider. $ABCD$ and $A'B'C'D'$ are tetrahedra such that $\overline{AB} = \overline{AC} = \overline{AD} = a$, $\overline{BC} = b$, $\overline{CD} = c$, $\overline{BD} = d$, $\overline{A'B'} = b$, $\overline{A'C'} = c$, $\overline{A'D'} = d$, $\overline{B'C'} = \overline{C'D'} = \overline{D'B'} = a$. For what values of a, b, c, d does the volume of $ABCD$ exceed the volume of $A'B'C'D'$?

Rider. More generally, establish that the volume of a tetrahedron $ABCD$ is given by

$$V^2 = \frac{(\overline{AB} \cdot \overline{AC} \cdot \overline{AD})^2}{36} \times$$
$$\{1 - \cos^2 \alpha - \cos^2 \beta - \cos^2 \gamma$$
$$+ 2 \cos \alpha \cos \beta \cos \gamma\},$$

where

$$\alpha = \angle BAC, \quad \beta = \angle CAD, \quad \gamma = \angle DAB.$$

Problem 308. *First solution.* Since the given polynomial is homogenous, any polynomial factor must also be homogeneous. If we set $(x, y, z, w) = (1, 1, 1, 1)$, we find that the polynomial vanishes. This suggest that we try linear factors of the form $x + y - z - w$ with an equal number of positive and negative terms. For this we need the Factor Theorem.

If we set $x = z + w - y$ we find that the polynomial, as a polynomial in y, (or z or w) has all its coefficients zero, so it vanishes identically. Hence $x + y - z - w$ is a factor. By symmetry, two other factors must be $x + w - z - y$ and $x + z - y - w$. There can be but one remaining factor of the first degree, and this must be symmetrical. Hence it is $x + y + z + w$.

Second solution. If there are four real linear factors, then by the symmetry and homogeneity, the factorization must be of either of the following forms:

(a) $a(x + y + z - bw)(y + z + w - bx) \times$
        $(z + w + x - by)(w + x + y - bz)$

(b) $c(x + y + z + w)(x + y + kz + kw) \times$
        $(x + z + kw + ky)(x + w + ky + kz).$

If there were a linear factor with at least three distinct coefficients, then, by the symmetry of the variables, there must be at least six such, which is too many (because of the degree).

Try a factorization of the form (a). Comparing coefficients of x^4, we have $-ab = 1$. Set $x = y = z = w = 1$ to get $a(3 - b)^4 = 0$, so $b = 3$ and $a = -\frac{1}{3}$. But then, setting $z = w = 0$ leads to

$$x^4 + y^4 - 2x^2 y^2 = -\frac{1}{3}(x + y)^2 (y - 3x)(x - 3y),$$

which is false when $x = y = 1$. Hence (a) is impossible.

Now try (b). From the coefficient of x^4, $c = 1$. Putting $x = y = z = w = 1$ yields $0 = 4 \cdot 2^3 (1 + k)^3$ whence $k = -1$. Expanding out, we verify that (b) works.

Here is a proof of the assertion at the beginning of this solution. (It holds for any number of variables.) If, for example, $P(x, y, z)$ is a polynomial of three variables, then P is homogeneous of degree n if and only if, for any t,

$$P(tx, ty, tz) = t^n P(x, y, z). \qquad (*)$$

Suppose $P(x, y, z) = Q(x, y, z) R(x, y, z)$. Replace x, y, z respectively by tx, ty, tz and use $(*)$ to obtain

$$t^n P(x, y, z) = Q(tx, ty, tz) R(tx, ty, tz).$$

Consider x, y, z to be fixed. Then both sides are polynomials in t and the equation gives a factorization of a constant times t^n. Since every factorization of t^n is of the form $t^m t^{n-m}$, we must have

$$Q(tx, ty, tz) = t^m H(x, y, z)$$

and

$$R(tx, ty, tz) = t^{n-m} K(x, y, z)$$

for some polynomials H and K. Now set $t = 1$ and argue that Q and R have to be homogeneous.

Rider. Show that the given polynomial equals

$$\text{determinant} \begin{pmatrix} x & y & z & w \\ y & x & w & z \\ z & w & x & y \\ w & z & y & x \end{pmatrix}$$

and use this to obtain the factorization.

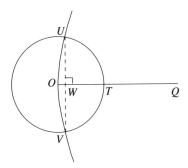

FIGURE 196

Problem 309. Suppose that the locus were a straight line segment.

Consider the extreme case that R and S both coincide with U; then P must coincide with U. Hence U is on the locus. Similarly, V is on the locus. Thus, the locus must be the line segment UV. Suppose UV intersects OQ in W. Since W is on the locus, W must be the midpoint of OT. Also, by symmetry (or by consideration of the congruent triangles UWQ and VWQ with corresponding sides equal) $UV \perp OT$.

Since UV is the right-bisector of OT,

$$\overline{TU} = \overline{OU} = \overline{OT} = \overline{OV} = \overline{TV},$$

so that O, U and V lie on a circle with center T. Thus, circles with centers Q and T have the three points O, U and V in common, and must therefore coincide. But this contradicts the fact that Q, lying outside the circle with center O must be distinct from T. Hence UV is not the required locus.

Problem 310. For $n = 1, 2, 3, 4$ we see in the nth row the square of the nth odd integer on the right (i.e., $(2n-1)^2$), and on the left $2n-1$ consecutive integers beginning with n and hence ending with $n + (2n - 1 - 1)1 = 3n - 2$. The suggested generalization is:

$$n + (n+1) + (n+2) + \cdots + (3n-2) = (2n-1)^2$$

for $n = 1, 2, 3, \ldots$. This is easy to prove by summing the left side as an arithmetic progression.

Problem 311. Since

$$b_0 + b_1 x + \cdots + b_{n+1} x^{n+1} + \cdots + b_{2n} x^{2n}$$
$$= (a_0 + a_1 x + a_2 x^2 + \cdots + a_n x^n)^2,$$

comparison of the coefficients of x^{n+1} shows that

$$b_{n+1} = a_1 a_n + a_2 a_{n-1} + \cdots + a_{n-1} a_2 + a_n a_1,$$

so that

$$b_{n+1} \le a_1 a_0 + a_2 a_0 + \cdots + a_{n-1} a_0 + a_n a_0.$$

Hence,

$$2b_{n+1}$$
$$\le a_1(a_0 + a_n) + a_2(a_0 + a_{n-1})$$
$$\quad + \cdots + a_{n-1}(a_2 + a_0) + a_n(a_0 + a_1)$$
$$\le a_1(a_0 + \cdots + a_n) + a_2(a_0 + \cdots + a_n)$$
$$\quad + \cdots + a_n(a_0 + \cdots + a_n)$$
$$= (a_1 + a_2 + \cdots + a_n)(a_0 + a_1 + \cdots + a_n)$$
$$\le (a_0 + a_1 + \cdots + a_n)^2 = (f(1))^2.$$

Problem 312. In effecting these summations, one employs the technique of "summing by differences," that is, expressing each term as a difference of two quantities so as to obtain a cancellation. Thus, in (i), one might write

$$\frac{1}{1 \cdot 3} + \frac{1}{3 \cdot 5} + \frac{1}{5 \cdot 7} + \cdots$$
$$= \frac{1}{2}\left(1 - \frac{1}{3}\right) + \frac{1}{2}\left(\frac{1}{3} - \frac{1}{5}\right) + \frac{1}{2}\left(\frac{1}{5} - \frac{1}{7}\right) + \cdots$$
$$= \frac{1}{2}\left(1 - \frac{1}{3} + \frac{1}{3} - \frac{1}{5} + \frac{1}{5} - \frac{1}{7} + \cdots\right) = \frac{1}{2}.$$

This computation is correct but needs some justification. On the face of it, one might just as easily write

$$1 + 1 + 1 + 1 + \cdots$$
$$= (2 - 1) + (3 - 2) + (4 - 3) + (5 - 4) + \cdots$$
$$= -1 + 2 - 2 + 3 - 3 + 4 - 4 + 5 + \cdots$$
$$= -1,$$

which seems strange. Consequently, we have to analyze what we mean by the sum of an infinite series. The definition adopted is that the sum of an infinite series is equal to the limit of the partial sum of the first n terms as n increases. This is illustrated in the present example.

The partial sums are:

$$\frac{1}{1\cdot 3} = \frac{1}{2}\left(1 - \frac{1}{3}\right)$$

$$\frac{1}{1\cdot 3} + \frac{1}{3\cdot 5} = \frac{1}{2}\left(1 - \frac{1}{3}\right) + \frac{1}{2}\left(\frac{1}{3} - \frac{1}{5}\right)$$

$$= \frac{1}{2}\left(1 - \frac{1}{5}\right)$$

$$\frac{1}{1\cdot 3} + \frac{1}{3\cdot 5} + \frac{1}{5\cdot 7} = \frac{1}{2}\left(1 - \frac{1}{3}\right) + \frac{1}{2}\left(\frac{1}{3} - \frac{1}{5}\right)$$

$$+ \frac{1}{2}\left(\frac{1}{5} - \frac{1}{7}\right) = \frac{1}{2}\left(1 - \frac{1}{7}\right)$$

and in general, for $n = 1, 2, 3, \ldots$,

$$\frac{1}{1\cdot 3} + \frac{1}{3\cdot 5} + \frac{1}{5\cdot 7} + \cdots + \frac{1}{(2n-1)(2n+1)}$$

$$= \frac{1}{2}\left(1 - \frac{1}{2n+1}\right).$$

(This illustrates the fundamental law of summation: $\sum_{k=1}^{n}\{F(k) - F(k-1)\} = F(n) - F(0)$.)
If $n = 100$, the value of the partial sum is $\frac{1}{2}\left(1 - \frac{1}{201}\right) = \frac{100}{200}$; if $n = 1{,}000{,}000$, it is $\frac{1000000}{2000001}$. Thus we can see that as n becomes larger, the partial sum becomes arbitrarily close to $\frac{1}{2}$. It is on this basis that we can give $\frac{1}{2}$ as the sum of the given series.

In (ii), we have

$$\frac{1}{1\cdot 2\cdot 3} = \frac{1}{6} = \frac{1}{2}\left(\frac{1}{2} - \frac{1}{6}\right),$$

$$\frac{1}{2\cdot 3\cdot 4} = \frac{1}{2}\left(\frac{1}{6} - \frac{1}{12}\right), \ldots$$

and generally

$$\frac{1}{n(n+1)(n+2)}$$

$$= \frac{1}{2}\left(\frac{1}{n(n+1)} - \frac{1}{(n+1)(n+2)}\right).$$

The partial sum of the first n terms is

$$\frac{1}{1\cdot 2\cdot 3} + \frac{1}{2\cdot 3\cdot 4} + \frac{1}{3\cdot 4\cdot 5}$$

$$+ \cdots + \frac{1}{n(n+1)(n+2)}$$

$$= \frac{1}{2}\left(\frac{1}{2} - \frac{1}{6}\right) + \frac{1}{2}\left(\frac{1}{6} - \frac{1}{12}\right) + \frac{1}{2}\left(\frac{1}{12} - \frac{1}{20}\right)$$

$$+ \cdots + \frac{1}{2}\left(\frac{1}{n(n+1)} - \frac{1}{(n+1)(n+2)}\right)$$

$$= \frac{1}{2}\left(\frac{1}{2} - \frac{1}{(n+1)(n+2)}\right)$$

$$= \frac{1}{4} - \frac{1}{2(n+1)(n+2)}.$$

As n increases, this number gets ever closer to $\frac{1}{4}$. Hence the sum of the series in (ii), according to our definition, is $\frac{1}{4}$.

Problem 313. *First solution.* Any positive integer b can be factored as follows:

$$b = 2^k \cdot a \quad k \in \{0, 1, 2, \ldots\}, \ a \in \{1, 3, 5, \ldots\},$$

i.e., a product of a power of 2 and an odd integer. Factor the $n + 1$ given integers in this way:

$$b_i = 2^{k_i} a_i, \qquad i = 1, 2, 3, \ldots, n.$$

Now the $n + 1$ odd integers a_i are all in $\{1, 3, 5, \ldots, 2n - 1\}$ (because $1 \le b_i \le 2n$), which has only n elements. Hence (by the Pigeonhole Principle) two of the a_i's are equal, say $a_l = a_m = a$. Then

$$b_l = 2^{k_l} a, \qquad b_m = 2^{k_m} a.$$

If $0 \le k_l \le k_m$ then b_l divides b_m; if $k_m \le k_l$ then b_m divides b_l.

Second solution (by induction). If $n = 1$, then the set must be $\{1, 2\}$. Hence the result holds for $n = 1$.

Suppose the result holds for $n = k$ $(k \ge 1)$. Let S be a subset of $\{1, 2, 3, \ldots, 2k + 2\}$ with $(k + 1) + 1 = k + 2$ members.

If $2k + 2 \notin S$, then at least $k + 1$ members of S must be chosen from $\{1, 2, \ldots, 2k\}$. By the induction hypothesis, one of the numbers must divide the other.

On the other hand, suppose $2k+2 \in S$. If $k+1 \in S$, then since $2k+2 = 2(k+1)$, we have found two numbers in S one of which divides the other. If $k+1 \notin S$, form the set T by removing $2k+2$ from S and replacing it by $k+1$. Then T has $k+2$ members, $k+1$ of which are in $\{1, 2, \ldots, 2k\}$. By the induction hypothesis, there are two numbers u, v in T for which v is a multiple of u. If u and v are both distinct from $k+1$, then $u, v \in S$. If one of u and v is $k+1$, it must be the larger v, and u must divide v. But then u divides $2v = 2k+2$, so that u and $2k+2$ are two members of S one of which divides the other.

Hence in every case S contains two numbers one of which divides the other.

Thus the result holds for $n = k+1$.

Problem 314. If X is situated so that $\overline{AX} = \overline{BX}$, then clearly t_X is parallel to AB. Also (since $\overline{AX} = \overline{BX}$, $\angle YXA = \angle ZXB = 90°$, $\angle YAX = \angle XBZ = 45°$) $\triangle AXY \cong \triangle BXZ$, so $\overline{AY} = \overline{BZ}$ and YZ is parallel to AB.

On the other hand, suppose that $\overline{AX} \neq \overline{XB}$. Then t_X meets AB, BZ and AY at points V, U and W respectively. Since $\angle AXB = 90°$, $\angle AXY = 90°$ and

$$\angle AYX + \angle YAX = 90° = \angle WXY + \angle WXA. \tag{1}$$

However, since t_X and t_A are tangents from the same point W, $\overline{WA} = \overline{WX}$, and hence $\angle YAX = \angle WAX = \angle WXA$. Now from (1),

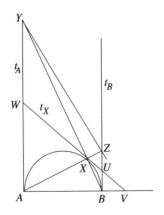

FIGURE 197

$\angle AYX = \angle WXY$, so that $\overline{WY} = \overline{WX} = \overline{WA}$ and W is the midpoint of AY.

Similarly U is the midpoint of ZB.

Now join VY and let it meet t_B at Z'. Since VW bisects AY and since $AY \parallel BZ'$, VW must bisect BZ'. Hence $\overline{BZ'} = 2\overline{BU} = \overline{BZ}$ so that $Z = Z'$. Thus YZ, WU and AB intersect at V.

Remark. The same result can be obtained using analytic geometry. Let $A = (-r, 0)$, $B = (r, 0)$ and $X = (u, v)$ where $u^2 + v^2 = r^2$. Then

$$Y = \left(-r, \frac{2rv}{r-u}\right), \quad Z = \left(r, \frac{2rv}{r+u}\right)$$

and, except when $u = 0$, then lines YZ, WU and AB intersect at $\left(\frac{r^2}{u}, 0\right)$.

Problem 315. Suppose, if possible, that a set of T of six consecutive integers be partitioned into disjoint subsets U and V whose least common multiples are both equal to m.

Suppose a prime p divides one of the numbers in T, say a number in U. Then p divides m, the least common multiple of V, so p must divide a number in V. Hence any prime p dividing a number in T must divide at least two numbers in T. Thus $p \leq 6$, so that p can be only 2, 3, or 5.

Now, distinct numbers divisible by 5 have a minimum difference of 5. Therefore there are four consecutive numbers in T none of which is divisible by 5. Among these are two odd numbers which can be divisible only by 3. But then we would have two consecutive odd numbers of the form $\pm 3^k$ ($k \geq 1$), which is impossible.

Therefore T cannot be partitioned as described above.

Problem 316. Consider the sequence $1, 2, 3, \ldots, f(n)$. In this sequence, the non-squares are $f(1), f(2), f(3), \ldots, f(n)$, while the squares are $1^2, 2^2, 3^2, \ldots, k^2$, where

$$k^2 < f(n) < (k+1)^2.$$

Thus

$$f(n) = n + k, \quad \text{where} \quad k^2 < f(n) < (k+1)^2.$$

Now we show that $k = \{\sqrt{n}\}$. We have

$$k^2 < n + k < (k+1)^2,$$

i.e.,

$$k^2 + 1 \leq n + k \leq (k+1)^2 - 1.$$

Hence

$$\left(k - \frac{1}{2}\right)^2 + \frac{3}{4} = k^2 - k + 1 \leq n$$

$$\leq k^2 + k = \left(k + \frac{1}{2}\right)^2 - \frac{1}{4}.$$

Therefore

$$k - \frac{1}{2} < \sqrt{n} < k + \frac{1}{2},$$

so that $\{\sqrt{n}\} = k$.

Problem 317. You should not. Since the cards neither of us gets are paired according to color, the number of cards in your pile at the conclusion of the game must equal the number of cards in mine. Thus you will lose one dollar every time you play.

Problem 318. *First solution.* The counterclockwise rotation about O through $90°$ takes A onto B and A' onto B'.

Since such a rotation preserves length, $\overline{AA'} = \overline{BB'}$. Since it takes any direction to a perpendicular direction, $AA' \perp BB'$.

Second solution. If $A'OA$ and $B'OB$ are straight line segments, the result is clearly valid. Suppose otherwise. Since $\overline{A'O} = \overline{B'O}$, $\overline{OA} = \overline{OB}$, $\angle A'OA = \angle B'OB = 90° \pm \angle A'OB$, the triangles $A'OA$ and $B'OB$ are congruent. Hence $\overline{AA'} = \overline{BB'}$ and $\angle OAA' = \angle OBB'$. From the latter, it follows that, if K is the point of intersection of AA' and BB' (produced, if necessary), then $\angle OAK = \angle OAA' = \angle OBB' = \angle OBK$. Thus, O, K, A, B are concyclic, so that $\angle AKB = \angle AOB = 90°$, as desired.

Problem 319. To motivate the solution, observe that if MPQ is a right isosceles triangle, then P, M, Q are three vertices of a square with diagonal PQ.

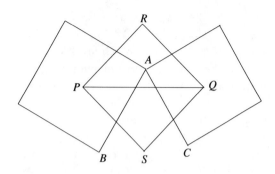

FIGURE 198

Draw the square $PRQS$ with diagonal PQ (see Figure 198). We show $S = M$. Since triangles RPS and APB are both right isosceles, by Problem 318 $\overline{BS} = \overline{RA}$ and $BS \perp RA$. Similarly, triangles RQS and AQC are right isosceles, so $\overline{RA} = \overline{SC}$ and $RA \perp SC$. Hence $\overline{BS} = \overline{SC}$ and $BS \parallel SC$. Thus, S must be M, the midpoint of BC.

Problem 320. (a) Let M be the midpoint of AC. By Problem 319, triangle PMR is isosceles and right-angled at M. Of course, triangle AMQ is isosceles and right-angled at M. By Problem 318 segments AP and QR are equal in length and lie on perpendicular lines.

(b) *First solution.* Let M be the midpoint of BD. By Problem 319 (applied to ABD) triangle PMS is isosceles and right-angled at M. Similarly, triangle QMR is isosceles and right-angled

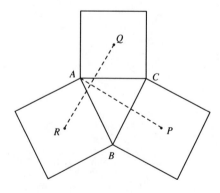

FIGURE 199

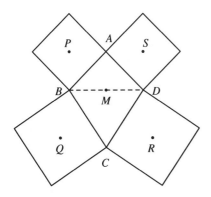

FIGURE 200

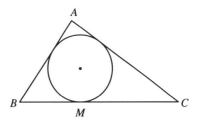

FIGURE 201

at M. By Problem 318, segments PR and QS are equal in length and lie on perpendicular lines.

Second solution. Let the vertices of A, B, C, D, correspond to the complex numbers $2Z_1$, $2Z_2$, $2Z_3$, $2Z_4$ respectively, i.e., if, in Cartesian coordinates, A is $(2x_1, 2y_1)$, we have $Z_1 = x_1 + iy_1$. Then we have

$$P = Z_1 + Z_2 + i(Z_1 - Z_2),$$

$$Q = Z_2 + Z_3 + i(Z_2 - Z_3),$$

$$R = Z_3 + Z_4 + i(Z_3 - Z_4),$$

$$S = Z_4 + Z_1 + i(Z_4 - Z_1).$$

Hence,

$$\overrightarrow{PR} = \overrightarrow{OR} - \overrightarrow{OP} = (Z_3 + Z_4 - Z_1 - Z_2)$$
$$+ i(Z_2 + Z_3 - Z_1 - Z_4)$$

and

$$\overrightarrow{QS} = \overrightarrow{OS} - \overrightarrow{OQ} = (Z_1 + Z_4 - Z_2 - Z_3)$$
$$+ i(Z_3 + Z_4 - Z_1 - Z_2),$$

and we see that $i\overrightarrow{PR} = \overrightarrow{QS}$, i.e., the segment QS is the image of segment PR when the plane is rotated about O through angle $90°$ counterclockwise.

Remark. Problems 319 and 320(a) are special cases of 320(b)!

Problem 321. (a) In Figure 201, we show the inscribed circle of $\triangle ABC$. If we keep the same

center, but increase the radius by any sufficiently small amount, we will obtain a circle which cuts each side in two distinct points. Thus the lower limit does not exceed the inradius. On the other hand, consider any circle whose radius is smaller than the inradius. If it intersects BC in two points straddling M (or coinciding with M), it cannot meet AB or AC. If it intersects BC in two points between M and C, it cannot meet AB. Thus, it cannot meet all three sides in two distinct points. Hence, the lower limit must be the inradius.

(b) It is easy to see that the upper limit must be at least R. To see how big a circle is possible, start with any circle $PQRSTU$ satisfying the condition (Figure 202). We can perform three processes which will continue to give us admissible circles of this size or larger:
(1) keeping $\overline{PQ}$ fixed, pull P and Q down towards B until the points R and S almost come together;
(2) rotate clockwise about P until the points R and S almost come together;
(3) keeping P and Q fixed, increase the size of the circle until S and T almost come together at C.

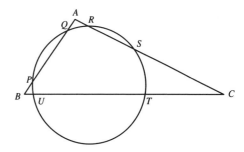

FIGURE 202

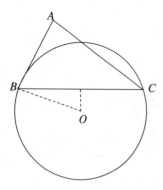

FIGURE 203

By these processes, we can find a bigger circle than the one we started with for which P and Q are as close as we like to B, and S and T are very close to C. Thus, every admissible circle is smaller than the circle tangent to AB at B passing through C, and we can find admissible circles as close to this in size as we like.

If a is the length of BC, the radius of this circle (OB in Figure 203) has length $\frac{a}{2} \csc B$. But $a = 2R \sin A$, where R is the radius of the circumcircle. Hence the upper limit of the radii of admissible circles is $\dfrac{R \sin A}{\sin B}$ which, except when $A = B$, exceeds R (since $A > B$ and $\pi - A > B$).

Problem 322. (a) $(0,0)$ is the only solution of $x + 2y = 0$, so $f(0) = 1$. $(1,0)$ is the only solution of $x + 2y = 1$, so $f(1) = 1$.

Now let $n \geq 2$. $(n, 0)$ is obviously a solution of $x + 2y = n$. Corresponding to any other solution (a, b), $b \geq 1$, there is a solution of $x + 2y = n - 2$, namely $(a, b - 1)$. Hence

$$f(n) = 1 + f(n - 2), \qquad n \geq 1.$$

Repeated use of this recurrence yields

$$f(n) = 1 + f(n - 2) = 2 + f(n - 4)$$

$$= 3 + f(n - 6) = \dots$$

$$= \begin{cases} \dfrac{n}{2} + f\left(n - 2\left(\dfrac{n}{2}\right)\right) = \dfrac{n}{2} + f(0) = \dfrac{n}{2} + 1 & \text{if } n \text{ is even} \\[2ex] \dfrac{n-1}{2} + f\left(n - 2\left(\dfrac{n-1}{2}\right)\right) = \dfrac{n-1}{2} + f(1) = \dfrac{n-1}{2} + 1 & \text{if } n \text{ is odd} \end{cases}$$

$$= \left[\dfrac{n}{2}\right] + 1.$$

This can also be deduced from the observation that there is a solution (a, b) of $x + 2y = n$ for every nonnegative integer b, $0 \leq b \leq \frac{n}{2}$.

(b) It is easy to verify that $g(0) = g(1) = 1$, $g(2) = 2$. When $n \geq 3$, the solutions (a, b, c) of $x + 2y + 3z = n$ with $c = 0$ satisfy $x + 2y = n$ and there are $f(n) = [\frac{n}{2}] + 1$ of these. Any solution (a, b, c) of $x + 2y + 3z = n$ with $c \geq 1$ corresponds to a solution $(a, b, c - 1)$ of $x + 2y + 3z = n - 3$. Hence

$$g(n) = g(n - 3) + \left[\frac{n}{2}\right] + 1, \qquad n \geq 3.$$

Problem 323. The equation is equivalent to

$$(x - y)^2 + (x - z)^2 + x^2 = 0,$$

so $x = y = z = 0$ is the only solution.

Problem 324. Let P, E, N denote the cost, in cents, of each pencil, eraser, notebook, respectively. P, E, N are positive integers such that

$$P + E + N = 100,$$

$$N > 2P,$$

$$3P > 4E,$$

$$3E > N.$$

We have $P < \frac{N}{2} < \frac{3E}{2}$ so

$$100 = P + E + N < \frac{3E}{2} + E + 3E = \frac{11}{2}E,$$

i.e.,

$$E > \frac{200}{11} > 18.$$

Also $N > 2P > \frac{8}{3}E$, so

$$100 = P + E + N > \frac{4E}{3} + E + \frac{8E}{3} = 5E,$$

i.e., $E < 20$ and thus $18 < E < 20$, so $E = 19$. Now we have

$$81 = N + P > 2P + P = 3P > 4E$$
$$= 4(19) = 76,$$

i.e., $81 > 3P > 76$, so $P = 26$. Finally,

$$N = 100 - P - E = 100 - 19 - 26 = 55.$$

Problem 325. If all row sums are equal to s, then all column sums must equal s as well (justify this), and no change is necessary. If some row sum is less than s, then some column sum is less than s as well; increase the entry in this row and column until the larger of the row and column sums equals s. Continue on.

In the example, if $s = 11$, the process yields

$$\begin{pmatrix} 3 & -5 & 2 \\ -6 & 4 & 1 \\ 1 & 0 & 8 \end{pmatrix} \rightarrow \begin{pmatrix} 14 & -5 & 2 \\ -6 & 4 & 1 \\ 1 & 0 & 8 \end{pmatrix}$$
$$\rightarrow \begin{pmatrix} 14 & -5 & 2 \\ -4 & 4 & 1 \\ 1 & 0 & 8 \end{pmatrix}$$
$$\rightarrow \begin{pmatrix} 14 & -5 & 2 \\ -4 & 14 & 1 \\ 1 & 0 & 8 \end{pmatrix}$$
$$\rightarrow \begin{pmatrix} 14 & -5 & 2 \\ -4 & 14 & 1 \\ 1 & 2 & 8 \end{pmatrix}$$

Problem 326. The problem is to determine a pattern for a sequence starting with $14, 38, 74, 122, \ldots$. Letting $g(n)$ denote the nth term of this sequence, we observe that successive differences are

$$g(2) - g(1) = 24,$$
$$g(3) - g(2) = 36,$$
$$g(4) - g(3) = 48,$$

and hence it seems that

$$g(n) - g(n-1) = 12n, \quad n = 2, 3, 4, \ldots.$$

Summing these equalities yields

$$g(n) - g(1) = 12(2 + 3 + \cdots + n), \ n \geq 2,$$

or

$$g(n) = 14 + 12(2 + 3 + \cdots + n)$$
$$= 14 + 12 \left\{ \frac{n(n+1)}{2} - 1 \right\}$$
$$= 2 + 6n(n+1) = 6n^2 + 6n + 2.$$

This suggests a general law, that is,

$$(2n)^2 + (2n+1)^2 + (2n+2)^2 + (6n^2 + 6n + 2)^2$$
$$= (6n^2 + 6n + 3)^2,$$

which is easily verified.

Problem 327. To get an idea of how to proceed, suppose the construction is completed. Let the circles, in decreasing order of radius, be α, β, γ and let ABC be the required triangle with A on α, B on β, C on γ. A rotation of the plane through $60°$ with center A in a suitable direction must carry B to C. Hence C must lie on the intersection of γ and the image of β under the rotation.

Construction. Let O be the common center of α, β, γ and let A be any point on α. Construct an equilateral triangle AOP, and let δ be the circle with center P whose radius is equal to that of β. Since $\overline{OP}$ equals the radius of α, by the condition imposed on the radii, the circles γ and δ must intersect. Let C be one of the intersection points. With center A and radius $\overline{AC}$, draw an arc cutting circle β in B such that triangles AOP and ABC are similarly directed. Then ABC is the required triangle.

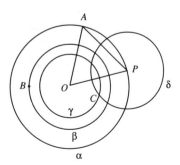

FIGURE 204

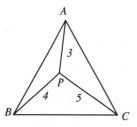

FIGURE 205

Proof. Consider the rotation with center A which takes P to O. This rotation through $60°$ takes circle δ onto circle β, and so takes the point C to a point B' on β. Hence $\overline{AB'} = \overline{AC} = \overline{AB}$ and $AB'C$ and AOP are similarly directed. Hence $B = B'$ so $\angle BAC = 60°$. It follows that triangle ABC is equilateral.

Rider. Given equilateral triangle ABC such that $\overline{PA} = 3$, $\overline{PB} = 4$, $\overline{PC} = 5$ for an interior point P, find $\overline{AB}$.

Hint. Rotate the figure about A through $60°$ so that C falls on B, B falls on B', P falls on P'. Show that $\angle BPP' = 90°$, $\angle APB = 150°$.

Problem 328. The problem is to find the highest power of 10 which divides 10000!. Since 2 divides 10000! more often than 5 does, the highest power of 10 dividing 10000! is equal to the highest power of 5 dividing 10000!. There is a factor 5 for every multiple of 5 not exceeding 10000 (there are 2000 such multiples), an additional 5 for every multiple of 25 (400 of them), yet another 5 for every multiple of 125 (80 of them), yet another 5 for every multiple of 625 (16 of them) and another 5 for every multiple of 3125 (3 of them). Therefore, the number of zeros is $2000 + 400 + 80 + 16 + 3 = 2499$.

In general, the number of zeros in which $n!$ ends with is given by

$$\left[\frac{n}{5}\right] + \left[\frac{n}{5^2}\right] + \left[\frac{n}{5^3}\right] + \cdots = \sum_{k=1}^{\infty}\left[\frac{n}{5^k}\right].$$

Problem 329. He can run $\frac{3}{8}$ of the length of the bridge in the time taken for the train to reach A.

If he runs towards B, he will be $\frac{3}{4}$ of the way towards B when the train arrives at A. Thus, he can run the remaining quarter of the distance from A to B in the time the train requires for the whole distance. Therefore his speed is a quarter that of the train, namely 20 kph.

Problem 330. *First solution.* Evidently, for $n \geq 2$,

$$\begin{aligned}
u_n &= 8(n-1) + u_{n-1} \\
&= 8(n-1) + 8(n-2) + u_{n-2} \\
&\;\;\vdots \\
&= 8[(n-1) + (n-2) + \cdots + 1] + u_1 \\
&= 4n(n-1) + 1 = (2n-1)^2.
\end{aligned}$$

Second solution (by induction). The result holds for $n = 1$. Assume $u_n = (2n-1)^2$. Then

$$\begin{aligned}
u_{n+1} &= u_n + 8n = (2n-1)^2 + 8n \\
&= (2n+1)^2 = (2(n+1)-1)^2.
\end{aligned}$$

Third solution (by Figure 206). The bottom corner square has $(2n-1)^2$ dots. The gnomon has $2(2n-1) + 2$ pairs of dots, i.e., $8n$ dots. Thus we can start with $u_1 = 1$ dot and build up squares of an odd number of dots by adding gnomons as indicated.

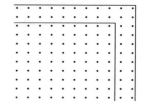

FIGURE 206

Problem 331. (a) By the Arithmetic-Geometric Mean inequality,

$$xyzw \leq \frac{1}{4}(x^2y^2 + y^2z^2 + z^2x^2 + w^4)$$

for all x, y, z, w. Suppose

$$x^2y^2 + y^2z^2 + z^2x^2 + w^4 - 4xyzw$$

$$= g_1(x, y, z, w)^2 + g_2(x, y, z, w)^2$$

$$+ \cdots + g_k(x, y, z, w)^2,$$

where $g_1, g_2, \ldots, g_k$ are polynomials. Since the left side is homogeneous of degree 4, each g_i must be homogeneous of degree 2 (see the comment after the solution for an argument of this).

No polynomial g_i can contain a term of the type ax^2, for then g_i^2 would contain the term a^2x^4, which would not be cancelled by a similar term from any other g_j^2 or from the left side. Similarly, no g_i contains nontrivial terms in y^2 nor z^2.

Can any g_i contain a term of the form bwx? If so, g_i^2 would contain $b^2x^2w^2$. Since such terms do not appear on the left side, this term would have to be cancelled by similar terms with negative coefficients.

These could only arise from the product of terms x^2 and w^2 appearing in some g_j, which we have excluded. Thus, no g_i contains terms in xw, nor likewise in yw or zw. But then each g_i would be of the form

$$axy + bxz + cyz + dw^2,$$

so that no g_i^2 would contain a term in $xyzw$. But this contradicts the appearance of $-4xyzw$ in the left side.

(b) Again, the inequality is a consequence of the arithmetic-geometric mean inequality. Suppose

$$x^4y^2 + y^4z^2 + z^4x^2 - 3x^2y^2z^2 = \sum_{i=1}^{m} h_i(x, y, z)^2.$$

Each h_i must be homogeneous of degree 3, and none can contain terms in x^3, y^3, z^3. In fact, the term xy^2 cannot appear. Otherwise, some h_i^2 would have a term ax^2y^4 with $a > 0$. This could only be cancelled by a term arising from a product $(x^2y)y^3$, which is impossible. Similarly, yz^2 and zx^2 do not appear in any h_i.

The left side contains the term $-3x^2y^2z^2$, which means that at least one h_i contains terms in x^2y, yz^2, or in y^2z, zx^2, or in z^2x, xy^2. But this gives a contradiction.

Hence, neither of the polynomials (a) nor (b) is the sum of squares of polynomials.

Remark. Part of the argument of both (a) and (b) depends on the following proposition:

Let $f(x_1, x_2, \ldots, x_n)$ be a homogeneous polynomial of degree $2r$ which can be written as a sum of squares of polynomials $f_i(x_1, x_2, \ldots, x_n)$. Then each f_i must be homogeneous of degree r.

In other words, we have to show that, if $f = f_1^2 + f_2^2 + \cdots + f_k^2$, then each term of each f_i is a monomial of degree r of the form $cx_1^{a_1} x_2^{a_2} \ldots x_n^{a_n}$ with $a_1 + a_2 + \cdots + a_n = r$.

Let p be the lowest degree of any of the terms in any f_i. Suppose $u(x_1, \ldots, x_n)$ is a monomial product of powers of x_i of degree p which appears in at least one f_i with nonzero coefficient. Then, for each i,

$$f_i = c_i u + v_i,$$

where c_i is a constant, not zero for at least one i, and v_i is a sum of terms of degree at least p (with no term in u). Then

$$\sum_{i=1}^{k} f_i^2 = \left(\sum_{i=1}^{k} c_i^2 \right) u^2 + \sum_{i=1}^{k} (2c_i u v_i + v_i^2).$$

Since none of the polynomials uv_i and v_i contain terms in u^2, the coefficient of u^2 in $f = \sum f_i^2$ must be positive. Hence u^2 must be of degree $2r$ and u of degree r.

The same type of argument shows that no f_i can contain terms of degree exceeding r.

Problem 332. (a) Let the roots be $n-1$, n, $n+1$. Then the sum of the roots is $3n = p$, the product is $n^3 - n = q$ and $3n^2 - 1 = (n-1)n + (n-1)(n+1) + n(n+1) = 11$. Thus, either $n = 2$, $p = 6$, $q = 6$ or $n = -2$, $p = -6$, $q = -6$.

(b) If the roots are all equal to r, the polynomial equals $(x - r)^3$, whence $p = 3r$, $q = r^3$ and $11 = 3r^2$. Thus $r = \pm\sqrt{\frac{11}{3}}$ and p and q can be found.

Problem 333. *First solution* (by induction). If $n = 1$, $2^{2n} + 24n - 10 = 18$. Suppose the assertion holds for some $n \geq 1$; then

$$2^{2(n+1)} + 24(n+1) - 10$$
$$= 4(2^{2n} + 24n - 10) - 18(4n - 3),$$

i.e., it holds for $n + 1$.

Second solution (direct). The quantity is obviously divisible by 2. Computing modulo 9 we have

$$2^{2n} + 24n - 10 \equiv (3-1)^{2n} + 6n - 1$$
$$\equiv (-2n \cdot 3 + 1) + 6n - 1 = 0,$$

and hence

$$2^{2n} + 24n - 10 \equiv 0 \pmod{18}.$$

Problem 334. Let $A = \{a_1, a_2, \ldots, a_n\}$ and $B = \{2n - a_1, 2n - a_2, \ldots, 2n - a_n\}$. Since all the elements of A are distinct, so are the elements of B. Furthermore,

$$1 \leq a_i \leq 2n - 1$$

implies

$$1 \leq 2n - a_i \leq 2n - 1 \quad \text{for} \quad i = 1, 2, \ldots, n.$$

Since

$$A \cup B \subseteq \{1, 2, \ldots, 2n - 1\},$$

we have that $\#A = \#B = n$ while $\#(A \cup B) \leq 2n - 1 < 2n$. Therefore $A \cap B \neq \emptyset$, that is, for some i and j,

$$a_i = 2n - a_j.$$

Problem 335. For each of the given integers, compute the absolute value of the difference between it and the nearest multiple of $2n$. The result is one of the $n + 1$ numbers $0, 1, 2, \ldots, n$. Since $n + 2$ integers are given, one of the results, say r, is obtained at least twice (by the Pigeonhole Principle). Consequently, there are integers u, v in the set such that one of the following holds (for integers p and q):

$$(1): \quad u = 2np + r, \quad v = 2nq + r;$$
$$(2): \quad u = 2np + r, \quad v = 2nq - r;$$
$$(3): \quad u = 2np - r, \quad v = 2nq - r.$$

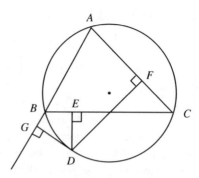

FIGURE 207

In cases (1) and (3), $u - v$ is divisible by $2n$; in case (2), $u + v$ is divisible by $2n$.

Problem 336. Let the perpendiculars meet BC, AC and AB produced in E, F, and G respectively, as in Figure 207.

Join GE, EF, BD, CD. Since $AGDF$ is concyclic, $\angle GDF + \angle GAF = 180°$. Since $ABCD$ is concyclic, $\angle BDC + \angle BAC = 180°$. Therefore $\angle GDF = \angle BDC$, whence $\angle BDG = \angle CDF$. Since $BGDE$ is concyclic, $\angle BDG = \angle BEG$. Since $DEFC$ is concyclic, $\angle CDF = \angle CEF$. Hence $\angle BEG = \angle CEF$, so G, E, F lie in a straight line. The line is called the Simson line but actually it was discovered in 1927 by William Wallace.

Problem 337. Without loss of generality we may take the leading coefficient of the cubic polynomial to be 1. We therefore have

$$x^3 - px^2 + qx - r$$
$$\equiv (x - u)(x - v)(x - uv),$$

with p, q, r rational. Expanding the right side and equating coefficients leads to:

$$p = u + v + uv,$$
$$q = uv + u^2v + uv^2, \qquad r = u^2v^2.$$

It is an easy exercise to deduce from these equations that, when $p \neq -1$,

$$uv = \frac{q + r}{1 + p},$$

which is rational. If $p = -1$, then $(1+u)(1+v) = 0$ so that either u or v is equal -1.

Problem 338. The cases for $n = 1$ and $n = 2$ show that either greatest common divisor is possible. Any common divisor d of $n^2 + 1$ and $(n+1)^2 + 1$ divides their difference, $2n+1$. Hence d divides

$$4(n^2 + 1) - (2n + 1)(2n - 1) = 5.$$

Rider. Show that the greatest common divisor is 5 if and only if $n \equiv 2 \pmod 5$.

Problem 339. If $x = \cos 36°$, then $\cos 72° = 2x^2 - 1$. Since

$$\cos 36° - \cos 72° = 2 \sin 54° \sin 18°$$

$$= 2 \cos 36° \cos 72°,$$

we have $x - 2x^2 + 1 = 4x^3 - 2x$, and hence

$$4x^3 + 2x^2 - 3x - 1 = 0.$$

Observing that -1 is a root permits us to factor:

$$(4x^2 - 2x - 1)(x + 1) = 0.$$

Since $x = \cos 36° \neq -1$, we have

$$4x^2 - 2x - 1 = 0.$$

Therefore,

$$\cos 36° - \cos 72° = x - (2x^2 - 1)$$

$$= -\frac{1}{2}(4x^2 - 2x - 1) + \frac{1}{2}$$

$$= \frac{1}{2}$$

and

$$\cos 36° \cos 72° = \frac{1}{2}(\cos 36° - \cos 72°) = \frac{1}{4}.$$

Problem 340.

$$7^{\frac{1}{2}} + 7^{\frac{1}{3}} + 7^{\frac{1}{4}} < 9^{\frac{1}{2}} + 8^{\frac{1}{3}} + 16^{\frac{1}{4}}$$

$$= 3 + 2 + 2 = 7$$

and

$$4^{\frac{1}{2}} + 4^{\frac{1}{3}} + 4^{\frac{1}{4}} > 2 + 1 + 1 = 4.$$

Rider. Can you find a simple way of deciding which is larger, $n^{\frac{1}{2}} + n^{\frac{1}{3}} + n^{\frac{1}{4}}$ or n, for $n = 5$, $n = 6$?

Problem 341. A player will be in a position to win if his opponent leaves him with any number from 45 to 55 inclusive. Thus the person who can attain 44 can win the game if he plays properly.

By a similar argument, the person who can attain 32 is in a position to win. Likewise for 20 and for 8.

Therefore, the first player has the advantage. If he plays 8 to begin with, and responds to his opponent by making the totals up to 20, 32, 44, 56 respectively in turn, he will win.

Riders. (a) Can 56 be replaced by a number which will give the second player an advantage? What are all the possibilities?

(b) Two players play the following game. The first player selects any integer from 1 to 55 inclusive. The second adds any positive integer not exceeding twice the integer chosen by the first. They continue, playing alternately, each adding an integer no more than twice that used by his predecessor. The player who reaches 56 wins. Which player has the advantage?

(c) The game is the same as (b), except that each player cannot add more than his predecessor. Who has the advantage now?

Problem 342. In Figure 208

$$\overline{AB} = 2a, \qquad \overline{AC} = \overline{BC} = R,$$

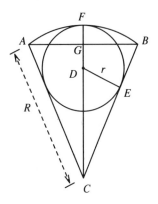

FIGURE 208

$$\overline{DE} = \overline{DF} = r, \qquad \overline{CD} = R - r.$$

By similar triangles GCB and DCE,

$$\frac{\overline{CD}}{\overline{DE}} = \frac{\overline{CB}}{\overline{BG}}$$

whence

$$\frac{R-r}{r} = \frac{R}{a}, \qquad \text{i.e.,} \qquad \frac{1}{r} = \frac{1}{R} + \frac{1}{a}.$$

Problem 343. Divide the parallelogram into four parallelograms by joining the midpoints of opposite sides, as indicated in Figure 209. We concentrate on the upper left parallelogram, $AEPH$. The quadrilateral $PQRS$ is that portion of the area of the figure lying inside $AEPH$.

First, observe that A, R, P are collinear and that $\overline{AP} = 3\overline{RP}$. This follows from the fact that R is the intersection of two medians BH and DE of $\triangle ADB$, and must therefore lie on the third median AP. Therefore, with $[\ldots]$ denoting area,

$$[APS] = 3[RPS] \qquad \text{and} \qquad [AQP] = 3[RQP].$$

Hence

$$[PQRS] = \frac{1}{3}[PQAS] = \frac{1}{6}[AEPH].$$

The same reasoning applies to the other three subparallelograms, with the result that the figure bounded by $AG, AF, BH, BG, CE, CH, DF,$ and DE has area

$$\frac{1}{6}[ABCD].$$

Problem 344. No. Indeed, we show by induction that

$$1 < u_n < \frac{1}{1-u} \qquad \text{for} \quad n = 1, 2, 3, \ldots.$$

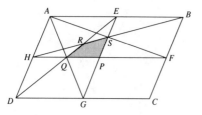

FIGURE 209

First note that $1 + u < \dfrac{1}{1-u}$. Clearly

$$1 < 1 + u = u_1, \qquad u_1 = 1 + u < \frac{1}{1-u}.$$

Now assume that for an $n \geq 2$

$$1 < u_n < \frac{1}{1-u}.$$

Then

$$1 = 1 - u + u < \frac{1}{u_n} + u = u_{n+1},$$

$$u_{n+1} = \frac{1}{u_n} + u < 1 + u < \frac{1}{1-u}.$$

Thus

$$1 < u_{n+1} < \frac{1}{1-u}$$

and the proof is complete.

Rider. Verify the result by determining u_n explicitly. One way of doing this is to define a sequence v_n for which $u_n = \dfrac{v_{n+1}}{v_n}$. Accordingly, take $v_1 = 1$, $v_2 = 1 + u$ and, generally,

$$v_{n+1} = u_n v_n, \qquad n = 1, 2, 3, \ldots.$$

Then v_n must satisfy

$$\frac{v_{n+2}}{v_{n+1}} = \frac{v_n}{v_{n+1}} + u,$$

or

$$v_{n+2} = u v_{n+1} + v_n, \qquad (n = 1, 2, 3, \ldots).$$

The general solution of this has the form

$$v_n = ar^n + bs^n,$$

where a, b are constants and r and s are roots of the polynomial $t^2 - ut - 1$.

Problem 345. (a) Any divisor of a and b divides $a - b$. If $b = a + 1$, any common divisor divides 1.

(b) $n! + 1$.

(c) Any prime p which divides any of the numbers $r+1, r+2, \ldots, 2r$ is obviously $\leq 2r$, and hence does not divide $(2r)!+1$. Thus $t = (2r)!+1$ does the trick.

Problem 346. *First Solution.* Yes. To multiply x by a negative number $-n^2$:

$$-n^2 x = n + \cfrac{1}{-\cfrac{1}{n} + \cfrac{1}{n + \cfrac{1}{x}}};$$

to multiply x by a positive number n^2:

$$n^2 x = n + \cfrac{1}{-\cfrac{n+1}{n} + \cfrac{1}{1 + \cfrac{1}{-(n+1) + \cfrac{1}{x}}}}.$$

Second Solution. If $u \neq 1$, we can find u^2 from $u^2 = u - \left(\dfrac{1}{u} + \dfrac{1}{1-u}\right)^{-1}$ and $\dfrac{1}{2}u$ from $\left(\dfrac{1}{u} + \dfrac{1}{u}\right)^{-1}$. Then we can find any product vw from

$$vw = \left(\frac{v+w}{2}\right)^2 - \left(\frac{v-w}{2}\right)^2.$$

Problem 347. It should be realized that the set has to have infinite extent, but need not be connected. A natural candidate to try is $\{(m, n) \mid m, n \text{ integers}\}$ which has, as axis of symmetry, each of the lines $x = 0$, $y = 0$, $x + y = 1$, and many more.

Rider. Suppose we asked for exactly three non-concurrent axes? exactly n non-concurrent axes?

Problem 348. $p + q$ is even, so $\frac{1}{2}(p + q)$ is an integer between p and q. Since p, q are *consecutive* primes, the integer $\frac{1}{2}(p + q)$ is not a prime, i.e., it is a product of two or more primes, and now

$$p + q = 2\left(\frac{p+q}{2}\right).$$

Problem 349. (a) will follow from a solution to (b).

Suppose $\sqrt{m} + \sqrt{n}$ is rational. Then

$$\sqrt{m} - \sqrt{n} = \frac{m - n}{\sqrt{m} + \sqrt{n}}$$

is rational.

Hence $\sqrt{m} = \frac{1}{2}(\sqrt{m} + \sqrt{n}) + \frac{1}{2}(\sqrt{m} - \sqrt{n})$ is rational. Therefore m is a perfect square. Similarly, n is a perfect square. Conversely, if m and n are squares then $\sqrt{m} + \sqrt{n}$ is an integer, hence rational.

Riders. (1) Show that if $m_1, m_2, \ldots, m_k$ are positive integers, then $\sum_{i=1}^{k} \sqrt{m_i}$ is irrational if at least one m_i is not a square of a rational.

(2) Show that $\sqrt[3]{3} + \sqrt[3]{9}$ is not rational.

(3) Let m and n be integers. When is $\sqrt[3]{m} + \sqrt[3]{n}$ rational? (Let this number be x; then $n = (x - \sqrt[3]{m})^3 = x^3 - m - 3x\sqrt[3]{mn}$.)

See Gregg N. Petruno, Sums of irrational square roots are irrational, *Math. Mag.* 61 (1988) 44–45.

Problem 350. Each choice of a pair of points P_i and a pair of points Q_j gives rise to one intersection point. By appropriately choosing the points on each line in turn, it can be arranged that the intersection points are all distinct. Hence the number of intersection points is

$$\binom{m}{2}\binom{n}{2}.$$

Rider. How many points of intersection are there of the diagonals of a regular n-gon?

Problem 351. Let

$$F(x, y, z) = (x + y + z)^5 - x^5 - y^5 - z^5.$$

Since

$$F(x, -x, z) = F(x, y, -x) = F(x, y, -y) = 0,$$

$F(x, y, z)$ has (by the Factor Theorem) the factors $x + y$, $x + z$ and $y + z$. Let

$$F(x, y, z) = (x+y)(x+z)(y+z)G(x, y, z). \quad (1)$$

Then $G(x, y, z)$ must be a symmetric, homogeneous polynomial of degree 2 and hence has the form

$$G(x, y, z) = a(x^2 + y^2 + z^2) + b(xy + xz + yz).$$

Setting $x = y = z = 1$ in (1) yields $3^5 - 3 = 8(3a + 3b)$, or $a + b = 10$. Setting $x = y = 1$

and $z = 0$ in (1) yields $2^5 - 2 = 2(2a + b)$, or $2a + b = 15$. We conclude that $a = b = 5$ and

$$G(x, y, z) = 5(x^2 + y^2 + z^2 + xy + xz + yz).$$

Problem 352. (a)

$112296^2 - 79896^2$

$= (112296 - 79896)(112296 + 79896)$

$= (32400)(192192) = 18^2 10^2 (192)(1001)$

$= (2 \cdot 9 \cdot 3 \cdot 6)(2 \cdot 5 \cdot 10)(2 \cdot 8 \cdot 12)(7 \cdot 11 \cdot 13)$

$= 13!$

(b) The equation suggests that we should try to pull a factor 10^4 out of $651^4 - 591^4$, so we calculate

$651^4 - 599^4$

$\quad = (651^2 - 599^2)(651^2 + 599^2)$

$\quad = (651 - 599)(651 + 599)(651^2 + 599^2)$

$\quad = 52 \cdot 1250((625 + 26)^2 + (625 - 26)^2)$

$\quad = 2^2 \cdot 13 \cdot 2 \cdot 5^4 \cdot 2(625^2 + 26^2)$

$\quad = 10^4 \cdot 13(25^4 + 26^2),$

deducing,

$$651^4 - 599^4 - 430^4 - 340^4 - 240^4 = 10^4 A,$$

where

$A = 13(25^4 + 26^2) - (43^4 + 34^4 + 24^4)$

$\quad = 12 \cdot 25^4 + 4 \cdot 13^3 - (34^4 + 24^4)$

$\quad \quad - (43^4 - 25^4)$

$\quad = 12 \cdot 25^4 + 4 \cdot 13^3 - 16(17^4 + 12^4)$

$\quad \quad - 18 \cdot 68 \cdot 2(34^2 + 9^2)$

$\quad = 4B$

with

$B = 3 \cdot 25^4 + 13^3 - 4(17^4 + 12^4)$

$\quad \quad - 17 \cdot 36(34^2 + 9^2).$

We have to show that $A = B = 0$. This can be done using the following lemma, the proof of which is left to the reader.

Lemma. *Let N be an integer which is divisible by positive integers $n_1, n_2, \ldots, n_k$. If the least common multiple of $n_1, n_2, \ldots, n_k$ exceeds N, then $N = 0$.*

This suggests that we show that $A \equiv 0 \pmod{m}$ for sufficiently many moduli m whose least common multiple exceeds A. Try various prime powers for m.

$B \equiv 3 \cdot 9^4 - 3^3 - 4(1 + 0) - 1 \cdot 4 \cdot (4 + 81)$

$\quad \equiv 3 \cdot 81^2 - 11 - 4 - 4 \cdot 1$

$\quad \equiv 3 - 11 - 4 - 4 = -16 \equiv 0 \pmod{16}$

so $A \equiv 0 \pmod{64}$.

$A \equiv 13(2^4 + 1^2) - (16^4 + 7^4 + 0)$

$\quad \equiv 13 \cdot 17 - (256^2 + 49^2)$

$\quad \equiv 13(-10) - ((-14)^2 + 49^2)$

$\quad \equiv -130 - 7^2(2^2 + 7^2)$

$\quad \equiv -130 - 49(-1) = -81 \equiv 0 \pmod{27}$

$A \equiv 13(0 + 1^2) - ((-7)^4 + 9^4 + 1)$

$\quad \equiv 13 - (49^2 + 81^2 + 1)$

$\quad \equiv 13 - (1 + 6^2 + 1) = -25 \equiv 0 \pmod{25}$

$A \equiv -1(4^4 + 2^2) - (1 + 1 + 4^4)$

$\quad \equiv -1(2^2 + 2^2) - (1 + 1 + 2^2)$

$\quad = -14 \equiv 0 \pmod{7}$

$A \equiv 2(3^4 + 4^2) - (1 + 1 + 2^4)$

$\quad \equiv 2(2^2 + 2^4) - 18 = 40 - 18 = 22$

$\quad \equiv 0 \pmod{11}$

$A \equiv 0 - (4^4 + 8^4 + 2^4) = -2^4(2^4 + 4^4 + 1)$

$\quad \equiv -2^4(4^2 + 3^2 + 1) \equiv -2^4(3 + 9 + 1)$

$\quad \equiv 0 \pmod{13}$

$B \equiv 3 \cdot 8^4 + (-4)^3 - 4(0 + 12^4) - 0$

$\quad \equiv 3 \cdot 4^6 - 4^3 - 3^4 \cdot 4^5$

$\quad = 4^3(3 \cdot 4^3 - 1 - 3^4 \cdot 4^2)$

$\quad \equiv 4^3(3(-4) - 1 - 3^4(-1))$

$\quad = 4^3(-12 - 1 + 81)$

$\quad = 4^3(68) \equiv 0 \pmod{17}$

so $A \equiv 0 \pmod{17}$. Hence, A is divisible by $64 \cdot 27 \cdot 25 \cdot 7 \cdot 11 \cdot 13 \cdot 17$, a number which exceeds $1600 \cdot 27 \cdot 1001 \cdot 17 > 10^6 (\frac{3}{2}) \cdot 27 \cdot 17 > 10^6 \cdot 40 \cdot 17 > 6 \cdot 10^8$.

However,

$$|A| < \max(13(25^4 + 26^2),\ 43^4 + 34^4 + 24^4)$$
$$< \max(13 \cdot 26^2(26^2 + 1),\ 3 \cdot 50^4)$$
$$< \max(13 \cdot 30^4, 3 \cdot 50^4) < 13 \cdot 50^4$$
$$= 13 \cdot 625 \cdot 10^4 < 13 \cdot 10^7 < 6 \cdot 10^8.$$

Since A is divisible by a number exceeding itself, A must be zero.

Problem 353. Let

$$S_n = \frac{1}{2} + \frac{3}{2^2} + \frac{5}{2^3} + \cdots + \frac{2n-1}{2^n}.$$

Multiply by $\frac{1}{2}$ to obtain

$$\frac{1}{2}S_n = \frac{1}{2^2} + \frac{3}{2^3} + \cdots + \frac{2n-5}{2^{n-1}} + \frac{2n-3}{2^n} + \frac{2n-1}{2^{n+1}}.$$

and now subtract the second equality from the first:

$$S_n - \frac{1}{2}S_n$$
$$= \frac{1}{2} + \frac{2}{2^2} + \frac{2}{2^3} + \cdots + \frac{2}{2^{n-1}} + \frac{2}{2^n} - \frac{2n-1}{2^{n+1}}.$$

Hence

$$\frac{1}{2}S_n$$
$$= \frac{1}{2} + \frac{1}{2} + \frac{1}{2^2} + \frac{1}{2^3} + \cdots + \frac{1}{2^{n-1}} - \frac{2n-1}{2^{n+1}}$$
$$= \frac{1}{2} + \frac{1}{2} \cdot \frac{1 - (1/2)^{n-1}}{1 - 1/2} - \frac{2n-1}{2^{n+1}}$$

and

$$S_n = 3 - \frac{2n+3}{2^n}.$$

Problem 354. By taking logarithms to base n we have to show that

$$2 = (\log_k m)(1 + \log_n k) \tag{1}$$
$$\text{or } 2 = \log_k m + \log_n m$$

(using the logarithm property $(\log_c b)(\log_b a) = \log_c a$). Now from the arithmetic progression condition, we have

$$2\log_m x = \log_n x + \log_k x.$$

Multiplying both sides by $\log_x m$ and using $(\log_a b)(\log_b a) = 1$ in addition to the previous "log property," we obtain (1).

Problem 355. Add the 100 equations to obtain

$$3(x_1 + x_2 + \cdots + x_{100}) = 0.$$

Therefore

$$0 = (x_1 + x_2 + x_3) + (x_4 + x_5 + x_6)$$
$$+ \cdots + (x_{97} + x_{98} + x_{99}) + x_{100}$$
$$= 0 + 0 + 0 + \cdots + 0 + x_{100},$$

i.e., $x_{100} = 0$, and similarly $x_1 = x_2 = \cdots = x_{99} = 0$.

Rider. What happens if we replace 100 by other positive integers?

Problem 356. Both (a) and (b) are obviously valid for $n = 1$, so we assume $n > 1$. x can be expressed in the form

$$x = a_0 + \frac{a_1}{n} + \frac{a_2}{n^2} + \cdots \tag{1}$$

where $a_0, a_1, a_2, \ldots$ are integers and

$$0 \le a_i \le n - 1 \quad \text{for} \quad i = 1, 2, 3, \ldots.$$

Then

$$[x] = a_0$$
$$nx = na_0 + a_1 + \frac{a_2}{n} + \frac{a_3}{n^2} + \cdots$$
$$[nx] = na_0 + a_1,$$

and

$$[nx] - n[x] = a_1,$$

establishing (a).

To prove (b), note that (using (1) and a similar expression for y) we can express x and y in the form

$$x = a_0 + \frac{a_1 + \alpha}{n},$$

$$y = b_0 + \frac{b_1 + \beta}{n}, \qquad 0 \le \alpha, \beta < 1.$$

and

$$x + y = a_0 + b_0 + \frac{a_1 + b_1 + \alpha + \beta}{n},$$

$$0 \le \alpha, \beta < 1.$$

Either (i) $a_1 + b_1 + \alpha + \beta < n$ or (ii) $n \le a_1 + b_1 + \alpha + \beta$.

In case (i), $[x + y] = a_0 + b_0$, so

$$[x] + [y] + (n - 1)[x + y]$$

$$= a_0 + b_0 + (n - 1)(a_0 + b_0)$$

$$= na_0 + nb_0 \le na_0 + a_1 + nb_0 + b_1$$

$$= [nx] + [ny].$$

In case (ii), since $\alpha + \beta < 2$ we have $a_1 + b_1 > n - 2$, so $a_1 + b_1 \ge n - 1$, while $n \le a_1 + b_1 + \alpha + \beta < 2n$; hence

$$[x + y] = a_0 + b_0 + 1$$

and

$$[x] + [y] + (n - 1)[x + y]$$

$$= a_0 + b_0 + (n - 1)(a_0 + b_0 + 1)$$

$$= na_0 + nb_0 + n - 1$$

$$\le na_0 + nb_0 + a_1 + b_1 = [nx] + [ny].$$

Problem 357. Let the segment be AB. Erect perpendiculars AX and BY to AB, on the same side of AB. Choose any two points P and Q on AX and draw through P and Q lines perpendicular to AX, meeting BY (produced if necessary) at R and S. Let segments AR and BP meet in M; let AS and BQ meet in N. Then line MN bisects AB.

Problem 358. Let ABC be the triangle, with medians AD, BE, and CF. Produce AD to H so that $\overline{AD} = \overline{DH}$. Then $\triangle BDH \equiv \triangle ADC$, so $\overline{BH} = \overline{AC}$. By the triangle inequality,

$$2\overline{AD} = \overline{AH} < \overline{AB} + \overline{BH} = \overline{AB} + \overline{AC}.$$

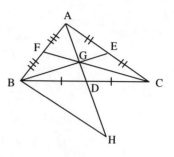

FIGURE 210

Similarly, $2\overline{BE} < \overline{AB} + \overline{BC}$ and $2\overline{CF} < \overline{AC} + \overline{BC}$, and adding these three inequalities, yields

$$\overline{AD} + \overline{BE} + \overline{CF} < \overline{AB} + \overline{BC} + \overline{CA}.$$

On the other hand,

$$\overline{BC} < \overline{BG} + \overline{GC} = \frac{2}{3}(\overline{BE} + \overline{CF}),$$

$$\overline{AC} < \frac{2}{3}(\overline{AD} + \overline{CF}),$$

$$\overline{AB} < \frac{2}{3}(\overline{AD} + \overline{BE})$$

and hence

$$\overline{AB} + \overline{BC} + \overline{CA} < \frac{4}{3}(\overline{AD} + \overline{BE} + \overline{CF}).$$

Problem 359. (a) 124 is larger, for each of the following inequalities is equivalent to the next one in the list:

$$29\sqrt{14} + 4\sqrt{15} < 124,$$

$$29\sqrt{14} < 124 - 4\sqrt{15},$$

$$29^2 \cdot 14 < 124^2 - 2 \cdot 4 \cdot 124\sqrt{15} + 4^2 \cdot 15,$$

$$(30 - 1)^2 \cdot 7 < 62 \cdot 124 - 4 \cdot 124\sqrt{15} + 8 \cdot 15,$$

$$4 \cdot 124\sqrt{15} < 62 \cdot 124 + 8 \cdot 15 - (901 - 60) \cdot 7,$$

$$496\sqrt{15} < 7688 + 120 - (6307 - 420) = 1921,$$

$$496^2 \cdot 15 < 1921^2,$$

$$3690240 < 3690241,$$

$$0 < 1.$$

(b) 2040 is larger, for each of the following inequalities is equivalent to the next one in the list:

$$2040 > 759\sqrt{7} + 2\sqrt{254},$$

$$2040 - 2\sqrt{254} > 759\sqrt{7},$$

$$(2040 - 2\sqrt{254})^2 > 7 \cdot 759^2,$$

$$4162616 - 8160\sqrt{254} > 4032567,$$

$$130049 > 8160\sqrt{254},$$

$$130049^2 > 254 \cdot 8160^2,$$

$$16912742401 > 16912742400,$$

$$1 > 0.$$

Problem 360. The differences of consecutive squares form an arithmetic progression with common difference 2:

$$1^2 - 0^2 = 1, \quad 2^2 - 1^2 = 3,$$

$$3^2 - 2^2 = 5, \quad 4^2 - 3^2 = 7, \ldots$$

Taking every second term (no two of which "involve" the same square) we have the two arithmetic progressions, both with common difference 4:

$$1^2 - 0^2 = 1, \quad 3^2 - 2^2 = 5, \quad 5^2 - 4^2 = 9, \ldots,$$

$$2^2 - 1^2 = 3, \quad 4^2 - 3^2 = 7, \quad 6^2 - 5^2 = 11, \ldots.$$

Thus

$$\left((q+3)^2 - (q+2)^2\right) - \left((q+1)^2 - q^2\right) = 4$$

for $q = 0, 1, 2, \ldots$, i.e.,

$$4 = q^2 - (q+1)^2 - (q+2)^2 + (q+3)^2$$

for $q = 0, 1, 2, \ldots$. Therefore, if n can be expressed in the form

$$n = \epsilon_1 1^2 + \epsilon_2 2^2 + \cdots + \epsilon_m m^2,$$

then so can $n + 4$:

$$n + 4 = \epsilon_1 1^2 + \epsilon_2 2^2 + \cdots + \epsilon_m m^2 + (m+1)^2$$
$$- (m+2)^2 - (m+3)^2 + (m+4)^2.$$

Since $1, 2, 3, 4$ can be expressed in the desired form, it follows that all integers can be so ex-

pressed. Now it is easy to write down such expressions. For $k \geq 1$,

$$1 + 4k = 1^2$$
$$+ \sum_1^{2k} (-1)^{j+1} \left((2j)^2 - (2j+1)^2\right),$$

$$2 + 4k = -1^2 - 2^2 - 3^2 + 4^2$$
$$+ \sum_2^{2k+1} (-1)^j \left((2j+1)^2 - (2j+2)^2\right),$$

$$3 + 4k = -1^2 + 2^2$$
$$+ \sum_1^{2k} (-1)^{j+1} \left((2j+1)^2 - (2j+2)^2\right),$$

$$4 + 4k = -1^2 - 2^2 + 3^2$$
$$+ \sum_2^{2k+1} (-1)^j \left((2j)^2 - (2j+1)^2\right).$$

Rider. Can you obtain a similar result for cubes in place of squares?

Problem 361. (a) Let AB and CM intersect at N (see Figure 211). Let $a = \overline{NM}$, $b = \overline{CD}$, $c = \overline{AN} = \overline{BN}$ and $r = \overline{AM} = \overline{DM} = \overline{BM}$. Since $\triangle ACM \sim \triangle NAM$, $\frac{CM}{AM} = \frac{AM}{NM}$ or $a(b+r) = r^2$. Now, with $[\ldots]$ denoting area,

$$[ABM] = ac, \quad [ACBM] = (b+r)c,$$
$$[ADBM] = rc = c\sqrt{a(b+r)},$$

which gives the desired result.

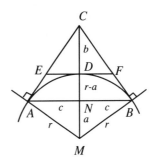

FIGURE 211

(b) The harmonic mean "h" of two numbers x and y is given by $\frac{2}{h} = \frac{1}{x} + \frac{1}{y}$. Here

$$x = [ADBM] = rc,$$

$$y = [ACBM] = (b+r)c$$

and

$$h = [AEDFBM] = (b+r)c - b\overline{ED}$$

$$= (b+r)c - \frac{b(bc)}{b+r-a}$$

since, from similar triangles CDE and CNA,

$$\frac{\overline{ED}}{c} = \frac{b}{b+r-a}.$$

Simplifying, we obtain

$$h = \frac{c((b+r)^2 - a(b+r) - b^2)}{b+r-a},$$

and using $a(b+r) = r^2$,

$$h = \frac{2brc}{b+r-a}.$$

Now

$$\frac{2}{h} - \frac{1}{x} - \frac{1}{y} = \frac{b+r-a}{brc} - \frac{1}{rc} - \frac{1}{(b+r)c}$$

$$= \frac{r-a}{brc} - \frac{1}{(b+r)c}$$

$$= \frac{(r-a)(b+r) - br}{brc(b+r)}$$

$$= \frac{r(b+r) - a(b+r) - br}{brc(b+r)}$$

$$= \frac{r(b+r) - r^2 - br}{brc(b+r)} = 0.$$

Problem 362. It is impossible. It does not change the problem if we divide either or both the numbers by a power of 10, so we assume for convenience that the two numbers are

$$a + r \quad \text{and} \quad b + s,$$

where a and b are integers and

$$1 \le a < b \le 9, \qquad 0 \le r, \ s < 1.$$

Suppose $(a+r)(b+s) < 10$. Then the first digit of the product is at least ab, a single-digit integer. Since $a \ge 1$, we have $a < b \le ab$, as desired.

Suppose $(a+r)(b+s) \ge 10$. Since

$$(a+r)(b+s) < (a+1)(9+1) = 10(a+1),$$

the product is a number between 10 and 90 (inclusive) whose first digit does not exceed a, again as desired.

Problem 363. By comparing the reading of his clock on departure and on return he knew how long he had been away. From his friend's clock he knew how long he was in his friend's house and thus deduced the time spent walking. Adding half of this time to the reading of his friend's clock when he departed the friend's home yields the correct time to set his clock.

Problem 364. Every time the gambler wins his fortune is multiplied by $\frac{3}{2}$; every time he loses it's multiplied by $\frac{1}{2}$. If in a sequence of $2n$ games he wins n times and loses n times then his fortune A dwindles to

$$A\left(\frac{3}{2}\right)^n \left(\frac{1}{2}\right)^n = \left(\frac{3}{4}\right)^n A.$$

It won't take long for him to be ruined.

Problem 365. The census taker calmly considered the sequence of squares

$$1, \ 4, \ 9, \ 16, \ 25, \ 36, \ 49, \ 64, \ 81, \ 100.$$

The age of the man was one of these, one which was the sum of three smaller numbers in the list. Being adept at mental arithmetic he soon found that

$$49 = 36 + 9 + 4.$$

In 1952, E. Teller (the father of the atomic bomb) was asked: "Will the thermonuclear device work?" He replied, "I don't know." "But if you didn't know five years ago, haven't you made any progress since then?" "Oh, yes," replied Teller, "Now I don't know on much better grounds."

The man's age was 49 and the wife's age 36. If the daughter's age were 9 the grandfather's age would be $49 + 36 + 9 = 94$, not a prime. So the daughter's age must be 4 and the grandfather's age $49 + 36 + 4 = 89$ a prime, leaving 9 as the son's age. The obvious remark is that the wife's age is the product of the ages of her children.

Problem 366. We consider (b) first. Let the vertices and the midpoints be denoted by $P_1, P_2, \ldots, P_n$ and $M_1, M_2, \ldots, M_n$, respectively, with M_i the midpoint of $P_i P_{i+1}$. Also, let $\overrightarrow{P_i}$ and $\overrightarrow{M_i}$ denote vectors from a common origin to P_i and M_i, respectively. Then we want to know when the following system of equations is uniquely solvable for the $\overrightarrow{P_i}$'s:

$$\overrightarrow{P_1} + \overrightarrow{P_2} = 2\overrightarrow{M_1}, \quad \overrightarrow{P_2} + \overrightarrow{P_3} = 2\overrightarrow{M_2},$$
$$\ldots, \quad \overrightarrow{P_n} + \overrightarrow{P_1} = 2\overrightarrow{M_n}.$$

Solving successively:

$$\overrightarrow{P_1} = 2\overrightarrow{M_1} - \overrightarrow{P_2} = 2\overrightarrow{M_1} - 2\overrightarrow{M_2} + \overrightarrow{P_3}$$
$$= \cdots = 2\overrightarrow{M_1} - 2\overrightarrow{M_2} + 2\overrightarrow{M_3}$$
$$- \cdots + 2\overrightarrow{M_n} - \overrightarrow{P_1}, \quad \text{if } n \text{ is odd,}$$

from which

$$\overrightarrow{P_1} = \overrightarrow{M_1} - \overrightarrow{M_2} + \overrightarrow{M_3} - \cdots + \overrightarrow{M_n}$$

uniquely, and similarly for the other $\overrightarrow{P_i}$'s. On the other hand, if n is even, we end up with

$$\overrightarrow{P_1} = 2\overrightarrow{M_1} - 2\overrightarrow{M_2} + 2\overrightarrow{M_3} - \cdots - 2\overrightarrow{M_n} + \overrightarrow{P_1}$$

which is consistent only if

$$\overrightarrow{M_1} + \overrightarrow{M_3} + \cdots + \overrightarrow{M_{n-1}}$$
$$= \overrightarrow{M_2} + \overrightarrow{M_4} + \cdots + \overrightarrow{M_n}.$$

When this condition is satisfied, the equations for the $\overrightarrow{P_i}$'s are not independent and there are infinitely many solutions.

For part (a), $\overrightarrow{P_1} = \overrightarrow{M_1} - \overrightarrow{M_2} + \overrightarrow{M_3} - \overrightarrow{M_4} + \overrightarrow{M_5}$ and is constructible. Then, $\overrightarrow{P_2} = 2\overrightarrow{M_1} - \overrightarrow{P_1}$, etc.

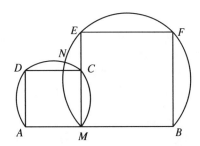

FIGURE 212

Problem 367. Clearly,

$$\angle ANM = \angle ACM = 45°,$$
$$\angle MNB = \angle MEB = 45°, \quad \angle BNE = 90°,$$

so

$$\angle ANE = \angle ANM + \angle MNB + \angle BNE = 180°,$$

and A, N, E are collinear. Furthermore,

$$\angle ANB = \angle ANM + \angle MNB$$
$$= 45° + 45° = 90°$$

and

$$\angle ANC = 90°,$$

so N, B, C are collinear.

Problem 368. (a) Any common divisor of $14n + 3$ and $21n + 4$ also divides

$$3(14n + 3) - 2(21n + 4) = 1.$$

(b) Any common divisor of $7an + 4$ and $7bn + 3$ must also divide

$$b(7an + 4) - a(7bn + 3) = 4b - 3a.$$

Thus if $4b - 3a = \pm 1$, then $7an + 4$ and $7bn + 3$ will be relatively prime. The general solution of $4b - 3a = 1$ is $b = 3t + 1$, $a = 4t + 1$; the general solution of $4b - 3a = -1$ is $b = 2 + 3t$, $a = 3 + 4t$. Setting $t = 0$ in the last case, i.e., $b = 2$, $a = 3$, answers part (a).

Problem 369. Let us agree that no person's set of acquaintances includes himself (otherwise, just add 1 to the number of each person's acquaintances). Suppose that there are n people at the concert, and that the kth person is acquainted with a_k people. Of course $0 \leq a_k \leq n-1$ for $k = 1, 2, \ldots, n$. However, it is not possible for one person to know no one else, while another knows everybody else, i.e., it is impossible that $a_i = 0$ and $a_j = n - 1$ for some i and j. Hence 0 and $n - 1$ are not simultaneously realizable as values of a_k's. Thus there are only $n - 1$ distinct possibilities for the n numbers $a_1, a_2, \ldots, a_n$ and, by the Pigeonhole Principle, two of the a_i's must be equal.

Problem 370. Observe that in the case $n = 7$, the 7 listed solutions have coordinate sums 1, 2, 3, 4, 5, 6, 7 (though not in this order). This suggests that, in the general case, we look at $x+y$ and try to pair off the solutions with the numbers $1, 2, \ldots, n$ by means of the sum $x + y$.

For any solution (x, y) of $x + 2y = n$, we have

$$\frac{n}{2} = \frac{1}{2}(x + 2y) \leq x + y \leq x + 2y \leq n;$$

for any integer k, $\frac{n}{2} \leq k \leq n$, the system

$$x + 2y = n, \qquad x + y = k$$

has the unique solution $(x, y) = (2k - n, n - k)$.

Similarly, for any solution of $2x + 3y = n - 1$ we have

$$\frac{n-1}{3} \leq x + y \leq \frac{n-1}{2};$$

for any k, $\frac{n-1}{3} \leq k \leq \frac{n-1}{2}$, the system

$$2x + 3y = n - 1, \qquad x + y = k$$

has the unique solution $(x, y) = (3k - (n - 1), n - 1 - 2k)$.

In general, any solution of $ix + (i + 1)y = n - i + 1$ satisfies

$$\frac{n+1-i}{i+1} \leq x + y \leq \frac{n+i-1}{i};$$

furthermore, for any k, $\frac{n+1-i}{i+1} \leq k \leq \frac{n+i-1}{i}$, the system

$$ix + (i + 1)y = n - i + 1 \qquad x + y = k$$

has the unique solution $(x, y) = ((i + 1)k - (n + 1 - i), (n + 1 - i) - ik)$.

We now show that for $k = 1, 2, \ldots, n$ there is a unique integer i for which

$$\frac{n+1-i}{i+1} \leq k \leq \frac{n+1-i}{i}.$$

This inequality is equivalent to

$$\frac{n+1-k}{k+1} \leq i \leq \frac{n+1}{k+1}.$$

Now

$$\frac{n+1-k}{k+1}, \frac{n+2-k}{k+1}, \ldots, \frac{n+(k+1)-k}{k+1}$$

$$= \frac{n+1}{k+1}$$

is a set of $k+1$ consecutive fractions with denominator $k + 1$, and exactly one of these must be an integer i. Hence for each $k = 1, 2, \ldots, n$ we can find exactly one equation which has exactly one solution for which $x + y = k$.

Problem 371. Starting with a sum counted in $a(n)$,

$$n = 1 + 2 + 2 + 1 + \cdots + 2 + 1 + 1 + 2$$

say, add a one at the beginning and at the end,

$$n + 2 = 1 + 1 + 2 + 2 + 1$$
$$+ \cdots + 2 + 1 + 1 + 2 + 1$$

and replace each 2 by $1 \big| 1$ (one stroke one)

$$n + 2 = 1 + 1 + 1 \big| 1 + 1 \big| 1 + 1$$
$$+ \cdots + 1 \big| 1 + 1 + 1 + 1 \big| 1 + 1$$

obtaining a sum

$$n + 2 = 3 + 2 + \cdots + 4 + 2$$

counted in $b(n + 2)$. This process, which can be reversed, sets up a one-to-one correspondence between the $a(n)$ ordered sums with "parts" 1 and 2 and the $b(n + 2)$ ordered sums with parts greater than 1. For example,

$$5 = 1+1+1+1+1$$
$$\rightarrow 7 = 1+1+1+1+1+1+1 \rightarrow 7 = 7$$

$$5 = 2+1+1+1$$
$$\rightarrow 7 = 1+2+1+1+1+1$$
$$\rightarrow 7 = 1+1\big|1+1+1+1+1 \rightarrow 7 = 2+5$$

$$5 = 1+2+1+1$$
$$\rightarrow 7 = 1+1+2+1+1+1$$
$$\rightarrow 7 = 1+1+1\big|1+1+1+1 \rightarrow 7 = 3+4$$

and so on.

Problem 372. Suppose that they each had x dollars at the completion of the fifth game. This means that at the beginning of the fifth game (i.e, the end of the fourth game) A, B, C, D, E had

$$\frac{x}{2}, \quad \frac{x}{2}, \quad \frac{x}{2}, \quad \frac{x}{2}, \quad 3x$$

dollars respectively; at the beginning of the fourth game they had

$$\frac{x}{4}, \quad \frac{x}{4}, \quad \frac{x}{4}, \quad \frac{11x}{4}, \quad \frac{3x}{2};$$

at the beginning of the third game they had

$$\frac{x}{8}, \quad \frac{x}{8}, \quad \frac{21x}{8}, \quad \frac{11x}{8}, \quad \frac{3x}{4};$$

at the beginning of the second game they had

$$\frac{x}{16}, \quad \frac{41x}{16}, \quad \frac{21x}{16}, \quad \frac{11x}{16}, \quad \frac{3x}{8};$$

at the beginning of the first game they had

$$\frac{81x}{32}, \quad \frac{41x}{32}, \quad \frac{21x}{32}, \quad \frac{11x}{32}, \quad \frac{3x}{16}.$$

The smallest integer x for which these "beginning" numbers are all integers is $x = 32$, and they could have started with as little as

$$81, \quad 41, \quad 21, \quad 11, \quad 6$$

dollars, respectively.

Problem 373. Since

$$\sum_i r_i = \sum_j c_j = \sum_{i,j} a_{ij}$$

it follows that $\sum_i (r_i - c_i) = 0$. Hence either $r_i - c_i = 0$ for all i or there are indices i and j for which $r_i - c_i > 0$ while $r_j - c_j < 0$. In both cases the desired inequality follows.

Problem 374. For any positive integer n and any number b

$$f(0) - f(b)$$
$$= f(0) - f\left(\frac{b}{n}\right) + f\left(\frac{b}{n}\right) - f\left(\frac{2b}{n}\right)$$
$$+ \cdots + f\left(\frac{(n-1)b}{n}\right) - f\left(\frac{nb}{n}\right)$$

and hence

$$|f(0) - f(b)|$$
$$\leq \left|f(0) - f\left(\frac{b}{n}\right)\right| + \left|f\left(\frac{b}{n}\right) - f\left(\frac{2b}{n}\right)\right|$$
$$+ \cdots + \left|f\left(\frac{(n-1)b}{n}\right) - f\left(\frac{nb}{n}\right)\right|$$
$$\leq \left(\frac{b}{n}\right)^2 + \left(\frac{b}{n}\right)^2 + \cdots + \left(\frac{b}{n}\right)^2$$
$$= n\left(\frac{b^2}{n^2}\right) = \frac{b^2}{n}.$$

With b fixed, and n increasing, this implies that

$$|f(0) - f(b)| = 0,$$

i.e., $f(b) = f(0)$.

Problem 375. Consider the graph of the velocity function $v(t)$ (t in hours, v in kph) of the car's motion (see Figure 213, where T is the time the car takes to travel the kilometer). The tangent line at any point has slope equal to the acceleration and,

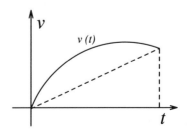

FIGURE 213

because the latter does not increase, the slope (of the tangent line) does not increase (as t increases), i.e, the tangent line rotates clockwise, never counterclockwise.

Alternatively, the function $v(t)$ is concave, or, the graph of $v(t)$ is concave down. The area under the curve is equal to the distance travelled (namely 1) and is not less than the area of the triangle with vertices $(0,0)$, $(T,0)$, $(T,240)$. Hence

$$\frac{1}{2}(240)T \le 1 \quad \text{(the area under the curve)}$$

and $T \le \frac{1}{120}$ i.e, the time is at most 30 seconds.

Problem 376. Let w_n denote the weight of the nth flea (from the bottom), so that

$$w_{n+1} = \sqrt{2 - w_n}, \quad n = 0, 1, 2, \ldots, w_1 = \sqrt{2}.$$

If the sequence $w_1, w_2, \ldots$ converges, to L say, then $L = \sqrt{2 - L}$, so L must be 1. We now show that the sequence does indeed converge to 1. It is not difficult to see that w_n is alternately greater than and less than 1, so we examine the relation between every second w_n and 1, as follows:

$$1 - w_{n+2}^2 = w_{n+1} - 1 = \sqrt{2 - w_n} - 1$$
$$= \frac{1 - w_n}{1 + \sqrt{2 - w_n}}$$

and hence

$$1 - w_{n+2} = \frac{1 - w_n}{(1 + w_{n+2})(1 + \sqrt{2 - w_n})}.$$

If $w_n < 1$, then

$$w_{n+2} < 1 \quad \text{and} \quad 1 - w_{n+2} < \frac{1 - w_n}{2}.$$

If $w_n > 1$, then

$$w_{n+2} > 1, \quad (1 + w_{n+2})(1 + \sqrt{2 - w_n}) > 2$$

and

$$w_{n+2} - 1 < \frac{w_n - 1}{2}.$$

Hence $|w_{n+2} - 1| < \frac{1}{2}|w_n - 1|$. In any case

$$|w_n - 1| < \frac{1}{2^{[n/2]}}, \quad \text{so} \quad \lim_{n \to \infty} |w_n - 1| = 0.$$

Problem 377. Observe that

$$x^3 + 1 = (x + 1)(x^2 - x + 1),$$

so $1 - r = -r^2$, $r^3 = -1$, $r^4 = -r$, $r^5 = -r^2$, and $r^6 = 1$. Now

$$S_n = r^n + (1 - r)^n = r^n + (-1)^n r^{2n},$$

and we have

$$S_{6m} = 1 + 1 = 2,$$
$$S_{6m+1} = r - r^2 = 1,$$
$$S_{6m+2} = r^2 + r^4 = -1,$$
$$S_{6m+3} = r^3 - r^6 = -2,$$
$$S_{6m+4} = r^4 + r^8 = -r + r^2 = -1,$$
$$S_{6m+5} = r^5 - r^{10} = -r^2 + r = 1.$$

This agrees with the desired result.

Problem 378. Mr. Laimbrain is wasting his time. He can never cover the whole area without cutting a sod. To obtain a neat proof of this, divide the area into one foot squares, and color every second square black. It is clear that in order to cover the whole area, each sod would have to cover two adjoining squares—one white and one black. Hence the 31 sods cover 31 white squares and 31 black squares. But our figure has 32 black squares and 30 white squares. Hence the task cannot be done.

FIGURE 214

> A student who has merely done mathematical exercises but has never solved a mathematical problem may be likened to a person who has learned the moves of the chess pieces but has never played a game of chess. The real thing in mathematics is to play the game.
> Stephen J. Turner

Problem 379. The proof of the first part is by induction. First note that

$$Q^2 = \begin{pmatrix} 1 & 1 \\ 1 & 0 \end{pmatrix} \begin{pmatrix} 1 & 1 \\ 1 & 0 \end{pmatrix}$$

$$= \begin{pmatrix} 1\cdot 1 + 1\cdot 1 & 1\cdot 1 + 1\cdot 0 \\ 1\cdot 1 + 1\cdot 0 & 1\cdot 1 + 0\cdot 0 \end{pmatrix}$$

$$= \begin{pmatrix} 2 & 1 \\ 1 & 1 \end{pmatrix}$$

$$= \begin{pmatrix} f_3 & f_2 \\ f_2 & f_1 \end{pmatrix}.$$

Now assume that for some $m \geq 2$,

$$Q^m = \begin{pmatrix} f_{m+1} & f_m \\ f_m & f_{m-1} \end{pmatrix}.$$

Then

$$Q^{m+1} = Q^m \cdot Q$$

$$= \begin{pmatrix} f_{m+1} & f_m \\ f_m & f_{m-1} \end{pmatrix} \begin{pmatrix} 1 & 1 \\ 1 & 0 \end{pmatrix}$$

$$= \begin{pmatrix} f_{m+1} + f_m & f_{m+1} \\ f_m + f_{m-1} & f_m \end{pmatrix}$$

$$= \begin{pmatrix} f_{m+2} & f_{m+1} \\ f_{m+1} & f_m \end{pmatrix},$$

and hence

$$Q^m = \begin{pmatrix} f_{m+1} & f_m \\ f_m & f_{m-1} \end{pmatrix},$$

$m = 2, 3, 4, \ldots$. Now for the second part. Since

$$Q^{3n} = \begin{pmatrix} f_{3n+1} & f_{3n} \\ f_{3n} & f_{3n-1} \end{pmatrix},$$

and

$$(Q^n)^3 = \begin{pmatrix} f_{n+1} & f_n \\ f_n & f_{n-1} \end{pmatrix}^3$$

$$= \begin{pmatrix} f_{n+1}^3 + 2f_{n+1}f_n^2 + f_n^2 f_{n-1} & f_{n+1}^2 f_n + f_{n+1}f_n f_{n-1} + f_n^3 + f_n f_{n-1}^2 \\ f_{n+1}^2 f_n + f_{n+1}f_n f_{n-1} + f_n^3 + f_n f_{n-1}^2 & f_{n+1}f_n^2 + 2f_n^2 f_{n-1} + f_{n-1}^3 \end{pmatrix}$$

we equate the entries in row 1, column 2 to find that

$$f_{3n}$$

$$= f_{n+1}^2 f_n + f_{n+1}f_n f_{n-1} + f_n^3 + f_n f_{n-1}^2$$

$$= f_{n+1}f_n(f_{n+1} + f_{n-1}) + f_n^3 + f_n f_{n-1}^2$$

$$= f_{n+1}(f_{n+1} - f_{n-1})(f_{n+1} + f_{n-1})$$

$$\quad + f_n^3 + f_n f_{n-1}^2$$

$$= f_{n+1}(f_{n+1}^2 - f_{n-1}^2) + f_n^3 + f_n f_{n-1}^2$$

$$= f_{n+1}^3 - f_{n+1}f_{n-1}^2 + f_n^3 + f_n f_{n-1}^2$$

$$= f_{n+1}^3 + f_n^3 - f_{n-1}^2(f_{n+1} - f_n)$$

$$= f_{n+1}^3 + f_n^3 - f_{n-1}^3.$$

Rider. Derive identities involving f_{3n+1} and f_{3n-1}.

Problem 380. For $n \geq 1$, the points of the sequence

$$L_n : (0, n), (1, n-1), (2, n-2), \ldots, (n, 0)$$

(the points on the line $x + y = n$) map onto the integers in the sequence

$$I_n : \frac{n(n+1)}{2}, \frac{n(n+1)}{2} + 1,$$

$$\frac{n(n+1)}{2} + 2, \ldots, \frac{n(n+1)}{2} + n.$$

It is easy to see that the sets $I_1, I_2, \ldots$ are disjoint and their union is the set of all positive integers.

Rider. More generally, let N be the sequence of positive integers. For each $n \in N$, establish the existence of a polynomial $F_n(x_1, x_2, \ldots, x_n)$ such that $F_n : N^n \to n$ is a bijection (one-to-one).

Remark. For a history of this problem and extensions, see *Crux Mathematicorum* 10 (1984) 300–303.

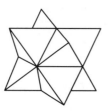

FIGURE 215

Problem 381. Each face of V is an equilateral triangle whose side length is half that of either tetrahedron. There are 8 such triangles, each contained in a face of one of the intersecting tetrahedra. Because of the symmetry, V must be an octahedron.

Let v be the volume of each tetrahedral "point" of the star U. The volume of each intersecting tetrahedron is $2^3 v = 8v$, so the volume of V is $8v - 4v = 4v$. Since U is made up of 8 tetrahedral "points" and the octahedron V, the volume of U is $4v + 8v = 12v$. Hence,

$$\frac{\text{volume } U}{\text{volume } V} = \frac{12v}{4v} = 3.$$

Problem 382. Since

$$(n^2 + n)^2 = n^4 + 2n^3 + n^2$$
$$< n^4 + 2n^3 + 2n^2 + 2n + 1$$
$$< n^4 + 2n^3 + 3n^2 + 2n + 1$$
$$= (n^2 + n + 1)^2,$$

we see that $n^4 + 2n^3 + 2n^2 + 2n + 1$ is between two consecutive squares and so cannot itself be a square.

Problem 383. The players of 1, 5, 8, 12 years of experience were on one side and those of 2, 3, 10, 11 years on the other. It is easy to check that

$$1 + 5 + 8 + 12 = 2 + 3 + 10 + 11,$$
$$1^2 + 5^2 + 8^2 + 12^2 = 2^2 + 3^2 + 10^2 + 11^2,$$
$$1^3 + 5^3 + 8^3 + 12^3 = 2^3 + 3^3 + 10^3 + 11^3.$$

Problem 384. The cheapest way would cost $290 and would require 4 large truckloads and 6 small truckloads. One way to solve this and

similar problems is to make a graph putting the number of large truckloads along the horizontal axis and the cost of the whole shipment along the vertical axis.

In the problem at hand, let x be the number of large trucks and y the number of small trucks. Then

$$18x + 13y \geq 150 \qquad (1)$$

and the total cost is

$$C = 35x + 25y.$$

For $x = 0, 1, 2, \ldots$ we compute the smallest y satisfying (1), and then the corresponding C.

x	smallest y for which $18x + 13y \geq 150$	C
0	12	300
1	11	310
2	9	295
3	8	305
4	6	290
5	5	300
6	4	310
7	2	295
8	1	305
9	0	315

Problem 385. For convenience we will count the time from 7:00 and use a rectangular coordinate system whose origin is the initial position of S and whose x-axis is the initial position of ray ST (Figure 216). If we let the x and y components of the velocity of S be a and b and those of T be c and d in kph, then the coordinates of S and T after t hours are

$$S = (at, bt), \qquad T = (m + ct, dt).$$

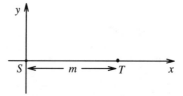

FIGURE 216

Then by the distance formula

$$(m + 3c - 3a)^2 + (3d - 3b)^2 = 25,$$

$$(m + 4c - 4a)^2 + (4d - 4b)^2 = 16,$$

$$(m + 6c - 6a)^2 + (6d - 6b)^2 = 100.$$

Letting $u = c - a$, $v = d - b$ and $u^2 + v^2 = x^2$, the above equations reduce to

$$m^2 + 6mu + 9x^2 = 25$$

$$m^2 + 8mu + 16x^2 = 16 \qquad (1)$$

$$m^2 + 12mu + 36x^2 = 100.$$

By subtraction,

$$2mu + 7x^2 = -9$$

$$4mu + 20x^2 = 84.$$

Solving, we find that $x^2 = 17$, $mu = -64$.

(a) From (1), $m^2 = 25 + 6(64) - 9(17) = 256$ or $m = 16$ km (and $u = -4$).

(b) Here,

$$m^2 + 2tmu + t^2x^2 = 26^2$$

or

$$16^2 - 2(64)t + 17t^2 = 26^2$$

or

$$(17t + 42)(t - 10) = 0.$$

Hence the two ships are 26 km. apart at 17:00, and also $\frac{42}{17}$ hours prior to 7:00 assuming they were sailing at that time.

(c) Here the distance apart $D(t)^2 = 256 - 128t + 17t^2$. By completing squares:

$$D(t)^2 = 17(t - \frac{64}{17})^2 + \frac{256}{17}.$$

Hence, the closest distance is $\frac{256}{17}$ km. occurring at $7:00 + \frac{64}{17}$ hours.

(d) Here we must use that the abscissae of S and T are the same, i.e., $at = m + ct$. Since $m = 16$ and $c - a = u = -4$ we have $t = 4$; S is due North of T at 11:00, provided $b > d$.

(e) Here we must have $T_x - S_x = T_y - S_y > 0$. Thus,

$$m + ut = vt > 0, \qquad u = -4.$$

Since $v^2 = x^2 - u^2 = 17 - 16 = 1$, or $v = \pm 1$. For $v = 1$, $t = \frac{16}{5}$. $v = -1$ is extraneous.

(f) Here, $a^2 + b^2 = c^2 + d^2$ and $c = 0$. Since $c - a = u = -4$, $a = 4$. Also, $(d - b)^2 + a^2 = 17$, giving $d - b = \pm 1$. Solving $d^2 - b^2 = 16$ with $d - b = \pm 1$ gives $d = -\frac{17}{2}$, $b = -\frac{15}{2}$ (note that d must be negative). Thus, S is steaming at a speed of $\sqrt{4^2 + (\frac{15}{2})^2} = \frac{17}{2}$ kph in a direction $\tan^{-1} \frac{8}{15}$ eastward of due south.

Problem 386. (a) The ratio of the area of the hexagon to that of the triangle is $\frac{6}{4} = \frac{3}{2}$.

(b) The ratio of the areas of the two hexagons is that of the areas of triangles OPQ and ORS, which is equal to $\frac{\overline{OA}^2}{\overline{OB}^2}$. If r is the radius of the circle, then $\overline{OA} = r$, $\overline{OB} = \frac{\sqrt{3}r}{2}$ and $\frac{\overline{OA}^2}{\overline{OB}^2} = \frac{4}{3}$.

An alternative quick solution to (b) can be seen from Figure 217.

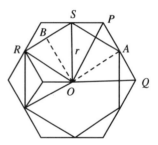

FIGURE 217

Rider. Show that the ratio of the areas of the circumscribed and inscribed regular n-gons of a circle is $\sec^2 \frac{\pi}{n}$.

Problem 387. Let D be the foot of the perpendicular from C to AP, so that $D \neq P$ and $\angle PCD = 30°$. See Figure 218. Then $\overline{PD} = \frac{1}{2}\overline{PC} = \overline{PB}$ so triangle BPD is isosceles and $\angle PBD = \angle PDB = 30°$. Furthermore

$$\angle DBA = \angle CBA - \angle PBD = 45° - 30° = 15°,$$

and

$$30° = \angle PDB = \angle DAB + \angle DBA$$

$$= \angle DAB + 15°.$$

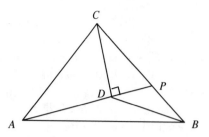

FIGURE 218

Then $\angle DAB = 15°$, $\overline{AD} = \overline{BD} = \overline{CD}$, so $\angle DAC = \angle DCA = 45°$ and $\angle ACB = 75°$.

Problem 388. Let PAB be a right-angled triangle, and angle θ as indicated in Figure 219. The area of triangle PAB is

$$\frac{1}{2}\left(\overline{AP}\right)\left(\overline{PB}\right) = \frac{1}{2}\left(\frac{a}{\cos\theta}\right)\left(\frac{b}{\sin\theta}\right)$$

$$= \frac{ab}{\sin 2\theta},$$

and this function assumes its minimum value, ab, when $\theta = 45°$.

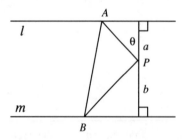

FIGURE 219

Rider. Redo the problem with $\angle APB = \alpha$ rather than $90°$; then $\theta = \frac{\pi-\alpha}{2}$.

Problem 389. If $3x^2+y^2-12$ and x^2-y^2+4 are both positive or both negative then the equation becomes

$$3x^2 + y^2 - 12 = x^2 - y^2 + 4 \quad \text{or} \quad x^2 + y^2 = 8.$$

Is every point on the circle $x^2 + y^2 = 8$ also on the given curve? The answer is yes, for if (x,y)

satisfies $x^2 + y^2 = 8$, then

$$|3x^2 + y^2 - 12| = |2x^2 - 4| = |x^2 - y^2 + 4|.$$

If $3x^2 + y^2 - 12$ and $x^2 - y^2 + 4$ have opposite signs, then the equation becomes

$$3x^2 + y^2 - 12 = -(x^2 - y^2 + 4) \quad \text{or} \quad |x| = \sqrt{2}.$$

Furthermore, it is easy to verify that if (x, y) lies on the curve $|x| = \sqrt{2}$, then it lies on the given curve. Hence the graph we seek is the union of the circle $x^2 + y^2 = 8$ and the lines $x = \sqrt{2}$ and $x = -\sqrt{2}$.

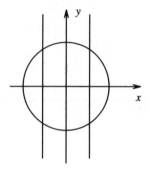

FIGURE 220

Problem 390. For any quadratic residue $r \pmod p$, there is an integer y, $1 \le y \le p-1$ whose square is congruent to r, i.e.,

$$y^2 \equiv r \pmod{p}, \quad 1 \le y \le p - 1;$$

of course $p - y$ satisfies the same conditions:

$$(p - y)^2 \equiv r \pmod{p}, \quad 1 \le p - r \le p - 1,$$

(and $y + (p - y) = p$). On the other hand, if

$$y_1^2 \equiv y_2^2 \pmod{p}, \quad 1 \le y_1 < y_2 \le p - 1,$$

then $p|(y_1^2 - y_2^2) = (y_1 - y_2)(y_1 + y_2)$; since $p \nmid (y_1 - y_2)$, we have $p|(y_1 + y_2)$ and hence $y_1 + y_2 = p$, i.e., $y_2 = p - y_1$. We conclude that, modulo p, the sequence 1^2, 2^2, $3^2, \ldots$, $(p-1)^2$ runs twice through the quadratic residues

$r_1, r_2, \ldots, r_q$ $(q = \frac{p-1}{2})$, and we have

$$2(r_1 + r_2 + \cdots + r_q)$$
$$\equiv 1^2 + 2^2 + \cdots + (p-1)^2 \pmod{p}$$
$$= \frac{(p-1)p(2p-1)}{6}.$$

Problem 391. It is easy to see that $\angle DAE = \alpha + \beta = 90°$, and hence $\angle ADE = \angle AED = 45°$ (Figure 221). Now $\angle ACB = 180° - (\angle ADE + \alpha) = 135° - \alpha$ and $45° = \alpha + \angle CBA$, so

$$\angle ACB - \angle CBA = (135° - \alpha) - (45° - \alpha)$$
$$= 90°.$$

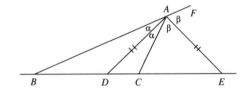

FIGURE 221

Problem 392. (a) A suggested generalization is

$$(n^3 + 1)^3 + (2n^3 - 1)^3 + (n^4 - 2n)^3$$
$$= (n^4 + n)^3, \quad n = 2, 3, 4, \ldots.$$

(b) A suggested generalization is

$$(3n^2)^3 + (6n^2 - 3n + 1)^3$$
$$+ (3n(3n^2 - 2n + 1) - 1)^3$$
$$= (3n(3n^2 - 2n + 1))^3 \quad n = 1, 2, 3, \ldots.$$

(c) A suggested generalization is

$$(3n^2)^3 + (6n^2 + 3n + 1)^3$$
$$+ (3n(3n^2 + 2n + 1))^3$$
$$= (3n(3n^2 + 2n + 1) + 1)^3 \quad n = 1, 2, 3, \ldots.$$

Problem 393. Every triangulation will dissect P into $n + 2m - 2$ triangles. We prove this by induction on m. The only way to triangulate the polygon when $m = 0$ is by drawing $n - 3$ diagonals, and this will always produce $n - 2$ triangles. This shows that the formula holds for $m = 0$. Now suppose the formula correct for a given value of m. If the new interior point is inside one of the triangles already formed, it must be connected to the vertices of this triangle; this replaces one triangle by three and so adds two to the total. If the new point lies on one of the interior segments already drawn, it must be connected to the opposite vertices of the two triangles which have this segment in common. This replaces two triangles by four triangles and so adds two to the total. Thus, in all cases, addition of an extra interior point adds two to the number of triangles. This completes the proof.

Another solution is to look at the sum of the angles of the triangles in two different ways. If there are k triangles, the angle sum is $k \cdot 180°$. The sum is also the sum of the angles of polygon P plus the sum of all the angles at the points Q_i i.e.,

$$k \cdot 180° = (n - 2)180° + m \cdot 360°,$$

and hence

$$k = n + 2m - 2.$$

For connections with Pick's theorem and Euler's formula, see the paper by R. W. Gaskell, M. S. Klamkin and P. Watson: Triangulations and Pick's Theorem, *Math. Mag.* 49 (1976) 35–37.

Problem 394. (Problem due to E. Just and B. Kabak; Solution by L. H. Cairoli.) The result follows immediately from the known relations:

$$\sin^2 A + \sin^2 B + \sin^2 C \qquad (1)$$
$$= 2 + 2\cos A \cos B \cos C,$$

$$\cos^3 A + \cos^3 B + \cos^3 C \qquad (2)$$
$$\geq 3\cos A \cos B \cos C.$$

Inequality (2) holds since

$$\cos^3 A + \cos^3 B + \cos^3 C - 3\cos A \cos B \cos C$$
$$= (\cos A + \cos B + \cos C)\frac{(\cos B - \cos C)^2 + (\cos C - \cos A)^2 + (\cos A - \cos B)^2}{2} \qquad (3)$$

and

$$\cos A + \cos B + \cos C$$
$$= 1 + 4\sin\frac{A}{2}\sin\frac{B}{2}\sin\frac{C}{2}.$$

Remark. A stronger inequality than the given one is

$$\cos^n A + \cos^n B + \cos^n C \geq \frac{3}{2^n} \qquad (4)$$

with equality only if the triangle is equilateral.

Riders. Establish identities (1) and (3). Prove (4), at least for the special case when the triangle is acute.

Problem 395. (a) It suffices to show that R, B, P are collinear and that $\overline{BP} = \overline{AQ}$. For then ΔPQR has $\angle PRQ = 60°$ and $\overline{RP} = \overline{RQ}$, from which the result follows. The counterclockwise rotation (of the plane) with center C and angle $60°$ takes A to B and D to P, and hence $\Delta CAD \cong \Delta CBP$. Thus $\overline{BP} = \overline{AD}$ yielding $\overline{BP} = \overline{AQ}$, and $\angle CBP = \angle CAD$ yielding R, B, P collinear.

(b) Let X, Y, Z, U, V be respectively the mid-points of PE, AQ, RD, QE, RA. Then $ZV \| DA$, $YU \| AE$ and

$$\overline{ZV} = \frac{1}{2}\overline{AD} = \frac{1}{2}\overline{AE} = \overline{YU}.$$

Also

$$\overline{VY} = \frac{1}{2}(\overline{RA} + \overline{AQ}) = \frac{1}{2}\overline{RQ} = \frac{1}{2}\overline{PQ} = \overline{UX}$$

and $\angle YVZ = 120° = \angle YUX$. Hence $\Delta YVZ \cong \Delta YUX$, so that $\overline{YZ} = \overline{YX}$ and $\angle VYZ = \angle YXU$. Thus

$$\angle ZYX = 180° - (\angle VYZ + \angle QYX)$$
$$= 180° - (\angle VYZ + 180°$$
$$- (\angle YQX + \angle YXQ)) = 60°.$$

It follows that ΔXYZ (being isosceles with vertex angle $60°$) is equilateral.

Problem 396. *First solution* (synthetic).

The circle with center O intersects the vertical axis in A and C. BC and OE intersect at D. We show that F (where the fixed circle intersects OE) is the midpoint of DE. Since $\angle ABC$ is subtended by a diameter,

$$\angle DBE = \angle CBA = 90°.$$

Since the angle subtended at the center is twice the angle subtended at the circumference, and since $\overline{OB} = \overline{OC}$,

$$\angle AOB = 2\angle ACB = 2\angle OBC.$$

Since $\angle OBF = 90° = \angle DBE$,

$$\angle OBC = \angle FBE.$$

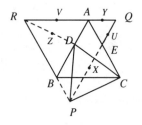

FIGURE 222

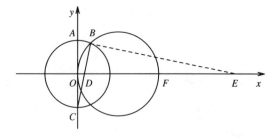

FIGURE 223

Since AOC is tangent to the fixed circle,

$$\angle AOB = \angle OFB.$$

Therefore,

$$\angle FBE + \angle FEB = \angle OFB = \angle AOB$$
$$= 2\angle OBC = 2\angle FBE,$$

so that $\angle FEB = \angle FBE$ and $\overline{BF} = \overline{FE}$. Also, $\angle FBD = 90° - \angle FBE = 90° - \angle ACB = \angle FDB$ so that $\overline{DF} = \overline{BF} = \overline{FE}$. Thus, F bisects DE. As the variable circle becomes smaller, B and C, and hence D move towards O. Thus E moves outwards to a point P for which F bisects OP.

Second solution (analytic). The circles

$$(x - 3)^2 + y^2 = 9 \quad \text{and} \quad x^2 + y^2 = h^2$$

intersect when $6x = h^2$, that is at

$$\left(\frac{h^2}{6}, \frac{h\sqrt{36 - h^2}}{6} \right).$$

The line joining

$$A(0, h) \quad \text{and} \quad B\left(\frac{h^2}{6}, \frac{h\sqrt{36 - h^2}}{6} \right)$$

is

$$y - h = \frac{(\sqrt{36 - h^2} - 6)x}{h},$$

whose intercept is $E(6 + \sqrt{36 - h^2}, 0)$. Hence, as $h \to 0$, the point E moves further out towards $(12, 0)$.

Problem 397. One way to approach this problem is to ask from where a positive rational v can come. Since, for $u > 0$, $u + 1 > 1$ and $\frac{u}{u+1} < 1$, we see that if $v > 1$, then v must have a unique "ancestor" u given by $u + 1 = v$ while, if $v < 1$, then v must have a unique ancestor given by $v = \frac{u}{u+1}$. It is easy to check that the sum of the numerator and denominator of v in either case exceeds the sum of the numerator and denominator of its unique ancestor. This suggests that an induction proof is appropriate.

The result is clearly true for $1 = \frac{1}{1}$, the only positive rational whose numerator and denominator add up to 2. Suppose it holds for all positive rationals whose numerator and denominator add up to $k \geq 2$. Let $v = \frac{p}{q}$ in lowest terms, where $1 \leq p, 1 \leq q, p + q = k + 1$. Then v is uniquely descended from $\frac{p-q}{q}$ if $p > q$, and from $\frac{p}{q-p}$ if $p < q$. But either of these ancestors, by the induction hypotheses, is uniquely descended from 1.

Problem 398. Let the sidelengths be a, b, c, d with $0 < a \leq b \leq c \leq d$. The sum of any three must strictly exceed the fourth. In particular, we have

$$d < a + b + c \leq 3d.$$

Since $a + b + c$ is a multiple of d, there are only two possibilities. One is that $a + b + c = 3d$, which requires that $a = b = c = d$, giving the result.

Thus, we may suppose that $a + b + c = 2d$. By hypothesis, there is an integer w for which $a + b + d = wc$. Hence,

$$2a + 2b + (a + b + c) = 2wc,$$

so that

$$(2w - 1)c = 3(a + b) \leq 6c.$$

Thus $w \leq 3$. Since $d \geq c$, $w \neq 1$.

Suppose $w = 2$. Then $3(a + b) + c = 4c$, so that $a + b = c$ and $2d = a + b + c = 2c$. Hence $c = d$ and the result follows.

On the other hand, let $w = 3$. Then $3(a + b) = 5c$ and

$$3d = a + b + c + d = (a + b + d) + c = 4c.$$

There are integers u and v for which

$$b + c + d = ua, \quad a + c + d = vb.$$

These, along with $a + b + d = 3c$, $a + b + c = 2d$, lead to

$$a - d = 2d - ua \quad \text{and} \quad b - d = 2d - vb.$$

Hence

$$a = \frac{3d}{u + 1} \quad \text{and} \quad b = \frac{3d}{v + 1}.$$

Thus

$$c = \frac{3}{5}(a+b) = \frac{9d}{5}\left(\frac{1}{u+1} + \frac{1}{v+1}\right)$$
$$= \frac{12c}{5}\left(\frac{1}{u+1} + \frac{1}{v+1}\right)$$

so that

$$\frac{1}{u+1} + \frac{1}{v+1} = \frac{5}{12}.$$

Clearly, $u \geq v$, so that either $u = 11$, $v = 2$ or $u = 5$, $v = 3$. The first of these is not possible, since $vb = a + c + d \geq a + 2b$. The second leads to $b = \frac{3d}{4} = c$. This establishes the result in all cases.

Problem 399. Look at two adjacent rows, A and B. Suppose row A has r red dots beside red dots in B, b blue dots beside blue dots in B, u red dots beside blue dots in B and v blue dots beside red dots in B. Then there are r red and b blue segments joining elements of A to elements of B. A has $r + u$ red dots and $b + v$ blue dots, while B has $r + v$ red dots and $b + u$ blue dots. Hence $r + u = b + v$ and $r + v = b + u$. Thus $r - b = v - u = -(v - u)$, so that $r = b$. Hence between adjacent rows, the number of red segments equals the number of blue segments. The same is true for adjacent columns. Hence the result follows.

Problem 400. The result is trivial if each person is acquainted with everyone else. Suppose A is not acquainted with B. Let X be any third person. If X fails to know someone, say Y, other than A or B, then $\{A, B, X, Y\}$ would be a foursome none of whom knows the other three. Hence X must know everyone, except possibly A and B.

Let C be someone apart from A and B who is not acquainted with everyone else and let D be any fourth person. Then, by the above argument, both C and D know everyone except A and B. Also, C is not acquainted with at least one of A and B. Therefore, D must be the one among $\{A, B, C, D\}$ who knows the other three. Hence D must know everyone and A, B, and C are the only three who are not acquainted with all of the others at the party.

Thus, there are no more than three who do not know everybody else.

Problem 401. Consider the circumcircle of the inscribed n-gon. It will have to intersect the given n-gon in each of its sides and its radius will be a minimum when it is the incircle of the given n-gon, and the rest then follows.

Problem 402. It follows from the first equation that

$$(z - x)(1 - S) = y - z,$$

where $S = x + y + z$. Similarly,

$$(x - y)(1 - S) = z - x,$$
$$(y - z)(1 - S) = x - y.$$

Thus,

$$(x - y)(y - z)(z - x)[(1 - S)^3 - 1] = 0,$$

and all solutions to this equation lead to $x = y$. Then,

$$x^2 + (1 - x - y)^2 + (1 - y)^2$$
$$= 2 - 6x(1 - x),$$

whence the minimum value is $2 - \frac{6}{4} = \frac{1}{2}$, taken on when $x = \frac{1}{2}$.

Comment. Geometrically, the problem is equivalent to finding the smallest equilateral triangle which can be inscribed in a given equilateral triangle. Intuitively, the vertices of the smallest such triangle will be the midpoints of the sides of the given triangle. (See Problem 401).

Problem 403. The equation can be rewritten as

$$(x^2 + 1)^2 = 2(x + 1)^2.$$

Thus,

$$x^2 + 1 = \pm\sqrt{2}(x + 1),$$

and the solutions of the these two quadratic equations give the desired roots.

Rider. Solve the more general quartic

$$x^4 - 4x = \frac{2}{k} - k^2.$$

Problem 404. We can assume without loss of generality that $P(x) > Q(x)$ for all real x. Then

$$P(P(x)) > Q(P(x))$$
$$= P(Q(x)) > Q(Q(x)),$$

which gives the desired result.

Problem 405. Clearing of fractions in (1) and factoring, we obtain

$$(b + c - a)(c + a - b)(a + b - c) = 0.$$

Thus, $a = b + c$ or $b = c + a$ or $c = b + a$. For any of these three cases, the three fractions on the left-hand side of (1) reduce to 2, 2, -2 in some order. Thus P has the constant value -8.

Problem 406. Let $\sqrt[3]{a}$, $\sqrt[3]{b}$, $\sqrt[3]{c}$ be the roots of the cubic equation

$$x^3 - px^2 + qx - r = 0. \tag{1}$$

Then,

$$S_1 = p,$$

$$S_2 = p^2 - 2q,$$

$$S_3 = pS_2 - qS_1 + 3r = p^3 - 3pq + 3r,$$

$$S_n = pS_{n-1} - qS_{n-2} + rS_{n-3} \text{ for } n > 3.$$

Thus, if p, q, r are integers, all the power sums S_n are also integers. It now remains to choose p, q, r so that the roots of (1) are non-integral and real. This is easily done. For example,

$$P(x) \equiv x^3 - 100x^2 + x + 1 = 0. \tag{2}$$

Since $P(0) = 1$, $P(1) = -97$, and $P(100) = 101$, there are real roots in the intervals $(0, 1)$ and $(1, 100)$. Since the sum of the three roots is 100, the third root is also real. Also, the roots are not integers since the only possible integer roots must be factors of the constant term 1 in (2). Therefore one possible triplet (a, b, c) is the cubes of the roots of equation (2), and there are infinitely many other such equations.

Problem 407. The result will follow from the stronger inequality

$$\frac{n+2}{S} \geq \sum_{i=1}^{n} \frac{1}{S - a_i}$$

or, equivalently,

$$2 \geq \sum_{i=1}^{n} \frac{a_i}{S - a_i} \tag{1}$$

Assume $a_1 \leq a_2 \leq \cdots \leq a_n$. Then since $S - a_1 \geq S - a_n \geq a_n$,

$$\sum_{i=1}^{n} \frac{a_i}{S - a_i} \leq \sum_{i=1}^{n-1} \frac{a_i}{S - a_n} + \frac{S - a_n}{S - a_n} = 2.$$

There is equality if and only if $a_1 = a_2 = \cdots = a_{n-2} = 0, a_{n-1} = a_n$ (for a degenerate polygon). The special case of (1) for $n = 3$ occurs in 0. Bottema, et al, *Geometric Inequalities,* Walters-Noordhoff, Gröningen, 1969, p. 15.

Problem 408. Consider a parallel projection onto another suitable plane which transforms the ellipse into a circle. This projection preserves parallelism and, consequently, the rectangle projects into a parallelogram. However, since this parallelogram is inscribed in a circle, it also must be a rectangle. Now if one considers the inverse transformation, the axes of the ellipse correspond to the diameters of the circle which were parallel to the sides of the rectangle inscribed in the circle.

Alternative solution with a generalization. Since the diagonals of a rectangle bisect each other and are equal, we will first show that the point of intersection two equal and mutually bisecting chords of an ellipse must be the center of the ellipse. This latter result is also valid for any strictly convex centrosymmetric region. Our proof is indirect.

Assume that AB and CD are two equal and mutually bisecting chords whose point of intersection is not the center of the region. Now reflect AB and CD through the center to produce two other mutually bisecting chords $A'B'$ and $C'D'$. Then,

$$\overline{AC} = \overline{DB} = \overline{B'D'} = \overline{C'A'},$$

$$AC \parallel DB \parallel B'D' \parallel C'A'.$$

This gives a contradiction since there are at most two equal and parallel chords of a strictly convex

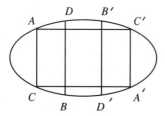

FIGURE 224

region (by convexity, the two inner chords must be longer than the two outer ones).

For the parallelogram to be an inscribed rectangle of the ellipse, the diagonals must have the same length and this requires that they be symmetrically placed in the ellipse.

More generally, if we are given a strictly convex centrosymmetric region with two perpendicular axes of symmetry such that the length of the radius vector from O decreases monotonically from C to B, then all inscribed rectangles must have their sides parallel to AB and CD.

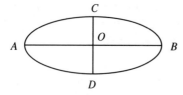

FIGURE 225

Problem 409. By Problem 408, the sides of the rectangle must be parallel to the axes of the ellipse. The coordinates of the four vertices will be of the form $(\pm h, \pm k)$. We now wish to maximize the area $A = 4hk$ subject to the constraint condition $\dfrac{h^2}{a^2} + \dfrac{k^2}{b^2} = 1$. By the Arithmetic-Geometric Mean inequality,

$$\frac{1}{2} = \frac{1}{2}\left\{ \frac{h^2}{a^2} + \frac{k^2}{b^2} \right\} \geq \frac{hk}{ab}$$

with equality if and only if $\dfrac{h^2}{a^2} = \dfrac{k^2}{b^2} = \dfrac{1}{2}$. Thus, maximum area is $4\dfrac{ab}{2} = 2ab$.

Comment. In many texts, it is assumed that the sides of the rectangle must be parallel to the axes of the ellipse. Although this is intuitive, it does require a proof.

Rider. Maximize $h^m k^n$, where m and n are given positive integers and $\dfrac{h^2}{a^2} + \dfrac{k^2}{b^2} = 1$.

Problem 410. Let $w = re^{i\alpha}$, $z = se^{i\beta}$, where r and s denote the absolute values of w and z, respectively. We now have to show that

$$2|re^{i\alpha} - se^{i\beta}| \geq (r+s)|e^{i\alpha} - e^{i\beta}| \quad (1)$$

or

$$4\{(r\cos\alpha - s\cos\beta)^2 + (r\sin\alpha - s\sin\beta)^2\}$$
$$\geq (r+s)^2 \times$$
$$\{(\cos\alpha - \cos\beta)^2 + (\sin\alpha - \sin\beta)^2\},$$

or equivalently (on simplifying)

$$2(r^2 + s^2 - 2rs\cos(\alpha - \beta))$$
$$\geq (r^2 + s^2 + 2rs)(1 - \cos(\alpha - \beta)),$$

or

$$(r - s)^2(1 + \cos(\alpha - \beta)) \geq 0.$$

This equality holds if and only if $r = s$ or $\alpha - \beta = \pm\pi$.

Alternative solution. In (1), let u, $-v$ equal $e^{i\alpha}$, $e^{i\beta}$, respectively, giving

$$\left| \frac{r}{r+s}u + \frac{s}{r+s}v \right| \geq \left| \frac{u+v}{2} \right|.$$

The latter immediately follows geometrically, i.e., the distance from the vertex of an isosceles triangle to any point on the base is at least the length of the altitude from that vertex.

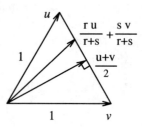

FIGURE 226

Problem 411. Multiply the equation by $(x-a)(x-b)$. The resulting equation is

$$(a-b)x^{n+2} - (a^{n+2}-b^{n+2})x \tag{2}$$
$$+ ab(a^{n+1}-b^{n+1}) = 0.$$

Without loss of generality, we assume that $a > b$.

Case 1: n even, $a > b \geq 0$. Equation (1) obviously has no positive roots, and we will show that it also has no negative roots. Since there are only 2 alternations of signs of the coefficients of (2), there are at most two positive roots of (2) and these correspond to $x = a$, $x = b$ which we introduced when we multiplied by $(x-a)(x-b)$. Changing x to $-x$, there are no alternations of signs in (2) and thus no negative roots in (2) or (1).

Case 2: n even, $a \geq 0 > b$. If $a^2 > b^2$, we get the same as the previous case. So we assume $a^2 < b^2$ and we get signs $\quad + \quad + \quad -$ which implies one positive root. Changing x to $-x$, the signs become $\quad + \quad - \quad -$ which implies one negative root. Thus, (1) has no real root.

Case 3: n even, $0 > a > b$. Here the coefficient signs are $+ \ + \ +$, implying no positive root. Changing x to $-x$, we get $+ \ - \ +$, implying two negative roots in (2). Again (1) has no real roots.

By now considering the same three above cases for n odd and proceeding similarly, we find that (1) has at most one real root.

Problem 412. First note that

$$(z-1)P(z)$$
$$= a_0 z^{n+1} - (a_0 - a_1)z^n - (a_1 - a_2)z^{n-1}$$
$$- \cdots - (a_{n-1} - a_n)z - a_n.$$

Since $a_0, a_0 - a_1, a_1 - a_2, \ldots, a_{n-1} - a_n, a_n$ are all nonnegative, the triangle inequality yields

$$|(z-1)P(z)| \geq a_0|z|^{n+1} \tag{1}$$
$$- \{(a_0 - a_1)|z|^n + (a_1 - a_2)|z|^{n-1} + \cdots + a_n\}.$$

Now note that if $|z| > 1$, the right-hand side of (1) is greater than or equal to

$$|a_0||z|^{n+1} - |z|^n\{(a_0 - a_1) + (a_1 - a_2)$$
$$+ \cdots + (a_{n-1} - a_n) + a_n\}$$
$$= a_0|z|^n(|z| - 1) > 0,$$

or $|P(z)| > 0$, and thus $|r| \leq 1$.

Problem 413. Noting that

$$[\sqrt{k}] = n \quad \text{for} \quad n^2 \leq k \leq (n+1)^2 - 1,$$

it is easy to see that

$$S_n = 1(2^2 - 1^2) + 2(3^2 - 2^2)$$
$$+ \cdots + (n-1)(n^2 - (n-1)^2)$$
$$= -1^2 - 2^2 - 3^3$$
$$- \cdots - (n-1)^2 + (n-1)n^2$$
$$= (n-1)n^2 - \frac{(n-1)(n)(2n-1)}{6}$$
$$= \frac{n(n-1)(4n+1)}{6},$$

completing the first part.

Noting that

$$[\sqrt[3]{k}] = n \quad \text{for} \quad n^3 \leq k \leq (n+1)^3 - 1,$$

it is easy to see that

$$T_n = 1(2^3 - 1^3) + 2(3^3 - 2^3)$$
$$+ \cdots + (n-1)(n^3 - (n-1)^3)$$
$$= -1^3 - 2^3 - 3^3$$
$$- \cdots - (n-1)^3 + (n-1)n^3$$
$$= (n-1)n^3 - \frac{(n-1)^2 n^2}{4}$$
$$= \frac{(n-1)n^2(3n+1)}{4}.$$

Problem 414. If the seven integers are a_1, $a_2, \ldots, a_7$, then

$$a_1 + a_2 + \cdots + a_7 + (a_1 + a_2)$$
$$+ (a_2 + a_3) + \cdots + (a_7 + a_1)$$
$$= 1 + 2 + \cdots + 14$$

or

$$a_1 + a_2 + \cdots + a_7 = 35.$$

1 must be selected. We now use some "trial and error". Since $5 = 1 + 4 = 2 + 3$, we try 4 next to 1. Then to get 14, we try 10 next to 4, etc. This leads to a correct solution:

$$\begin{array}{ccccc} & & 1 & & \\ & 4 & & 6 & \\ 10 & & & & 2 \\ & 3 & & 9 & \end{array}.$$

Rider. Determine all solutions in distinct positive integers.

Problem 415. Any arrangement will do! First note that $3168 = (32)(99)$.

The remainder on division by 32 depends only on the last five digits of the dividend (since $2^5 | 10^5$). The remainder on division by 99 is unaltered by any interchange of a pair of even spaced digits or any pair of odd spaced digits, since $a10^n + b10^{n+2m} \equiv b10^n + a10^{n+2m} \pmod{99}$. Finally, $39997 \equiv 1981 \pmod{3168}$.

Problem 416. Since
$$\tan(A - B) = \frac{\tan A - \tan B}{1 + \tan A \tan B},$$
etc., we obtain by expanding out and replacing $\tan A$, $\tan B$, $\tan C$ by u, v, w respectively,

$$\frac{u - v}{1 + uv} + \frac{v - w}{1 + vw} + \frac{w - u}{1 + wu} = 0.$$

On combining fractions and factoring, we find that $(u - v)(v - w)(w - u) = 0$.

Hence the triangles must be isosceles.

Problem 417. Our proof is indirect. Assume three such numbers exist. Then $cr^{n_1} = \sqrt[m]{p_1}$, $cr^{n_2} = \sqrt[m]{p_2}$, $cr^{n_3} = \sqrt[m]{p_3}$, where n_1, n_2, n_3 are distinct integers. Eliminating c gives

$$r^{n_1 - n_2} = \sqrt[m]{\frac{p_1}{p_2}}, \quad r^{n_2 - n_3} = \sqrt[m]{\frac{p_2}{p_3}}.$$

Eliminating r then gives

$$\left(\frac{p_1}{p_2}\right)^{n_2 - n_3} = \left(\frac{p_2}{p_3}\right)^{n_1 - n_2},$$

and this is impossible by the unique factorization theorem. For a related problem concerning terms in an arithmetic progression, see Problem 182.

Problem 418. Our proof is indirect. Assume there are two parallel chords of equal maximum length. The endpoints of these two chords are vertices of a parallelogram, at least one of whose diagonals is longer than each of the sides. (Note that if the sides of the parallelogram are a and b, the squares of the diagonals are $a^2 + b^2 + \pm 2ab \cos \theta$.) This contradicts our initial assumption and gives the desired result.

Problem 419. Consider two corresponding maximum-length parallel chords* AB and $A'B'$ of the figures. We can only have the two correspondences $A \leftrightarrow B'$, $B \leftrightarrow B'$ or $A \leftrightarrow B'$, $B \leftrightarrow A'$ between them. The two homothetic centers are the intersections of AA' with BB' and AB' with $A'B$. Note that if the figures are also congruent, then one of the centers is the "point at infinity".

Problem 420. Choose the three vertices to have coordinates $(a, 0, 0)$, $\left(-\frac{a}{2}, a\frac{\sqrt{3}}{2}, 0\right)$ and $\left(-\frac{a}{2}, -a\frac{\sqrt{3}}{2}, 0\right)$, and that of the projectile to be (x, y, h). Then

$$R_1^2 = (x - a)^2 + y^2 + h^2,$$

$$R_2^2 = \left(x + \frac{a}{2}\right)^2 + \left(y - a\frac{\sqrt{3}}{2}\right)^2 + h^2,$$

and

$$R_3^2 = \left(x + \frac{a}{2}\right)^2 + \left(y + a\frac{\sqrt{3}}{2}\right)^2 + h^2.$$

Expanding out:

$$R_1^2 = b^2 - 2ax,$$
$$R_2^2 = b^2 + ax - ay\sqrt{3},$$
$$R_3^2 = b^2 + ax + ay\sqrt{3},$$

---

* By the preceding problem, there is only one pair of maximum-length parallel chords of the two figures. Also, for the figures to have two centers of homotheticity, they must have some symmetry, e.g., two homothetic ellipses or parallelograms.

where $b^2 = x^2 + y^2 + h^2 + a^2$. Adding: $R_1^2 + R_2^2 + R_3^2 = 3b^2$. Then

$$ax = \frac{1}{2}(b^2 - R_1^2)$$

and

$$ay\sqrt{3} = \frac{1}{2}(3b^2 - 2R_2^2 - R_1^2),$$

from which,

$$x^2 + y^2 = \frac{(b^2 - R_1^2)^2}{4a^2} + \frac{(3b^2 - 2R_2^2 - R_1^2)^2}{12a^2}.$$

Finally,

$$h^2 = b^2 - a^2 - \frac{(b^2 - R_1^2)^2}{4a^2}$$
$$- \frac{(3b^2 - 2R_2^2 - R_1^2)^2}{12a^2},$$

where $b^2 = \frac{1}{3}(R_1^2 + R_2^2 + R_3^2)$.

Problem 421. If we choose a rectangular coordinate system whose axes are along the three given chords, then B is the point $(2b, 0, 0)$ and (by symmetry) the center O of the sphere is the point $(b - a, d - c, f - e)$ where we are assuming, without loss of generality, that $b \geq a$, $d \geq c$, $f \geq e$. Now

$$R^2 = \overline{OB}^2$$
$$= (b - a - 2b)^2 + (d - c)^2 + (f - e)^2$$

and, because $ab = cd = ef$,

$$R^2 = a^2 + b^2 + c^2 + d^2 + e^2 + f^2 - 2ef.$$

Problem 422. If at some stage there are r sheets, and s of them are left intact while $r - s$ of them are each cut into 7 pieces, the number of pieces will then be

$$s + 7(r - s) = 7r - 6s \equiv r \pmod 6.$$

At the beginning $r = 7$, so at the end the number of pieces will be the integer x, $1988 < x < 1998$, for which $x \equiv 7 \pmod 6$, i.e., $x = 1993 = 6(331) + 7$.

Problem 423. Consider a line of length $r = 2^s + t$, where s and t are positive integers and

$0 < t \leq 2^s$. We desire to cut (with piling permitted) the line into r unit pieces. With one cut we can obtain pieces of length 2^s and t; then s more obvious pile-and-cuts will give us r pieces all of unit length. Fewer than $s + 1$ cuts cannot do the job, for each cut cannot more than double the number of pieces.

It follows that if

$$a = 2^m + d \quad (0 < d \leq 2^m),$$
$$b = 2^n + e \quad (0 < e \leq 2^n),$$
$$c = 2^p + f \quad (0 < f \leq 2^p),$$

then the minimum number of plane cuts required to cut a block $a \times b \times c$ into abc unit cubes is

$$m + n + p + 3.$$

Problem 424. Let vectors from one of the vertices to the remaining vertices be denoted by $\vec{A}$, $\vec{B}$, $\vec{C}$, $\vec{D}$, $\vec{E}$ where $\vec{A}$, $\vec{B}$, $\vec{C}$ are linearly independent. (See Figure 227.) Since opposite sides are parallel,

$$\vec{D} - \vec{C} = -u\vec{A},$$
$$\vec{E} - \vec{D} = v(\vec{A} - \vec{B}),$$
$$-\vec{E} = w(\vec{B} - \vec{C}),$$

and, on adding,

$$-\vec{C} = -w\vec{C} + (w - v)\vec{B} + (v - u)\vec{A}.$$

Since $\vec{A}$, $\vec{B}$, $\vec{C}$ are linearly independent, $w = 1 = v = u$. Thus, the opposite sides of the

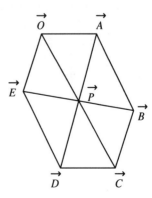

FIGURE 227

hexagon are both congruent and parallel. Since the diagonals of a parallelogram bisect each other, the three diagonals OC, AD, BE are concurrent (at point P, say) and the hexagon is centrosymmetric about P. It now suffices to show that the midpoints F, G, H, I of OA, AB, BC, CD, respectively are coplanar. This follows since $GH \parallel AC \parallel FI$.

Problem 425. Let the left-hand circle have equation $x^2 + y^2 = R^2$. Then if the x-axis is along the line of centers, the other circle has the equation $(x - 2R)^2 + y^2 = R^2$. The coordinates of the particle moving on the left-hand circle and starting at the point of tangency is given by

$$x = R\cos\omega t, \qquad y = R\sin\omega t,$$

where ω denotes the angular velocity. The coordinates of the other particle are given by

$$x' - 2R = R\cos(\pi + \omega t) = -R\cos\omega t,$$

$$y' = R\sin(\pi + \omega t) = -R\sin\omega t.$$

The motion (x'', y'') of (x', y') relative to (x, y) is given by

$$x'' = x' - x = 2R - 2R\cos\omega t,$$

$$y'' = y' - y = -2R\sin\omega t.$$

Thus, $(x'' - 2R)^2 + y''^2 = (2R)^2$. If the angular speed is not constant then ωt is replaced by $F(t) = \int_0^t \omega dt$, which yields the same result.

Problem 426. Let the equation of the parabola be $y = ax^2$ and the fixed interior point be (h, k). Then the family of lines of the chords have the equation $y - k = m(x - h)$. Solving for the points of intersection (x_1, y_1) and (x_2, y_2): $ax^2 - k = m(x - h)$. Thus, the midpoints of the chords are given by

$$X = \frac{x_1 + x_2}{2} = \frac{m}{2a},$$

$$Y = \frac{y_1 + y_2}{2} = m\frac{x_1 + x_2}{2} + k - mh$$

$$= \frac{m^2}{2a} + k - mh.$$

Eliminating m gives $Y = 2aX^2 - 2ahX + k$ or

$$Y - k + \frac{ah^2}{2} = 2a\left(X - \frac{h}{2}\right)^2. \qquad (1)$$

Thus the locus is a translated affine image of the original parabola.

To complete the problem, we must show that every point (r, s) on the locus (1) has the property: the midpoint of the chord joining the points where the line through (r, s) and (h, k) meets the parabola $y = ax^2$ is the point (r, s). The slope of the line through (r, s) and (h, k) is

$$\frac{k - s}{h - r} = \frac{k - \{2ar^2 - 2ahr + k\}}{h - r}$$

$$= 2ar.$$

From the first part of the solution, the line through (h, k) with this slope determines a chord with midpoint abscissa $\dfrac{2ar}{2a} = r$, and we are done.

Problem 427. First note that $\dfrac{(n + 3)^2}{12}$ is never of the form $k + \frac{1}{2}$. For if it were, then $(n + 3)^2 = 12k + 6$. But this is impossible since the left side is congruent to 0 or 1 (mod 4) and the right side is congruent to 2 (mod 4). Now suppose $g(n) = k$. Then,

$$k - \frac{1}{2} < \frac{(n + 3)^2}{12} < k + \frac{1}{2}$$

or

$$k - \frac{1}{2} < \frac{[(n - 3)^2 + 12n]}{12} = \frac{(n - 3)^2}{12} + n$$

$$< k + \frac{1}{2}.$$

Thus,

$$(k - n) - \frac{1}{2} < \frac{(n - 3)^2}{12} < (k - n) + \frac{1}{2}$$

and

$$G(n - 6) = k - n = G(n) - n.$$

Problem 428. (J. L. Selfridge) If $1, 2, \ldots, [\frac{m}{2}]$ are selected from $1, 2, \ldots, m$, then none of those remaining divides another of them. This shows that $n < [\frac{m}{2}]$. Now we show that for any $n \leq [\frac{m}{2}] - 1$, some a and b will remain unselected with $b = 2^k a$. Suppose that no such a and b exist. Then we must select all but one from each sequence $r, 2r, \ldots, 2^s r$. These sequences are disjoint if r is odd, and we count the numbers selected by noticing that selecting all but the largest in each sequence means selecting $1, 2, \ldots, [\frac{m}{2}]$.

Problem 429. Consider the six numbers

$$N_1 = 1 - f(1) - g(1) - h(1),$$

$$N_2 = 1 - f(1) - g(1) - h(1),$$

$$N_3 = 0 + f(0) + g(1) + h(1),$$

$$N_4 = 0 + f(1) + g(0) + h(1),$$

$$N_5 = 0 + f(1) + g(1) + h(0),$$

$$N_6 = 0 - f(0) - g(0) - h(0).$$

Since $\sum N_i = 2$, one of the numbers N_i is at least $\frac{1}{3}$. Thus the given relation holds for at least one of the points

$$(x, y, z) = (1, 1, 1), \ (0, 1, 1), \ (1, 0, 1),$$

$$(1, 1, 0), \ (0, 0, 0).$$

To show $\frac{1}{3}$ is best possible, we exhibit three functions f, g, h for which we get exactly $\frac{1}{3}$.

Let $f(x) = g(x) = h(x) = \dfrac{3x - 1}{9}$. Then

$$|xyz - f(x) - g(y) - h(z)|$$

$$= \left| xyz - \frac{x + y + z}{3} + \frac{1}{3} \right|.$$

It suffices now to show that

$$\frac{1}{3} \geq xyz - \frac{x + y + z}{3} + \frac{1}{3} \geq -\frac{1}{3}$$

for all $0 \leq x, y, z \leq 1$. Since the expression is linear in each variable, its extreme values are taken on at the end points. The rest follows.

For a generalization, see *Mathematics Magazine,* 62 (1989) 198; 63 (1990) 194–197. A simpler version appeared in the 1959 Putnam Competition. Problem 1325.

Problem 430. Two natural cases to look at are the two limiting cases of a regular triangular pyramid for which the altitude to the base approaches zero or becomes unbounded. In the former case, the dihedral angles will approach $0, 0, 0, \pi, \pi, \pi$,

while in the latter case they will approach $\frac{\pi}{2}, \frac{\pi}{2}, \frac{\pi}{2}, \frac{\pi}{3}, \frac{\pi}{3}, \frac{\pi}{3}$.

Problem 431. Let

$$F(t) = \frac{x_n t^n + x_{n-1} t^{n-1} + \cdots + x_1 t + x_0}{x_{n-1} t^{n-1} + x_{n-2} t^{n-2} + \cdots + x_1 + x_0}$$

$$= 1 + \frac{x_n}{G},$$

where

$$G = \frac{x_{n-1}}{t} + \frac{x_{n-2}}{t^2} + \cdots + \frac{x_1}{t^{n-1}} + \frac{x_0}{t^n}.$$

Since each $x_i \geq 0$, G decreases as t increases so that $F(t)$ is an increasing function of t. Hence,

$$\frac{A}{C} = F(a) > F(b) = \frac{B}{D}$$

if and only if $a > b$.

Problem 432. (D. Duncan) Three great circles intersecting in distinct points divide a spherical surface into 8 spherical triangles occurring in symmetric pairs which lie in opposite hemispheres. Let 2 great circles intersect at O and O', and let them meet a third great circle in points A, A' and B, B', respectively. On the hemisphere bounded by the great circle through A, B, A', B' and containing point O, examine the configuration obtained by drawing a fourth great circle. This fourth great circle meets the third great circle at the diametrically opposite points C, C', the semicircle AOA' at P, and the semicircle BOB' at Q. Thus a triangle OPQ is formed and is surrounded cyclically by, say, triangle AOB, quadrilateral $BOPC$, triangle CPA', quadrilateral $A'PQB'$, triangle $B'QC'$, and quadrilateral $C'QOA$. Hence 4 great circles having all intersection points distinct must always yield 8 spherical triangles and 6 spherical quadrilaterals on a sphere.

Consider again the hemisphere bounded by the original third great circle and containing the aforesaid configuration. Let the fifth great circle cut the third great circle in points D and D', one of which must lie on a side of a triangle, the other on a side of a quadrilateral. Let D lie on side AB of AOB and D' lie on side $A'B'$ of $A'PQB'$. Observe that the sides of the triangle OPQ are also sides of quadrilaterals. Now DD' will or will not enter triangle OPQ.

If DD' does enter triangle OPQ it must cut 2 of its sides, thereby dividing triangle OPQ into a triangle and a quadrilateral. Semicircle DD' must also divide AOB into a triangle and a quadrilateral, the quadrilateral from which it enters OPQ into a triangle and a pentagon, and $A'PQB'$ into 2 quadrilaterals. Hence the fifth circle yields a configuration on the hemisphere consisting of 5 triangles, 5 quadrilaterals and one pentagon.

If DD' does not enter OPQ it must pass through four consecutive peripheral polygons, thereby dividing AOB into a triangle and a quadrilateral, the adjacent quadrilateral into 2 quadrilaterals, the next adjacent triangle into a triangle and a quadrilateral, and $A'PQB'$ into a triangle and a pentagon. Again, we have 5 triangle, 5 quadrilaterals and one pentagon.

Hence, the configuration formed on a sphere by 5 great circles, no 3 of which are concurrent, consists of exactly 10 triangles, 10 quadrilaterals and 2 pentagons. Furthermore, a sixth great circle will either leave the pentagon intact, will divide it into a pentagon and a quadrilateral, or will divide it into a triangle and a hexagon. Hence, if on the sphere there are n great circles ($n \geq 5$), no 3 of which are concurrent, there will always be at least one spherical polygon with 5 or more sides.

Problem 433. One obvious solution is any triplet satisfying $x + y = 0$. Then by symmetry and the factor theorem,

$$(x + y + z)^3 - x^3 - y^3 - z^3$$

$$= 3(x + y)(y + z)(z + x).$$

Thus all the solutions are given by $x + y = 0$, or $y + z = 0$, or $z + x = 0$.

Problem 434. Either $\sin x + \cos x = 0$, in which case $x = \frac{3\pi}{4}, \frac{7\pi}{4}$, or we can divide out the factor $\sin x + \cos x$ to give

$$11 = 16(\sin^4 x - \sin^3 x \cos x + \sin^2 x \cos^2 x$$

$$- \sin x \cos^3 x + \cos^4 x)$$

$$= 16(1 - \sin^2 x \cos^2 x - \sin x \cos x)$$

or, equivalently,

$$(4 \sin x \cos x - 1)(4 \sin x \cos x + 5) = 0.$$

The only nonextraneous solutions of this equation satisfy

$$\sin x \cos x = \frac{1}{4}$$

or

$$\sin 2x = \frac{1}{2}, \qquad 0 \leq x \leq 2\pi.$$

Thus all the required values of x are $\frac{\pi}{12}, \frac{5\pi}{12}, \frac{9\pi}{12}, \frac{13\pi}{12}, \frac{17\pi}{12}, \frac{21\pi}{12}$.

Problem 435. Looking at the first few terms $1, 1, 1, 2, 3, 7, 11, 26, \ldots$ of the sequence, we are led to conjecture

$$a_{n+2} = 4a_n - a_{n-2} \quad (n = 3, 4, 5, \ldots).$$

This can be established by induction. Assuming its truth for $n - 1$, we have

$$a_{n-1}a_{n+2} = 1 + a_{n+1}a_n$$

$$= 1 + (4a_{n-1} - a_{n-3})a_n$$

$$= 4a_{n-1}a_n - a_{n-1}a_{n-2},$$

whence the result. From this, we see that, if $a_1, a_2, \ldots, a_{n+1}$ are integers, so also is a_{n+2}.

Rider. It can be conjectured also that, for $k = 1, 2, \ldots, a_{2k+1} = 2a_{2k} - a_{2k-1}$ and $a_{2k+2} = 3a_{2k+1} - a_{2k}$. Establish these equations and give an alternative solution to the problem.

Problem 436. Let m be the gcd of a and c, and suppose that $a = mu$ and $c = mv$. Then $ub = vd$. Since gcd $(u, v) = 1$, u must divide d. If $d = nu$, then $b = nv$ and

$$a^2 + b^2 + c^2 + d^2 = (m^2 + n^2)(u^2 + v^2).$$

Rider. Show that $a^2 + b^2 + c^2 + d^2 \not\equiv 3 \pmod{4}$.

Problem 437. Consider first the same problem with 5 cards. A shuffle is equivalent to a permutation of the 5 cards and can be represented by

$$\begin{pmatrix} 1 & 2 & 3 & 4 & 5 \\ a_1 & a_2 & a_3 & a_4 & a_5 \end{pmatrix},$$

which means that card r is replaced by card a_r. For example, consider the shuffle represented by the permutation

$$\begin{pmatrix} 1 & 2 & 3 & 4 & 5 \\ 2 & 1 & 4 & 5 & 3 \end{pmatrix}.$$

In the following table, line 0 gives the original order of the cards and line i, $1 \le i \le 6$, gives the result of the ith shuffle.

0	1	2	3	4	5
1	2	1	4	5	3
2	1	2	5	3	4
3	2	1	3	4	5
4	1	2	4	5	3
5	2	1	5	3	4
6	1	2	3	4	5

In this case, it takes 6 shuffles to get the cards back to their original position. Now every permutation can be "factored" into a product of cycles. Here we have

$$\begin{pmatrix} 1 & 2 & 3 & 4 & 5 \\ 2 & 1 & 4 & 5 & 3 \end{pmatrix} = (1,2)(3,4,5)$$

where the cycle $(3,4,5)$, for example, means $3 \to 4$, $4 \to 5$, $5 \to 3$. Since the periods of the two cycles are 2 and 3, respectively, the period of the permutation is $2 \cdot 3$, which is the least common multiple (lcm) of 2 and 3.

For our original problem, we have to partition 13 into a sum of positive integers whose lcm is a maximum. A little trial and error yields

$$13 = 5 + 4 + 3 + 1$$

giving a maximum period of $5 \cdot 4 \cdot 3 = 60$.

For 52 cards, the maximum period 180180 can be found from

$$52 = 13 + 11 + 9 + 7 + 5 + 4 + 2 + 1.$$

As a research problem, the reader may wish to consider devising an algorithm to determine the maximum cycling period for n cards.

Problem 438. We have to show that

$$P_1 = 0 \quad \Longrightarrow \quad P_2 = 0 \qquad (1)$$

where

$$P_n \equiv \frac{a}{(bc - a^2)^n} + \frac{b}{(ca - b^2)^n} + \frac{c}{(ab - c^2)^n}, \quad n = 1, 2.$$

If we set

$$P_0 \equiv \frac{1}{bc - a^2} + \frac{1}{ca - b^2} + \frac{1}{ab - c^2},$$

then straightforward multiplication (and subsequent simplification) shows that $P_0 \cdot P_1 = P_2$, and (1) follows at once, as well as

$$P_0 = 0 \quad \Longrightarrow \quad P_2 = 0 \qquad (2)$$

Observe now that the converse of either (1) or (2) is not true, since

$$P_2 = 0 \quad \Longrightarrow \quad P_0 = 0 \ \text{ or } \ P_1 = 0. \qquad (3)$$

Moreover, the disjunction in (3) cannot be avoided since neither equation on the right of (3) implies the other. For example, if ω is an imaginary cube root of unity, then $(a, b, c) = (0, \omega, -1)$ is a solution of $P_0 = 0$ but not of $P_1 = 0$, while the reverse is true for $(a, b, c) = (0, 1, -1)$.

Problem 439. It will be seen that the proof of (ii) implies (i); nevertheless we include a direct proof of (i).

(i) Let $p = a + a'$, $q = b + b'$, $r = c + c$. We assume that the edges of the tetrahedron have been labeled so that $p \ge q \ge r$. Then (see Figure 228)

$$b + c > a', \qquad b' + c' > a',$$

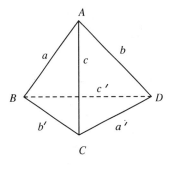

FIGURE 228

$$b + c' > a, \qquad b' + c > a,$$

and adding these inequalities yields $q + r > p$. Thus there is a triangle PQR with sides of lengths p, q, r.

(ii) From the law of cosines involving the largest angle P,

$$p^2 = q^2 + r^2 - 2qr \cos P,$$

we see that triangle PQR is acute-angled if and only if $\cos P > 0$ or

$$q^2 + r^2 - p^2 > 0. \qquad (1)$$

We will establish the inequalities

$$b^2 + b'^2 + c^2 + c'^2 - a^2 - a'^2 \geq 0 \qquad (2)$$

and

$$2bb' + 2cc' - 2aa' > 0, \qquad (3)$$

from which (1) follows by addition.

To prove (2), let $\vec{A}, \vec{B}, \vec{C}, \vec{D}$ denote vectors from some origin to the vertices A, B, C, D of the tetrahedron (see Figure 228). With the usual notation, $\vec{v}^2 = \vec{v} \cdot \vec{v} = |\vec{v}|^2 \geq 0$ and other properties of the dot product of two vectors, the left side of (2) becomes

$$(\vec{A} - \vec{D})^2 + (\vec{B} - \vec{C})^2 + (\vec{A} - \vec{C})^2$$
$$+ (\vec{B} - \vec{D})^2 - (\vec{A} - \vec{B})^2 - (\vec{C} - \vec{D})^2$$
$$= (\vec{A} + \vec{B} - \vec{C} - \vec{D})^2 \geq 0.$$

To prove (3), suppose the framework of the tetrahedron in our figure is considered to be hinged along CD and flattened to form a convex quadrilateral $ADBC$. The lengths of all edges remain the same except that of AB which is increased. For the quadrilateral, the *Ptolemaic inequality* assures us that, in magnitude only,

$$\overline{AD}\,\overline{BC} + \overline{AC}\,\overline{BD} \geq \overline{AB}\,\overline{CD}, \qquad (4)$$

with equality if and only if $ADBC$ is a cyclic quadrilateral (Ptolemy's Theorem). Hence (4) holds a fortiori, with strict inequality for the tetrahedron, where AB is shorter, and thus (3) is established.

Problem 440. Note that

$$16 - 24x + 9x^2 = (4 - 3x)^2 > x^3$$

for $x < 1$ (sketch graphs!), so the expression makes sense. Let the first radical be denoted by m, the second radical by n and the sum by s. Then

$$s^3 = (m + n)^3 = m^3 + n^3 + 3mns$$

and

$$s^3 = 8 - 6x + 3sx,$$

or

$$(s - 2)(s^2 + 2s + 4 - 3x) = 0.$$

When $x < 1$ the second factor has complex roots, so s must be constant and equal to 2. This is true also for $x = 1$.

Problem 441. Writing $u_n = 2n - t_n$, we see that the sequence $\{t_n\}$ is

$$1; \ 2, 2; \ 3, 3, 3; \ 4, 4, 4, 4; \ 5, 5, \dots.$$

For each t_n, we have

$$1 + 2 + 3 + \cdots + (t_n - 1)$$
$$\leq n - 1 < 1 + 2 + 3 + \cdots + t_n,$$

with equality on the left exactly when n is the first index for which t_n assumes a specific value. The left inequality yields $t_n^2 - t_n - 2(n - 1) \leq 0$, from which $t_n \leq \frac{1}{2}(1 + \sqrt{8n - 7})$. Since t_n is the largest integer satisfying this inequality, the result follows.

Problem 442. It follows easily that $F = \frac{1}{2}ef \sin \theta$ (see Figure 229). Then $e^2 + f^2 \geq 2ef \geq 2ef \sin \theta$.

There is equality if and only if $e = f$ and $\theta = 90°$. The result is also valid for non-convex

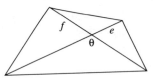

FIGURE 229

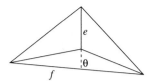

FIGURE 230

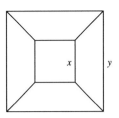

FIGURE 231

quadrilaterals (see Figure 230) since we still have $F = \frac{1}{2}ef\sin\theta$.

Problem 443. Divide the cube into $13^3 = 2197$ small cubes of side $\frac{15}{13}$ units. The space diagonal of each small cube has length $\frac{15\sqrt{3}}{13}$ units. Now $\left(\frac{15\sqrt{3}}{13}\right)^2 = \frac{675}{169} < 4$, or $\frac{15\sqrt{3}}{13} < 2$.

Therefore each small cube can be contained within a sphere of unit radius. Since there are 11000 points and 2197 small cubes, and $5 \cdot 2197 = 10985 < 11000$, at least one cube must contain at least 6 points (pigeonhole principle). Therefore there is a unit sphere which contains at least 6 points.

Problem 444. Consider a $6 \times 6 \times 6$ "board." Each winning line of the $4 \times 4 \times 4$ central cube intersects 2 distinct $1 \times 1 \times 1$ cubes in the outer shell, and each $1 \times 1 \times 1$ cube is the continuation of exactly one winning line. Hence the number of ways of winning is $\frac{1}{2}$ of the number of outer $1 \times 1 \times 1$ cubes, or $\frac{1}{2}(6^3 - 4^3) = 76$.

If the game were to be played on an n-dimensional board, k to a side, the corresponding number is

$$\frac{1}{2}\{(k+2)^n - k^n\}.$$

Problem 445. This problem is a "trickie." A little reflection, or algebra, reveals that the number of red cards in the top 26 must always equal the number of black cards in the bottom 26. Hence, by the rules of logic, the whole assertion is correct no matter what follows the word *then*.

Problem 446. Although the area A, of Figure 231 can be calculated, in terms of x and y, directly from the given diagram, the easiest way of finding

A is to note that 4 identical pieces of this form can be put together to form a hollow square or square frame as in the figure. (Indeed this was what the carpenter used the pieces for.) Thus

$$A = \frac{y^2 - x^2}{4} = \left(\frac{y+x}{2}\right)\left(\frac{y-x}{2}\right).$$

If we set $m = \frac{y+x}{2}$ and $n = \frac{y-x}{2}$ then $m + n = y$ and $m - n = x$, and also $A = mn$. From this and the conditions of the problem it is evident that A must be a number which can be factored into two factors in at least four ways. It is easily found that 30 is the only number under 40 having this property. Thus $30 = 1 \cdot 30 = 2 \cdot 15 = 3 \cdot 10 = 5 \cdot 6$. The corresponding values of y and x are then found to be

$$(31, 29), \quad (17, 13), \quad (13, 7), \quad (11, 1).$$

Problem 447. The cases $m = 1$ or $n = 1$ are obvious. Let $\sqrt[n]{m} = 1 + u$, $\sqrt[m]{n} = 1 + v$, where $u, v > 0$. Then, from the binomial expansion, $m = (1+u)^n > 1+nu$, $n = (1+v)^m > 1+mv$, so that

$$u < \frac{m-1}{n} \quad \text{and} \quad v < \frac{n-1}{m}.$$

Then

$$\frac{1}{\sqrt[n]{m}} + \frac{1}{\sqrt[m]{n}} = \frac{1}{1+u} + \frac{1}{1+v}$$
$$> \frac{n}{n+m-1} + \frac{m}{n+m-1}$$
$$= \frac{n+m}{n+m-1} > 1.$$

Remark. More generally, $x^y + y^x > 1$ whenever $x, y > 0$. A short calculus proof appears in *Mathematics Magazine* 52 (1979) 118. Can you give a non-calculus proof?

Problem 448. (a) The statement is false. Observe that $2N = 1052631578947368420$. Thus

$$19N = 20N - N = 9999999999999999990,$$

which does not contain all the ten digits.

(b) There are no persistent numbers! Our proof is indirect. Assume that N is persistent. We can express N in the form $N = 2^a 5^b M$, where M is relatively prime to both 2 and 5. Since

$$2^b 5^a N = 10^{a+b} M$$

must also be persistent, all multiples of M must contain the nine nonzero digits. We now invoke Euler's generalization of Fermat's Theorem: *If a is relatively prime to n, then*

$$a^{\varphi(n)} - 1 = kn,$$

where $\varphi(n)$ (Euler's φ-function) is the number of positive integers less than or equal to n that are relatively prime to n. Now we have

$$10^{\varphi(M)} - 1 = kM, \tag{1}$$

and this gives a contradiction, for kM should contain all nine nonzero digits, but the left side of (1) contains only nines.

For an alternative, more elementary, solution, assume again that N is persistent and consider the remainders obtained by dividing the following N numbers by N:

$$1, \ 11, \ 111, \ \ldots, \ 111\ldots 1,$$

where the last number has N digits. Since at most $N - 1$ different nonzero remainders can result, either one of the above numbers is divisible by N, in which case N is not persistent, or else two of them, say

$$R = \underbrace{111\ldots 1}_{r} \ \text{ and } \ S = \underbrace{111\ldots 1}_{s}, \quad s > r,$$

give the same remainder, in which case their difference

$$S - R = \underbrace{11\ldots 1}_{s-r}\underbrace{00\ldots 0}_{r}$$

is divisible by N, and N is not persistent.

Problem 449. The result follows immediately from the identity

$$(a^2 + b^2 - 2)^2 + (c^2 + d^2 - 2)^2$$
$$+ 2(ac - bd)^2$$
$$= (a^2 + c^2 - 2)^2 + (b^2 + d^2 - 2)^2$$
$$+ 2(ab - cd)^2.$$

The result can be generalized. It is then equivalent to stating that if the row vectors of an $n \times n$ matrix are orthonormal, so are the column vectors.

Problem 450. Let S, S_1 and S_2 denote the sets of permutations of $1, 2, \ldots, n$ with respectively no restriction, with $a_1 = 1$ and $a_2 = 2$. Then (with $|X|$ denoting the number of elements in set X)

$$|S| = n!,$$
$$|S_1| = |S_2| = (n-1)!,$$
$$|S_1 \cap S_2| = (n-2)!.$$

Hence

$$|S_1 \cup S_2| = |S_1| + |S_2| - |S_1 \cap S_2|$$
$$= 2(n-1)! - (n-2)!,$$

and the number of permutations with $a_1 \neq 1$ and/or $a_2 \neq 2$ is

$$n! - 2(n-1)! + (n-2)!$$
$$= (n-2)!(n^2 - 3n + 3).$$

Problem 451. Let the man's age at the time be n and that of his grandson be k. Then we must have

$$(k+i) \text{ divides } (n+i) \text{ for } i = 0, 1, 2, 3, 4, 5.$$

Thus, $k + i$ divides $n - k$ and so $n - k$ is a multiple of the least common multiple of $k, k + 1, \ldots, k + 5$. If $k = 1$, then $n - 1$ is a multiple of $\text{lcm}(1, 2, 3, 4, 5, 6) = 60$. Here the man is 61 and the boy is 1. Any other k leads to an unrealistic solution. For example, if $k = 2$, then $n - 2$ is a multiple of $\text{lcm}(2, 3, 4, 5, 6, 7) = 420$.

Problem 452. Assume that $\log_{10} 2 = \frac{a}{b}$, where a, b are positive integers. Then $10^{\frac{a}{b}} = 2$, so

$10^a = 2^b$. This gives a contradiction, since the left side is divisible by 5 whereas the right side is not. Similarly, for integral m, $\log_{10} m$ is irrational unless m is a power of 10.

Problem 453. Each four points are the vertices of a quadrilateral which contains at least one angle of at least $90°$. Therefore at least one of the four triangles formed from the four points is obtuse or right-angled.

 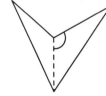

FIGURE 232

Consider any five points. Each four of them gives rise to at least one obtuse or right-angled triangle. There are five ways of choosing four points, so there are at least five obtuse or right-angled triangles. But any triangle is counted twice, so we can be assured only that at least three of these triangles are distinct. Thus, from the $\binom{5}{3} = 10$ possible triangles from five points, at least three are obtuse or right-angled. Now look at all $\binom{75}{5}$ possible choices of 5 points from the 75. For each choice, catalogue all the triangles obtainable; 30% at least will be obtuse or right-angled. In the catalogue of triangles for all possible choices, each triangle appears exactly $\binom{72}{2}$ times, and at least 30% of them are obtuse or right-angled. Therefore, in the list of triangles without repetitions, at least 30% are obtuse or right-angled.

(This problem with 100 points appeared in the 12th International Mathematics Olympiad.)

Problem 454. Let the angles of T_1 and T_2 be respectively A_1, B_1, C_1 and A_2, B_2, C_2. Since $A_1 + B_1 + C_1 = 180° = A_2 + B_2 + C_2$, it cannot happen that each angle of one exceeds the corresponding angle of the other. Consequently, by relabelling if necessary, we may suppose that

$$A_1 \geq A_2 \qquad \text{and} \qquad B_1 \leq B_2.$$

Then $\Delta_1 = \frac{1}{2} b_1 c_1 \sin A_1 \geq \frac{1}{2} b_2 c_2 \sin A_2 = \Delta_2$ (since A_1 and A_2 are acute). Also (Tool Chest, E 45),

$$R_1 = \frac{b_1}{2 \sin B_1} \geq \frac{b_2}{2 \sin B_2} = R_2.$$

The formula for the inradius of the triangle is $r = \frac{\Delta}{s}$, where s is the semiperimeter (Tool Chest, E 45). The easiest cases for calculating area are right-angled and isosceles triangles. Consequently, we try to make T_1 a right-angled triangle and T_2 an isosceles triangle with one side shorter for which $r_1 < r_2$. If $(a_1, b_1, c_1) = (3, 4, 5)$ and $(a_2, b_2, c_2) = (3, 4, 4)$ then $\Delta_1 = 6$, $s_1 = 6$, $r_1 = 1$ while

$$\Delta_2 = \frac{3\sqrt{55}}{4}, \quad s_2 = \frac{11}{2}, \quad r_2 = \sqrt{\frac{45}{44}} > 1.$$

Note. If just T_1 is acute, it still follows that $\Delta_1 \geq \Delta_2$.

Problem 455. Suppose the given skew lines are a and b, respectively, containing the points A and B. If C is the center of the sphere, then $a \perp AC$, $b \perp BC$. Also, C is equidistant from A and B. Hence C lies on three planes:

σ: through A, perpendicular to a;

μ: through B, perpendicular to b;

τ: right-bisecting the segment s joining A and B.

Since a and b are skew, σ and μ are neither parallel nor coincident, and hence intersect in a line m. Suppose, if possible, that τ is parallel to, or contains m. Then m is perpendicular to s as well as to a and b. Therefore, m is perpendicular to the planes determined by a and s (through A) and by b and s (through B). Since both planes contain s, they must coincide, so that a and b lie in a common plane, contrary to hypothesis. Hence τ meets m in exactly one point, and only this can be (and is) the center of the required sphere.

Problem 456. Except for 2 and 3, each prime is of the form $6u + 1$ or $6u - 1$. If a prime $p = 6u + 1$ is in a chain, its successor, if any, must be

$2p - 1 = 12u + 1 = 2(6u) + 1$, since $2p + 1$ is divisible by 3. Therefore successive elements of the chain must be (insofar as primes occur)

$$\{6u + 1, \ 2(6u) + 1,$$
$$2^2(6u) + 1, \ldots, 2^i(6u) + 1, \ldots\}.$$

This sequence cannot go on forever giving primes. For (Tool Chest, B 8), there exists a number k such that $2^k \equiv 1 \pmod{6u + 1}$. For such k,

$$2^k(6u) + 1 \equiv 6u + 1 \equiv 0 \pmod{6u + 1},$$

so that $2^k(6u) + 1$ must be composite.

A similar argument can be given for primes $p = 6u - 1$ in the chain.

Rider. Is there a maximum length for such chains of primes?

Problem 457. We have, for triangle ABC with circumradius R,

$$r = 4R \sin \frac{\angle A}{2} \sin \frac{\angle B}{2} \sin \frac{\angle C}{2},$$

$$t = 4R \cos \frac{\angle A}{2} \cos \frac{\angle B}{2} \sin \frac{\angle C}{2},$$

and hence

$$\frac{r}{t} = \tan \frac{\angle A}{2} \tan \frac{\angle B}{2}.$$

Likewise

$$r_1 = 4R_1 \sin \frac{\angle A}{2} \sin \frac{\angle AMC}{2} \sin \frac{\angle ACM}{2},$$

$$t_1 = 4R_1 \cos \frac{\angle A}{2} \cos \frac{\angle AMC}{2} \sin \frac{\angle ACM}{2},$$

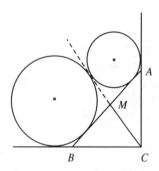

FIGURE 233

$$r_2 = 4R_2 \sin \frac{\angle B}{2} \sin \frac{\angle BMC}{2} \sin \frac{\angle BCM}{2},$$

$$t_2 = 4R_2 \cos \frac{\angle B}{2} \cos \frac{\angle BMC}{2} \sin \frac{\angle BCM}{2}.$$

Since

$$\sin \frac{\angle AMC}{2} = \cos \frac{\angle BMC}{2}$$

and

$$\cos \frac{\angle AMC}{2} = \sin \frac{\angle BMC}{2}$$

($\frac{\angle AMC}{2} + \frac{\angle BMC}{2} = 90°$), the result follows.

Problem 458. Since the largest (smallest) number in any ring is not more (less) than the largest (smallest) number in its predecessor ring, the absolute value of the difference between the largest and smallest numbers in a ring must decrease (or remain the same) as the process is repeated. Eventually, any such sequence of nonnegative integers must have two adjacent entries the same. With no loss of generality, we may assume the smallest element of the ring R at this stage is 0. If a is the largest, then there are three possibilities for R:

(i) $\begin{pmatrix} a & 0 \\ a & b \end{pmatrix}$

(ii) $\begin{pmatrix} a & 0 \\ b & 0 \end{pmatrix}$

(iii) $\begin{pmatrix} a & 0 \\ b & b \end{pmatrix}$

Applying the process to each, we have:

(i)

$$\begin{pmatrix} a & 0 \\ a & b \end{pmatrix} \rightarrow \begin{pmatrix} a & b \\ 0 & a - b \end{pmatrix}$$

$$\rightarrow \begin{pmatrix} a - b & |a - 2b| \\ a & a - b \end{pmatrix}$$

$$\rightarrow \begin{pmatrix} c & c \\ b & b \end{pmatrix} \rightarrow \begin{pmatrix} 0 & |c - b| \\ |c - b| & 0 \end{pmatrix}$$

$$\rightarrow \begin{pmatrix} |c - b| & |c - b| \\ |c - b| & |c - b| \end{pmatrix} \rightarrow \begin{pmatrix} 0 & 0 \\ 0 & 0 \end{pmatrix},$$

where $c = |a - b - |a - 2b||$,

(ii)

$$\begin{pmatrix} a & 0 \\ b & 0 \end{pmatrix} \rightarrow \begin{pmatrix} a & 0 \\ a-b & b \end{pmatrix}$$

$$\rightarrow \begin{pmatrix} a & b \\ b & |a-2b| \end{pmatrix} \rightarrow \begin{pmatrix} a-b & c \\ a-b & c \end{pmatrix}$$

$$\rightarrow \begin{pmatrix} d & 0 \\ 0 & d \end{pmatrix} \rightarrow \begin{pmatrix} d & d \\ d & d \end{pmatrix} \rightarrow \begin{pmatrix} 0 & 0 \\ 0 & 0 \end{pmatrix},$$

where $c = |b - |a - 2b||$ and $d = |a - b - c|$,

(iii)

$$\begin{pmatrix} a & 0 \\ b & b \end{pmatrix} \rightarrow \begin{pmatrix} a & b \\ a-b & 0 \end{pmatrix} \rightarrow \begin{pmatrix} a-b & b \\ b & a-b \end{pmatrix}$$

$$\rightarrow \begin{pmatrix} |a-2b| & |a-2b| \\ |a-2b| & |a-2b| \end{pmatrix} \rightarrow \begin{pmatrix} 0 & 0 \\ 0 & 0 \end{pmatrix}.$$

It can be shown that if one started a similar process with n (≥ 3) numbers instead of four numbers, we will always end up with all zeros provided $n = 2^m$ (see Problem 69-1, *SIAM Review* 12 (1970) 297–300). Also, Lotan (*Amer. Math. Monthly* 56 (1949) 535–541) showed that if one starts with four real numbers whose first difference set is not of the form 1, q, q^2, q^3, where q is the positive solution of $q^3 = q^2 + q + 1$, then apart from trivial transformations, one will always get a set of all zeros.

Problem 459. The only inscribed ellipse in the regular pentagon is the inscribed circle. If there were another inscribed ellipse, it would have to intersect the inscribed circle in at least five points. But this is impossible since the maximum number of points of intersection between two distinct ellipses is four. To prove this, we can eliminate y^2 between the two second-degree equations for the ellipses and solve for y. Then, substituting back, we obtain a fourth degree polynomial equation in x.

More generally, for n-dimensional Euclidean space, if p algebraic surfaces have degrees $d_1, d_2, \ldots, d_p$ and have only a finite number of common points, then that number is at most $d_1 d_2 d_3 \cdots d_{p-1} d_p$. This is Bezout's Theorem.

Problem 460. The inequality is true if $x, y, z \leq 0$. Henceforth, with no loss of generality, assume

$z > 0$. By considering the degenerate triangles $(1, 0, 1)$ and $(0, 1, 1)$ we must have $x < 0$ and $y < 0$. Let $x = -u$, $y = -v$, $z = w$, with u, v, w positive. The condition becomes

$$wc^2 \leq ua^2 + vb^2 \text{ for all triangles } (a, b, c). \quad (1)$$

Suppose (1) holds. Then it holds in the degenerate case $c = a + b$, so that

$$(u - w)a^2 - 2wab + (v - w)b^2 \geq 0. \quad (2)$$

Now this holds for any positive ratio $\frac{a}{b}$, so $u \geq w$ and $v \geq w$. Since $w \geq 0$, we see that (2) holds also for any nonpositive ratio $\frac{a}{b}$, so (2) must hold for *all* pairs (a, b). Hence the discriminant of the left side is negative, i.e., $w^2 \leq (u - w)(v - w)$ or

$$\frac{1}{u} + \frac{1}{v} \leq \frac{1}{w} \quad (3)$$

On the other hand, let (3) hold. Then $u \geq w$, $v \geq w$ and the discriminant of the left side of (2) is negative. If (a, b, c) is any triangle, then

$$ua^2 + vb^2 - wc^2$$
$$\geq ua^2 + vb^2 - w(a + b)^2$$
$$= (u - w)a^2 - 2wab + (v - w)b^2$$
$$\geq 0.$$

The solution of the problem is the set all (x, y, z) for which one of the following four sets of conditions holds:

(i) $x \leq 0$, $y \leq 0$, $z \leq 0$;

(ii) $x < 0$, $y < 0$, $z > 0$, $-\frac{1}{x} - \frac{1}{y} \leq \frac{1}{z}$;

(iii) $x < 0$, $y > 0$, $z < 0$, $-\frac{1}{x} - \frac{1}{z} \leq \frac{1}{y}$;

(iv) $x > 0$, $y < 0$, $z < 0$, $-\frac{1}{y} - \frac{1}{z} \leq \frac{1}{x}$.

Problem 461. Letting $\dfrac{\overline{BA'}}{\overline{A'C}} = \dfrac{v}{u}$, where $u + v = 1$, we have $\overrightarrow{A'} = u\overrightarrow{B} + v\overrightarrow{C}$ and the square of the cevian c_a is given by

$$c_a^2 = (\overrightarrow{A'} - \overrightarrow{A}) \bullet (\overrightarrow{A'} - \overrightarrow{A})$$
$$= \{u(\overrightarrow{B} - \overrightarrow{A}) + v(\overrightarrow{C} - \overrightarrow{A})\}^2$$
$$= u^2(\overrightarrow{B} - \overrightarrow{A})^2 + v^2(\overrightarrow{C} - \overrightarrow{A})^2 \quad (1)$$
$$+ uv\left[2(\overrightarrow{B} - \overrightarrow{A}) \bullet (\overrightarrow{C} - \overrightarrow{A})\right]$$
$$= u^2c^2 + v^2b^2 + uv(b^2 + c^2 - a^2).$$

(Incidentally, this is Stewart's Theorem.) By cyclic interchange of a, b, c, we obtain the other cevians:

$$c_b^2 = u^2 a^2 + v^2 c^2 + uv(c^2 + a^2 - b^2), \quad (2)$$

$$c_c^2 = u^2 b^2 + v^2 a^2 + uv(a^2 + b^2 - c^2). \quad (3)$$

We now show that $c_a = c_b = c_c$ if and only if $a = b = c$. Clearly, if $a = b = c$, then $c_a = c_b = c_c$ (from (1),(2),(3)). To show that $c_a = c_b = c_c$ implies $a = b = c$, we have to show that the system of linear equations in a, b, c ((1), (2), (3)) has a unique solution. Rewriting this system, we have

$$-uva^2 + vb^2 + uc^2 = c_a^2,$$

$$ua^2 - uvb^2 + vc^2 = c_b^2 = c_a^2,$$

$$va^2 + ub^2 - uvc^2 = c_c^2 = c_a^2.$$

Equivalently, we have to show that the coefficient determinant

$$D = \det \begin{pmatrix} -uv & v & u \\ u & -uv & v \\ v & u & -uv \end{pmatrix}$$

is not zero (nonsingular). Expanding out the determinant, we get

$$D = u^3 + v^3 + 3u^2v^2 - u^3v^3$$
$$= u^2 - uv + v^2 + 3u^2v^2 - u^3v^3$$
$$= 1 - 3uv + 3u^2v^2 - u^3v^3 = (1 - uv)^3.$$

However, $1 - uv = 1 - u(1 - u)$ cannot be zero for real u.

Problem 462. We have

$$\sin A_1 \sin A_2 \ldots \sin A_n$$

$$= \cos A_1 \cos A_2 \ldots \cos A_n$$

$$= (2^{-n} \sin 2A_1 \sin 2A_2 \ldots \sin 2A_n)^{\frac{1}{2}}$$

$$\leq 2^{-\frac{n}{2}},$$

with equality if and only if

$$\sin 2A_1 = \sin 2A_2 = \ldots = \sin 2A_n = 1.$$

This occurs when each A_i equals $\frac{\pi}{4}$, in which case the condition is satisfied.

Problem 463. Let $0 \leq \theta < \pi$. The quadratic form

$$x^2 + y^2 - 2xy \cos \theta$$
$$= (x - y \cos \theta)^2 + y^2 \sin^2 \theta \qquad (1)$$

is always nonnegative and vanishes only if $\theta \neq 0$, $x = y = 0$ or $\theta = 0$, $x = y$. Now set $x = b_1 c_2$, $y = b_2 c_1$ and $\theta = A_1 - A_2$, where A_i is the angle opposite side a_i. Then

$$2b_i c_i \cos A_i = b_i^2 + c_i^2 - a_i^2 \,,$$

$$b_i c_i \sin A_i = 2\Delta_i, \text{ for } i = 1, 2.$$

Hence from (1) we get

$$b_1^2 c_2^2 + b_2^2 c_1^2 - \frac{1}{2}(b_1^2 + c_1^2 - a_1^2) \times$$
$$(b_2^2 + c_2^2 - a_2^2) - 8\Delta_1 \Delta_2 \geq 0,$$

which is equivalent to the desired inequality. For equality, we must have $\frac{b_1}{b_2} = \frac{c_1}{c_2}$ and $A_1 = A_2$. Thus the two triangles have to be similar. For other proofs and discussion, see: D. Pedoe, Thinking geometrically, *Amer. Math. Monthly* 77 (1970) 711–721; O. Bottema, M.S. Klamkin, Joint triangle inequalities, *Simon Steven* 48 (1974/5) 3–8.

Problem 464. We can assume, without loss of generality, that $A_1 B_1 \geq A_2 B_2$ which implies that $B_1 A_2 \leq B_2 A_3$. Since also $B_1 B_2 = B_2 B_3$, we get $A_2 B_2 \geq A_3 B_3$ and then $B_2 A_3 \leq B_3 A_4$. (Note that $\angle A_2 = \angle A_3$.) Continuing, we finally obtain

$$A_1 B_1 \geq A_2 B_2 \geq \cdots \geq A_n B_n \geq A_1 B_1.$$

Thus $A_i B_i = $ constant.

Problem 465. Since 0 is obviously good, the largest good number is ≥ 0. Furthermore, if x is good then (1) holds with $a = m$ and $b = 0$, showing that $x \leq \frac{n}{m} \leq 1$. Thus we may restrict our attention to good numbers in the interval $[0,1]$. For any x in this interval there is a unique angle θ, $\frac{\pi}{3} \leq \theta \leq \frac{\pi}{2}$, for which $\cos \theta = \frac{x}{2}$. Hence $x = 2\cos \theta$ is good if

$$m^2 + n^2 - 2mn \cos \theta \geq a^2 + b^2 - 2ab \cos \theta$$
$$\text{for all } \quad 0 \leq a \leq m, \ 0 \leq b \leq n. \qquad (2)$$

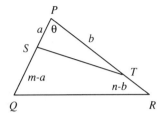

FIGURE 234

This suggests that we consider the following geometric version of (2). See Figure 234.

Let PQR be a triangle with $\overline{PQ} = m$, $\overline{PR} = n$, $\angle QPR = \theta$. For arbitrary a, b let S be on PQ with $\overline{PS} = a$ and T on PR with $\overline{PT} = b$. By the Cosine Law for triangles, (2) is equivalent to

$$\overline{QR}^2 \geq \overline{ST}^2 \tag{3}$$

for all S on PQ and T on PR,

and this is equivalent to $\overline{QR}^2 \geq \overline{PQ}^2$ (for (3) holds if and only if QR is the longest side of $\triangle PQR$). Thus x is good if and only if

$$m^2 + n^2 - mnx = m^2 + n^2 - 2mn\cos\theta \geq m^2,$$

i.e., $0 \leq x \leq \frac{n}{m}$, so the largest good number is $\frac{n}{m}$.

Problem 466. Consider any $(n+1)$-gon with sides $a_1, a_2, \ldots, a_{n+1}$. Then by power mean inequality

$$\left(\frac{a_1^m + a_2^m + \cdots + a_n^m}{n} \right)^{\frac{1}{m}} \geq \frac{a_1 + a_2 + \cdots + a_n}{n}$$

for $m \geq 1$, with equality if and only if $a_1 = a_2 = \cdots = a_n$. Also, by the triangle inequality

$$a_1 + a_2 + \cdots + a_n \geq a_{n+1}.$$

Thus,

$$a_1^m + a_2^m + \cdots + a_n^m \geq \frac{a_{n+1}^m}{n^{m-1}},$$

with equality if and only if $a_1 = a_2 = \cdots = a_n = \frac{1}{n} a_{n+1}$, i.e., the polygon is degenerate. The desired result corresponds to the special case $m = 4$, $n = 3$. (See also Problem 275.)

Problem 467. First note that by the Arithmetic-Geometric Mean inequality, we have

$$x + y = \frac{x}{2} + \frac{x}{2} + y \geq 3 \sqrt[3]{\frac{x^2 y}{4}}$$

with equality if and only if $x = 2y$. Also

$$2x^2 + 2xy + 3y^2$$
$$= \frac{2x^2}{8} + \cdots + \frac{2x^2}{8} + \frac{2xy}{4}$$
$$+ \cdots + \frac{2xy}{4} + y^2 + y^2 + y^2$$
$$\geq 15 \left\{ \left(\frac{2x^2}{8} \right)^8 \left(\frac{2xy}{4} \right)^4 (y^2)^3 \right\}^{\frac{1}{15}}$$
$$= 15 \left(\frac{x^2 y}{4} \right)^{\frac{2}{3}}$$

with equality if and only if $\frac{2x^2}{8} = \frac{2xy}{4} = y^2$ or $x = 2y$. Thus,

$$k = x + y + \left(2x^2 + 2xy + y^2 \right)^{\frac{1}{2}}$$
$$\geq (3 + \sqrt{15}) \sqrt[3]{\frac{x^2 y}{4}}$$

and

$$\max x^2 y = \frac{4k^3}{(3 + \sqrt{15})^3}.$$

Problem 468. Our proof is by induction. Assume the left side of (1) is valid for $m = n$. Then

$$\binom{2n+2}{n+1} = \frac{2(2n+1)}{n+1} \binom{2n}{n}$$
$$> \frac{2(2n+1)4^n}{(n+1)2\sqrt{n}}$$
$$= \frac{2(2n+1)4^n}{\sqrt{4n(n+1)}\sqrt{n+1}}$$
$$> \frac{4^{n+1}}{2\sqrt{n+1}}.$$

Thus (1) is also valid for $n + 1$, and since it is valid for $m = 2$ it is valid for all m. The right-hand inequality of (1) is proven in a similar way.

Problem 469. Suppose (x, y) is a solution. If $x = 0$, we get $y = 0$ or 1 and the solutions $(0, 0)$,

$(0, 1)$. If $x \neq 0$, there is a unique rational number $m \neq -1$ such that $y = mx$, and substitution yields

$$x = \frac{1 + m^2}{1 + m^3}, \quad y = \frac{m(1 + m^2)}{1 + m^3}. \quad (1)$$

Conversely, it is easily verified that (1) yields a solution for any rational $m \neq -1$.

Problem 470. For $0 \leq k \leq n$, the probability of k sixes turning up in a random toss of n fair dice is

$$\binom{n}{k}\left(\frac{5}{6}\right)^{n-k}\left(\frac{1}{6}\right)^k;$$

hence, with $a = \frac{5}{6}$ and $b = \frac{1}{6}$, the required probability is

$$P = \binom{n}{1}a^{n-1}b + \binom{n}{3}a^{n-3}b^3$$

$$+ \binom{n}{5}a^{n-5}b^5 + \cdots$$

= sum of the even-ranked terms in the

expansion of $(a + b)^n$

$$= \frac{1}{2}\{(a + b)^n - (a - b)^n\}$$

$$= \frac{1}{2}\left\{1 - \left(\frac{2}{3}\right)^n\right\}.$$

Problem 471. Let Ma's age be M, Pa's age P, brother's B and mine I. Then

$$M + P + B + I = 83, \quad 6P = 7M, \quad M = 3I.$$

and hence

$$\frac{5}{2}M + B = 83.$$

Since P and M are integers, M must be a multiple of 6. The only values of M satisfying this condition and which make B positive and less than M are $M = 30$ (and then $B = 8$) or $M = 24$ (and then $B = 23$). The second solution would make Ma only one year older than my brother, so that's out. Hence $M = 30$, $B = 8$, $P = 35$ and $I = 10$. ("Me" is fairly bright to be composing such puzzles at 10 years of age!)

Problem 472. Let $a = \prod p_i^{a_i}$, $b = \prod p_i^{b_i}$, $c = \prod p_i^{c_i}$, where the p_i are the prime factors of a, b, c (some of the exponents may be zero). Since

$$(a, b) = \prod p_i^{\min(a_i, b_i)},$$

$$[a, b] = \prod p_i^{\max(a_i, b_i)}, \text{ etc.,}$$

we have to show that

$$2\max(a_i, b_i, c_i) - \max(b_i, c_i)$$

$$- \max(c_i, a_i) - \max(a_i, b_i)$$

$$= 2\min(a_i, b_i, c_i) - \min(b_i, c_i)$$

$$- \min(c_i, a_i) - \min(a_i, b_i).$$

Without loss of generality, let $a_i \geq b_i \geq c_i$ for any particular index i. We then obtain

$$2a_i - b_i - a_i - a_i = 2c_i - c_i - c_i - b_i.$$

It is easy to give extensions using min and max.

Problem 473. Let $n = (abc)$ and N be the desired product. Clearly, none of a, b, c is 5. Further, $c \neq 1$. Otherwise $(cba) < 200$, so $N < 200 \times 10^3 \times 10^3 = 2 \times 10^8$, which is false. We can apply two criteria:

(a) $a \times b \times c$ must end in 6;

(b) since, modulo 3, any number is congruent to the sum of its digits, $N \equiv 35 \equiv 2 \pmod 3$, whence $n \equiv a + b + c \equiv 2 \pmod 3$.

Criterion (a) leaves as possibilities $n = 876, 964, 843, 763, 983, 632, 742, 862, 972$.

Criterion (b) eliminates all but 983 and 632. We must have $n = 983$ and $N = 328, 245, 326$.

Problem 474. We rotate the coordinate system $45°$ (see Figure 235) by letting

$$x + y = \sqrt{2}X,$$

$$y - x = \sqrt{2}Y,$$

and the equation becomes

$$(5X + Y)^2Y = 2X.$$

This is unchanged if we put $-X$, $-Y$ for X, Y. Also Y and X must be of the same sign, and (solving as a quadratic for X) $Y^2 \leq \frac{1}{10}$. The

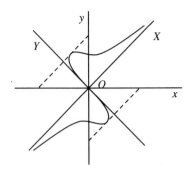

FIGURE 235

only real asymptote is $Y = 0$, and $X = 0$ is an inflexional tangent at the origin.

Problem 475. We have $\dfrac{1}{A_{n+1}} - \dfrac{1}{A_n} = n$. Thus,

$$\frac{1}{A_1} - \frac{1}{A_0} = 0,$$

$$\frac{1}{A_2} - \frac{1}{A_1} = 1,$$

$$\vdots$$

$$\frac{1}{A_{1990}} - \frac{1}{A_{1989}} = 1989,$$

Adding them all up, we obtain

$$\frac{1}{A_{1990}} - \frac{1}{A} = 1 + 2 + \cdots + 1989$$

$$= (995)(1989)$$

and hence

$$A_{1990} = \frac{A}{1 + (995)(1989)A}.$$

Problem 476. Our proof is indirect. Assume all positive remainders $1, 2, \ldots, 100$ are obtained. Let the pairs be denoted by $a_1 b_1, a_2 b_2, \ldots, a_{100} b_{100}$ and let $a_i b_i \equiv c_i \pmod{101}$. Then, by Wilson's Theorem,

$$\prod c_i = 100! \equiv -1 \pmod{101}.$$

Also

$$\prod c_i \equiv \prod a_i b_i \equiv (100!)^2 \equiv 1 \pmod{101},$$

giving a contradiction.

Problem 477. Let $x = y = a$ where a is arbitrary. Then $f(0) = ba^2$. Since $f(0)$ has a single value and a is arbitrary, $b = 0$. A real function exists because $f(x) = mx$ satisfies $f(x - y) = f(x) - f(y)$.

Problem 478. Since any polygon with four or more sides can be subdivided into triangles, the problem reduces to showing how an obtuse triangle can be subdivided into acute triangles. In Figure 236, $\angle BAC$ is obtuse and X is the incenter. The angles marked α, β, γ, δ, ϵ are clearly acute, and angles α, β, γ each exceed $45°$. It follows that all angles at X are acute.

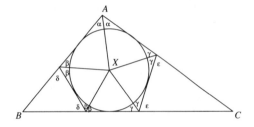

FIGURE 236

Problem 479. The converse is false. We will show that if a tetrahedron is merely isosceles (opposite edges congruent in pairs), hence not necessarily regular, then O, I, and G coincide. Let $ABCD$ be an isosceles tetrahedron with centroid G (see Figure 237) and let

$$a = \overrightarrow{GA}, \quad b = \overrightarrow{GB}, \quad c = \overrightarrow{GC}, \quad d = \overrightarrow{GD}.$$

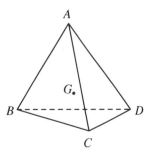

FIGURE 237

Because the tetrahedron is isosceles, we have (using the notation $\vec{V}^2 = \vec{V} \bullet \vec{V} = |\vec{V}|^2$)

$$(\vec{a} - \vec{b})^2 = (\vec{c} - \vec{d})^2,$$
$$(\vec{a} - \vec{c})^2 = (\vec{b} - \vec{d})^2, \qquad (1)$$
$$(\vec{a} - \vec{d})^2 = (\vec{b} - \vec{c})^2.$$

Because G is the centroid, $\vec{a} + \vec{b} + \vec{c} + \vec{d} = \vec{O}$, and hence

$$(\vec{a} + \vec{b})^2 = (\vec{c} + \vec{d})^2,$$
$$(\vec{a} + \vec{c})^2 = (\vec{b} + \vec{d})^2, \qquad (2)$$
$$(\vec{a} + \vec{d})^2 = (\vec{b} + \vec{c})^2.$$

Expanding the squares and adding corresponding equations in (1) and (2) yield

$$\vec{a}^2 + \vec{b}^2 = \vec{c}^2 + \vec{d}^2,$$
$$\vec{a}^2 + \vec{c}^2 = \vec{b}^2 + \vec{d}^2,$$
$$\vec{a}^2 + \vec{d}^2 = \vec{b}^2 + \vec{c}^2,$$

from which $\vec{a}^2 = \vec{b}^2 = \vec{c}^2 = \vec{d}^2$. Thus $|\vec{GA}| = |\vec{GB}| = |\vec{GC}| = |\vec{GD}|$ and G coincides with the circumcenter O.

It is obvious that in an isosceles tetrahedron the four faces are congruent and hence so are the four altitudes (consider the volume!). Since G divides each median (segment from a vertex to the centroid of the opposite face) in the ratio $3 : 1$, the distance of G from any face is $\frac{1}{4}$ the corresponding altitude. Hence G is equidistant from the four faces and thus coincides with the incenter I.

Problem 480. We wish to show that, for any finite set S with the given property, we can find another number of the form $2^r - 3$ which is coprime with every integer in S. Let $p_1, p_2, \ldots, p_k$ be all the primes which divide at least one number in S, and define $r = (p_1 - 1)(p_2 - 1) \cdots (p_k - 1) + 1$. Consider the number $2^r - 3$. Since by Fermat's theorem $2^{p_1 - 1} \equiv 1 \pmod{p_1}$, we have

$$2^r - 3 = 2(2^{p_1 - 1})^{(p_2 - 1) \cdots (p_k - 1)} - 3$$

$$\equiv 2 - 3 = -1 \pmod{p_1},$$

so $p_1 \nmid 2^r - 3$. Similarly $p_i \nmid (2^r - 3)$ for $i \geq 2$, and the result follows.

Problem 481. It follows easily that (with S denoting the sum)

$$4R^2 S = \sum_{r=1}^{n-1} \csc^2 \left(\frac{r\pi}{n} \right).$$

We now give a method which can be used to find S as well as sums of other trigonometric functions, such as $\sum \sec^2 \frac{r\pi}{n}$, $\sum \csc^4 \frac{r\pi}{n}$. Our starting point is the expansion

$$\tan(\theta_1 + \theta_2 + \cdots + \theta_n)$$
$$= \frac{S_1 - S_3 + S_5 - \cdots}{1 - S_2 + S_4 - \cdots},$$

where S_r denotes the sum of the products of $\tan \theta_1, \tan \theta_2, \ldots, \tan \theta_n$, taken r at a time. This can be proved by induction. Now let $\theta_1 = \theta_2 = \cdots = \theta_n = \theta$ to give

$$\tan n\theta = \frac{\binom{n}{1}x - \binom{n}{3}x^3 + \binom{n}{5}x^5 - \cdots}{1 - \binom{n}{2}x^2 + \binom{n}{4}x^4 - \cdots},$$

where $x = \tan \theta$.

Since $\tan n\theta$ vanishes for $\theta = \frac{r\pi}{n}$, $r = 0, 1, \ldots, n - 1$, the roots of

$$\binom{n}{1} - \binom{n}{3}x^2 + \binom{n}{5}x^4 - \cdots = 0$$

are $\tan(\frac{r\pi}{n})$, $r = 1, 2, \ldots, n - 1$. Thus

$$\sum_{r=1}^{n-1} \cot(\frac{r\pi}{n}) = 0,$$

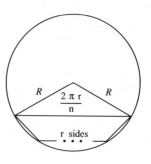

FIGURE 238

$$\sum_{r=1}^{n-1} \cot^2\left(\frac{r\pi}{n}\right) = 0 + 2\frac{\binom{n}{3}}{\binom{n}{1}}$$

$$= \frac{1}{3}(n-1)(n-2).$$

Finally,

$$\sum_{r=1}^{n-1} \csc^2\left(\frac{r\pi}{n}\right) = \sum_{r=1}^{n-1}\left\{1 + \cot^2\left(\frac{r\pi}{n}\right)\right\}$$

$$= \frac{1}{3}(n^2-1)$$

and

$$S = \frac{(n^2-1)}{12R^2}.$$

Rider. Determine $\sum \csc^4\left(\frac{r\pi}{n}\right)$.

Problem 482. We are not concerned with the order of the coins in any row; it is simply a case of distribution into rows. n^2 coins can be distributed into n rows of n each in $\frac{(n^2)!}{(n!)^n}$ ways; but if the silver coins are to be one in each row the number of ways is $n!\frac{(n^2-n)!}{\{(n-1)!\}^n}$. The probability that there should be one silver coin in each row is therefore

$$\frac{n^{n-1}(n-1)!(n^2-n)!}{(n^2-1)!},$$

and the required probability is the complement of this.

Problem 483. (Solution by E.P. Starke, *Math. Mag.* 23 (1949)) Let the given integer be

$$k = a_{n-1}b^{n-1} + a_{n-2}b^{n-2} + \cdots + a_1b + a_0,$$

where b is the base of the system of numeration. Consider the n integers

$$k_1 = a_0, \ k_2 = a_1b + a_0, \ldots,$$

$$k_{n-1} = a_{n-2}b^{n-2} + \cdots + a_1b + a_0, \ k_n = k.$$

If some k_j is divisible by n, obviously the problem is solved by replacing the first $n-j$ digits of k with zeros, thus obtaining k_j. If no $k_{j'}$ is divisible by n, then at least two of the k_j's must be congruent to each other, say

$$k_j \equiv k_i \pmod{n}, \quad i < j.$$

Then the problem is solved by replacing the first $n-j$ digits and the last i digits of k with zeros. By hypothesis, the remaining number has at least one nonzero digit.

If the number of digits of k is less than n, the proposition need not be true. Consider the $b-2$ digit number $111\ldots111$ and take $n = b-1$. If any j digits ($0 \le j < n-1$) are replaced by zeros, the resulting number is congruent (mod n) to $n-j-1$ which cannot be zero.

Problem 484. Let O be the center of the parallelogram, and let OA, OB be drawn parallel to the sides, the angle between them being ω. Any inscribed rhombus must have diagonals intersecting at right angles and the same center O (why?). Let P and Q be two adjacent vertices of the rhombus and suppose $\angle POA = \theta$. Since the area of the rhombus is $2\overline{OP}\cdot\overline{OQ}$, we minimize $\overline{OP}\cdot\overline{OQ}$.

Now $\angle OPA = \pi - (\omega + \theta)$ and $\angle OQB = \frac{\pi}{2} \pm \theta$. (2 cases) By the Law of Sines,

$$\frac{\overline{OP}}{\overline{OA}} = \frac{\sin\omega}{\sin(\omega+\theta)} \quad \text{and} \quad \frac{\overline{OQ}}{\overline{OB}} = \frac{\sin\omega}{\cos\theta},$$

whence

$$\frac{\overline{OP}\cdot\overline{OQ}}{\overline{OA}\cdot\overline{OB}} = \frac{\sin^2\omega}{\sin(\omega+\theta)\cos\theta}$$

$$= \frac{2\sin^2\omega}{\sin(\omega+2\theta) + \sin\omega}.$$

This is minimum when $\omega + 2\theta = \frac{\pi}{2}$, i.e., $\theta = \frac{\pi}{4} - \frac{\omega}{2}$. Hence if OM and ON are drawn perpen-

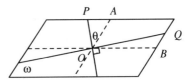

FIGURE 239

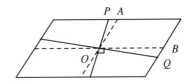

FIGURE 240

dicular to AP and BQ respectively, the diagonals of the minimum rhombus will bisect the angles AOM, BON respectively.

Remarks. The solution assumes that $\overline{AP}$ will be less than $\overline{OB}$, and $\overline{BQ}$ less than $\overline{OA}$. If this is not the case the minimum rhombus will have for one of its diagonals the shorter diagonal of the parallelogram.

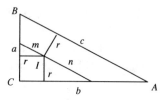

FIGURE 241

Problem 485. From similar triangles in Figure 241, $a = \frac{rc}{n}$, $b = \frac{rc}{m}$, $c^2 = a^2 + b^2$, and by twice the area $ab = r(a + b + c)$. Thus $r = \dfrac{mn}{\sqrt{m^2 + n^2}}$ and

$$\frac{r^2 c^2}{mn} = ab = rc\left(\frac{r}{n} + \frac{r}{m} + 1\right),$$

so $c = m + n + \sqrt{m^2 + n^2}$ and the area is

$$\frac{1}{2} ab = \frac{r^2 c^2}{2mn} = \frac{mn\left(m + n + \sqrt{m^2 + n^2}\right)^2}{2(m^2 + n^2)}.$$

Problem 486. Let p_r stand for the sum of the products of the four letters c, d, e, f, taken r at a time. Then $p_1 = -(a + b)$. Also

$$p_1(p_1^2 - 3p_2) = \sum c^3 - 3p_3$$
$$= -(a^3 + b^3) - 3p_3;$$

therefore

$$p_3 = \frac{1}{3}\{(a + b)^3 - (a^3 + b^3)\} - p_2(a + b)$$
$$= (a + b)(ab - p_2).$$

Hence

$$(a + c)(a + d)(a + e)(a + f)$$
$$= a^4 - a^3(a + b) + a^2 p_2$$
$$+ a(a + b)(ab - p_2) + p_4$$
$$= a^2 b^2 - abp_2 + p_4$$
$$= (b + c)(b + d)(b + e)(b + f).$$

Problem 487. Since $2739726 = 9999 \times 274$, it will be sufficient to prove that every even multiple of 9999 up to the 20000th has the sum of its digits equal to 36.

(i) Consider any multiple of 9999 up to the 10000th. If $x < 10000$, the number $9999(x + 1)$, or $10^4 x + (9999 - x)$, has not more than 8 digits; and if a, b, c, d are the digits of x, those of $9999(x + 1)$ are

$$a,\ b,\ c,\ d,\ 9 - a,\ 9 - b,\ 9 - c,\ 9 - d,$$

so that their sum is 36.

(ii) Consider any even multiple of 9999 from the 10002th to the 20000th. Then since $(9999)(10001) = 10^8 - 1$, if we add to this any odd multiple of 9999 up to the 9999th, i.e., if we add a number whose digits are

$$a,\ b,\ c,\ d,\ 9 - a,\ 9 - b,\ 9 - c,\ 9 - d,$$

($9 - d$ being odd) we get a number whose digits are

$$1,\ a,\ b,\ c,\ d,\ 9 - a,\ 9 - b,\ 9 - c,\ 8 - d,$$

and the sum of these is 36.

Problem 488. Since $|z_i| = |\overline{z_i}|$ and $z_i \overline{z_i} = 1$,

$$|z_2 z_3 + z_3 z_1 + z_1 z_2|$$
$$= |z_1 z_2 z_3|\left|\frac{1}{z_1} + \frac{1}{z_2} + \frac{1}{z_3}\right|$$
$$= \left|\frac{1}{z_1} + \frac{1}{z_2} + \frac{1}{z_3}\right| = |\overline{z_1} + \overline{z_2} + \overline{z_3}|$$
$$= |z_1 + z_2 + z_3|.$$

Thus $k = 1$.

Problem 489. If the bettor backs all the horses, viz., if he bets $\dfrac{b_r}{a_r + b_r}$ to $\dfrac{a_r}{a_r + b_r}$ that the rth horse will win, then, if the rth horse wins, the bettor wins $\dfrac{a_r}{a_r + b_r}$ and loses $\sum \dfrac{b}{a+b} - \dfrac{b_r}{a_r + b_r}$. Therefore his net winnings are

$$\frac{a_r}{a_r + b_r} - \sum \frac{b}{a+b} + \frac{b_r}{a_r + b_r}$$

$$= 1 - \sum \frac{b}{a+b},$$

which is positive and the same whatever horse wins.

Problem 490. The arithmetic mean of any number of positive and unequal quantities is greater than their geometric mean, i.e.,

$$p_1 + p_2 + \cdots + p_m$$
$$> m \sqrt[m]{p_1 p_2 \ldots p_{m-1} p_m}.$$

Applying the same theorem to their reciprocals, we find on multiplication

$$\sum (p) \cdot \sum (\frac{1}{p}) > m^2.$$

Let the p_i be the products of the roots of the given equation taken r at a time, so that

$$\sum p_i = \binom{n}{r} \frac{a_r}{a_0}, \quad \sum \frac{1}{p_i} = \binom{n}{r} \frac{a_{n-r}}{a_n}.$$

Also m, the number of terms in each sum, is $\binom{n}{r}$. Therefore

$$\binom{n}{r} \frac{a_r}{a_0} \binom{n}{r} \frac{a_{n-r}}{a_n} > \binom{n}{r}^2 \quad \text{or} \quad a_r a_{n-r} > a_0 a_n.$$

Problem 491. The given inqualities are equivalent to

$$(u-v)(w-1) \geq 0 \quad \text{and} \quad (w-v)(1-u) \geq 0.$$

If $u = w = 1$, then the inequalities hold for all values of v.

If $u = 1$, $w > 1$, then the inequalities hold if and only if $v \leq u = 1$.

If $w = 1$, $u < 1$, then the inequalities hold if and only if $v \leq w = 1$.

If $u < 1 < w$, then the inequalities hold if and only if $v \leq u$.

Problem 492. Let (u, v, w) and (p, q, r) be the points on the ellipsoid and the plane, respectively. Then we wish to minimize

$$D^2 = (u - p)^2 + (v - q)^2 + (w - r)^2.$$

By Cauchy's inequality

$$D^2(A^2 + B^2 + C^2)$$
$$\geq \{A(u - p) + B(v - q) + C(w - r)\}^2$$
$$= \{1 - (Au + Bv + Cw)\}^2,$$

with equality when (p, q, r) is of the form $(u + \lambda A, v + \lambda B, w + \lambda C)$. Since the plane and the ellipsoid do not intersect, and since the ellipsoid is on the origin side of the plane,

$$1 > Au + Bv + Cw.$$

So, to minimize D, we maximize $Au + Bv + Cw$. Again by Cauchy's inequality

$$\left(\frac{u^2}{a^2} + \frac{v^2}{b^2} + \frac{w^2}{c^2} \right) (a^2 A^2 + b^2 B^2 + c^2 C^2)$$
$$\geq (Au + Bv + Cw)^2,$$

with equality when (u, v, w) is of the form $(\mu A a^2, \mu B b^2, \mu C c^2)$. Hence

$$\max(Au + Bv + Cw)$$
$$= \sqrt{a^2 A^2 + b^2 B^2 + c^2 C^2}$$

and the shortest distance is

$$D_{m,n} = \frac{1 - \sqrt{a^2 A^2 + b^2 B^2 + c^2 C^2}}{\sqrt{A^2 + B^2 + C^2}}.$$

This distance is attained when

$$(u, v, w) = (\mu A a^2, \mu B b^2, \mu C c^2),$$

$$\mu^2 (A^2 a^2 + B^2 b^2 + C^2 c^2) = 1$$

$$(p, q, r) = (u + \lambda A, v + \lambda B, w + \lambda C),$$

$$\lambda = \frac{1 - \sqrt{a^2 A^2 + b^2 B^2 + c^2 C^2}}{A^2 + B^2 + C^2}.$$

Problem 493. We can find the equation of a normal line by first finding the equation of a tangent line. If $y = mx + b$ is tangent to $y^2 = 2ax$, there can only be a double root for the polynomial $(mx + b)^2 - 2ax$. Thus its discriminant must vanish, i.e., $b = \frac{a}{2m}$. Then, if (h, k) is the point of tangency, $k^2 = 2ah$ and $k = mh + \frac{a}{2m}$, leading to $0 = (km - a)^2$. Thus $m = \frac{a}{k}$ and the normal line through (h, k) is

$$y - k = -\frac{k}{a}(x - h).$$

Solving this simultaneously with $y^2 = 2ax$ yields

$$k^2(x - h)^2 - 2ak^2(x - h) + a^2k^2 - 2a^3x = 0,$$

which (using $k^2 = 2ah$) leads to $k^2(x - h) = 2a(a^2 + k^2)$ for the abscissa of the other point of intersection of chord and parabola.

The endpoints of the chord are (h, k) and $\left(\frac{(2a^2 + k^2)^2}{2ak^2}, \frac{-(2a^2 + k^2)}{k}\right)$ and the square L^2 of its length is $\frac{4}{k^4}(k^2 + a^2)^3$. Then

$$\left(\frac{L^2}{4}\right)^{\frac{1}{3}} = \frac{(t^6 + a^2)}{t^4},$$

where $t^3 = k$. Finally, by the Arithmetic-Geometric Mean Inequality

$$\frac{t^6}{2} + \frac{t^6}{2} + a^2 \geq 3\left[\frac{t^{12}a^2}{4}\right]^{\frac{1}{3}}.$$

Thus,

$$\left(\frac{L^2}{4}\right)^{\frac{1}{3}} \geq 3\left[\frac{a^2}{4}\right]^{\frac{1}{3}}$$

(with equality if and only if $t^6 = 2a^2$, or $k^2 = 2a^2$) and

$$L_{\min} = 3a\sqrt{3}.$$

Problem 494. When Alice has n counters, let u_n be her chance of success. Then, as Alice either wins or loses the next game,

$$u_n = \frac{1}{2}u_{n+1} + \frac{1}{2}u_{n-1}$$

or

$$u_{n+1} - 2u_n + u_{n-1} = 0.$$

Hence, observing that $u_0 = 0$, we have

$$u_n - u_{n-1} = u_{n-1} - u_{n-2} = u_{n-2} - u_{n-3}$$

$$= \cdots = u_2 - u_1 = u_1.$$

Therefore $u_n = nu_1$ and, since $u_{a+b} = 1$, $u_1 = \frac{1}{a+b}$. Thus $u_a = \frac{a}{a+b}$.

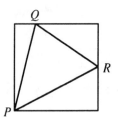

FIGURE 242

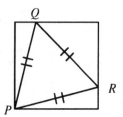

FIGURE 243

Problem 495. If any of three points are inside the square, it can be moved to the boundary without decreasing the distances. With all three points on the boundary, one can move one of the points to a vertex, if not already there, again without decreasing the distances. Thus we are led to consider the configuration of Figure 242, and the problem now is to locate Q and R such that the shortest side of triangle PQR is as large as possible. This occurs when PQR is equilateral (Figure 243). It follows that $\alpha = 15°$ and

$$\overline{PQ} = \overline{QR} = \overline{RP} = 2\sqrt{2 - \sqrt{3}}.$$

Problem 496. Divide the rectangle into two unit squares. By the Pigeon Hole Principle, one of the squares contains 3 of the points. The minimum distance determined by these three points is maximized by the configuration in Figure 243 of Problem 495 and hence the solution is given by Figure 244.

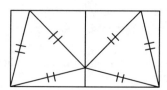

FIGURE 244

Problem 497. A general solution is given by the recursive formula

$$x_r = x_1 x_2 \cdots x_{r-1} + 1,$$

$$1 < r \le n, \quad x_1 = 2.$$

We note that the equation is satisfied for $n = 1$. Assuming that the set $\{x_1, x_2, \ldots, x_r\}$ satisfies the equation for $n = r$, then, for $n = r + 1$ we have

$$\frac{1}{x_1} + \frac{1}{x_2} + \cdots + \frac{1}{x_{r+1}} + \frac{1}{x_1 x_2 \cdots x_{r+1}}$$

$$= 1 - \frac{1}{x_1 \cdots x_r} + \frac{1}{x_{r+1}} + \frac{1}{x_1 \cdots x_{r+1}}$$

$$= 1 + \frac{1}{x_{r+1}} - \frac{x_{r+1} - 1}{x_1 \cdots x_{r+1}} = 1,$$

completing the induction.

The solution is unique (except for permutation of the x_i) for $n = 1, 2$. For each $n > 2$ there exists an infinite set of solutions. (See also Problem 179.)

Problem 498. Cut the tetrahedron $ABCD$ along AB, AC, and AD and lay out flat as in Figure 245. Since ABA, ACA and ADA are all straight, B, D and C are the midpoints of the larger outer triangle. Then $\overline{AB} = \overline{CD}$, $\overline{BC} = \overline{DA}$ and $\overline{CA} = \overline{BD}$.

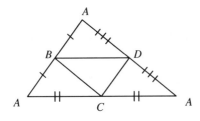

FIGURE 245

News item in the Bozeman (Montana) Chronicle:
Average American is growing older.

Problem 499. Let the ith term of the given sequence

$$\{3, -2, 12, -8, 48, -32, 192, \ldots\}$$

be a_i. We have

$$1 = a_1 + a_2, \quad 2 = a_2 + a_3 + a_4, \quad 3 = a_1.$$

Suppose $n \ge 4$ and all integers $1, 2, \ldots, n - 1$ can be written as the sum of distinct $a_i's$. There are four cases:

(1) $n = 4k$ where $1 \le k < n$. If $k = \sum_{i \in S} a_i$, where $S \subset \{1, 2, 3, \ldots\}$, then $n = \sum_{i \in S} a_{i+2}$ is the required representation;

(2) $n = 4k + 1$ where $1 \le k < n$. If $k = \sum_{i \in S} a_i$, then $n = a_1 + a_2 + \sum_{i \in S} a_{i+2}$;

(3) $n = 4k - 2$ where $2 \le k < n$. If $k = \sum_{i \in S} a_i$, then $n = a_2 + \sum_{i \in S} a_{i+2}$;

(4) $n = 4k + 3$ where $1 \le k < n$. If $k = \sum_{i \in S} a_i$, then $n = a_1 + \sum_{i \in S} a_{i+2}$.

500. For E^3, consider the 4 rays from the center of a regular tetrahedron to the vertices. By symmetry all the angles between the pairs of rays are equal. For E^n, we consider the $n + 1$ rays from the center of a regular simplex (n-dimensional tetrahedron). Let $\vec{V_1}, \vec{V_2}, \ldots, \vec{V}_{n+1}$, denote vectors from the center to the vertices. By symmetry (show!) the circumcenter and centroid of the simplex are the same point. Thus $|\vec{V_i}| = R$, $i = 1, 2, \ldots, n + 1$ and

$$\vec{V_1} + \vec{V_2} + \cdots + \vec{V}_{n+1} = \vec{0}.$$

Then

$$(\vec{V_1} + \vec{V_2} + \cdots + \vec{V}_{n+1})$$
$$\bullet (\vec{V_1} + \vec{V_2} + \cdots + \vec{V}_{n+1}) = 0$$

or

$$(n + 1)R^2 + 2\binom{n+1}{2} \vec{V_i} \bullet \vec{V_j} = 0.$$

Finally, if θ is the common angle between any vector pair $(\vec{V_i}, \vec{V_j})$, then $\cos \theta = -\frac{1}{n}$ and $\theta = \arccos(-\frac{1}{n})$.

THE TOOL CHEST

While the ingenuity of the solver alone will unravel many of the problems in this collection, many others require elementary knowledge in some mathematical area. Herewith is a brief list of "tools" that may be needed; your textbook should provide fuller details of those facts you do not wish to (or cannot) verify yourself.

A. Combinatorics

A 1. "r factorial"

$$r! = r(r-1) \cdot \ldots \cdot 2 \cdot 1$$

is the number of ways of listing (in order) r distinct objects.

A 2. "n choose r"

$$\binom{n}{r} = \frac{n(n-1)\ldots(n-r-1)}{1 \cdot 2 \cdot \ldots \cdot r}$$

is the number of ways of choosing r objects from among n distinct objects, order not taken into account.

A 3. Set notation: Let A and B be sets. Then:
"A union B"

$$A \cup B = \{x \mid x \in A \quad \text{or} \quad x \in B\};$$

"A intersection B"

$$A \cap B = \{x \mid x \in A \quad \text{and} \quad x \in B\};$$

"complement of A"

$$\backslash A = \{x \mid x \notin A\};$$

Note that $A \cap B \subseteq A \cup B$ ($\subseteq$ means "is contained in").

A 4. The empty set, which contains no elements, is denoted by $\varnothing$. This set is sometimes called the null or void set.

A 5. The number of elements in a set A is denoted by $\#A$, and also by $|A|$.

A 6. Pigeonhole Principle (Dirichlet box principle): If n objects are distributed among fewer than n boxes, some box must contain at least two of the objects. (With an inspired choice of objects and boxes, several problems are rendered completely innocuous by this principle.)

A 7. The Principle of Inclusion-Exclusion. Let S be a set of n objects, and let $P_1, P_2, \ldots, P_m$ be m properties such that for each object $x \in S$ and each property P_i, either x has property P_i or x does not have property P_i. Let $f(i, j, \ldots, k)$ denote the number of elements of S each of which has properties $P_i, P_j, \ldots, P_k$ (and possibly others as well). Then the number of elements of S each having none of the properties $P_1, P_2, \ldots, P_m$ is

$$n - \sum_{1 \leq i \leq m} f(i) + \sum_{1 \leq i < j \leq m} f(i, j)$$

$$+ \sum_{1 \leq i < j < \ell \leq m} f(i, j, \ell)$$

$$+ \cdots + (-1)^m f(1, 2, \ldots, m).$$

B. Arithmetic

B 1. The division algorithm: If a and b are any positive integers, there are nonnegative integers

q (the quotient) and r (the remainder) such that $a = bq + r$ and $0 \leq r < b$.

B 2. Greatest common divisor: If a and b are any integers, we say that "a divides b" or "b is a multiple of a" if there is an integer c with $ac = b$. An integer a is a common divisor of m and n if a divides both m and n. The greatest common divisor of m and n, denoted by (m, n) or by $\gcd(m, n)$ is the largest integer which divides both m and n; any common divisor (of m and n) divides (m, n).

B 3. Two positive integers are coprime (or relatively prime) if their greatest common divisor is 1.

B 4. Let b be a positive integer greater than 1. Every positive integer a can be written $a = a_0 + a_1 b + a_2 b^2 + \cdots + a_n b^n$ where $0 \leq a_i \leq b - 1$ for each i. The digital representation of a in base b is

$$a = a_n a_{n-1} \ldots a_1 a_0.$$

(The context will in general preclude confusion with the notation for multiplication.)

B 5. $[x]$ is the greatest integer not exceeding the real number x, e.g., $[-1.5] = -2$, $[3] = [\pi] = 3$. Note that, always, $x - 1 < [x] \leq x$. If n, k and r are positive integers with $1 \leq r \leq k-1$, $\left[\frac{n}{k}\right]$ is the number of positive multiples of k not exceeding n; $\left[\frac{n+r}{k}\right]$ is the number of positive integers not exceeding n which when divided by k leave a remainder $k - r$.

B 6. $a \equiv b \pmod{m}$, "a is congruent to b modulo m", is the notation which expresses the fact that a and b leave the same remainder when divided by m (i.e., $a - b$ is a multiple of m). This relation satisfies the following laws:

If $a \equiv b \pmod{m}$ and $c \equiv d \pmod{m}$ then $a + c \equiv b + d \pmod{m}$ and $ac \equiv bd \pmod{m}$.

B 7. Squares:

$$n^2 \equiv \begin{cases} 0 \pmod{4} & \text{when } n \text{ is even,} \\ 1 \pmod{4} & \text{when } n \text{ is odd.} \end{cases}$$

$$n^2 \equiv \begin{cases} 0 \pmod{8} & \text{when } n \equiv 0 \pmod{4}, \\ 4 \pmod{8} & \text{when } n \equiv 2 \pmod{4}, \\ 1 \pmod{8} & \text{when } n \text{ is odd.} \end{cases}$$

B 8. Fermat's Little Theorem. Let a and n be two relatively prime integers. Then $a^{\varphi(n)} \equiv 1 \pmod{n}$ where $\varphi(n)$ is the number of integers among $\{1, 2, 3, \ldots, n - 1\}$ which are relatively prime to n. If $n = p$ a prime, then

$$a^{p-1} \equiv 1 \pmod{p} \text{ whenever } \gcd(a, p) = 1$$

$$a^p \equiv a \pmod{p} \text{ for every integer } a.$$

If k is the smallest positive integer for which $a^k \equiv 1 \pmod{n}$, then k divides $\varphi(n)$.

B 9. Figurate numbers. If dots are arranged in an array in the shape of a regular polygon, their number is a figurate number. The triangular numbers $1, 3, 6, 10, 15, \ldots$, square numbers $1, 4, 9, 16, 25, \ldots$ and pentagonal numbers $1, 5, 12, 22, \ldots$ are illustrated in Figure 246.

B 10. Linear diophantine equation. Let a, b, c be integers and let g be the greatest common divisor of a and b (B2). Then the equation $ax + bx = c$ has a solution in integers x and y if and only if g divides c. If (x_0, y_0) is one solution, then

$$\left\{ (x_0 + \frac{bt}{g}, y_0 - \frac{at}{g}) \mid t \text{ an integer} \right\}$$

is a complete set of integer solutions.

In particular, if a and b are coprime (relatively prime, B4), then $ax + by = 1$ is solvable for integers x and y. Conversely, if there are integers x and y for which $ax + by = 1$ then a and b must be coprime.

B 11. Quadratic residue. Let m be a positive integer and a be an integer coprime with m. Then a is a quadratic residue (modulo m) if and only if for some integer x, $x^2 \equiv a \pmod{m}$. If a is not a quadratic residue, then a is called a quadratic nonresidue. For the first few primes p, we indicate the positive quadratic residues mod (p) not exceeding p:

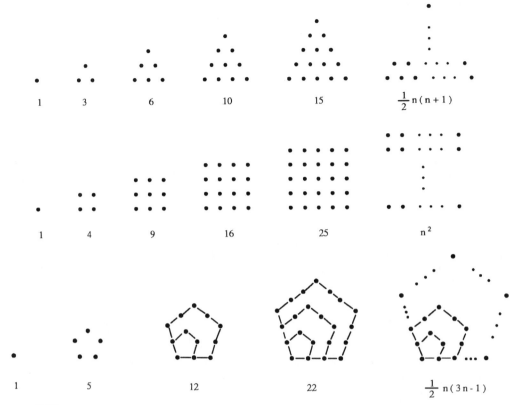

FIGURE 246

p	quadratic residues
2	1
3	1
5	1, 4
7	1, 2, 4
11	1, 3, 4, 5, 9
13	1, 3, 4, 9, 10, 12
17	1, 2, 4, 8, 9, 13, 15, 16
19	1, 4, 5, 6, 7, 9, 11, 16, 17

B 12. Casting out nines. Let $a_r a_{r-1} \ldots a_1 a_0$ be the digital representation of a positive integer to base 10. Then the remainder of this integer under division by 9 is equal to that of $a_r + a_{r-1} + \cdots + a_1 + a_0$. In particular, a number is divisible by 9 if and only if the sum of its digits is divisible by 9.

B 13. Divisibility by 11. Let $a_r a_{r-1} \ldots a_1 a_0$ be the digital representation of a positive integer.

Then the integer is congruent to $a_0 - a_1 + a_2 - a_3 + \ldots + (-1)^r a_r \pmod{11}$.

B 14. Wilson's Theorem. The positive integer n is prime if and only if $(n-1)! \equiv -1 \pmod{n}$.

C. Algebra

C 1. Summation

$$\sum_{k=m}^{n} a_k = a_m + a_{m+1} + \cdots + a_n$$

$$= \sum_{k=m+1}^{n+1} a_{k-1} = \sum_{i=m}^{n} a_i;$$

$$\sum_{k=1}^{n} k = \frac{n(n+1)}{2};$$

$$\sum_{k=1}^{n} k^2 = \frac{n(n+1)(2n+1)}{6}.$$

Sum of a Geometric Progresion:

$$\sum_{k=0}^{n} ar^k = \frac{a(1 - r^{n+1})}{1 - r}.$$

Sum of an Arithmetic Progression:

$$\sum_{k=0}^{n} (a + kd) = \frac{n+1}{2}\{2a + nd\}.$$

$$\sum_{k=1}^{n} \frac{1}{k(k+1)} = \sum_{k=1}^{n} \frac{1}{k} - \frac{1}{k+1}$$

$$= 1 - \frac{1}{n+1}.$$

If the quantity a is independent of the index of summation, then

$$\sum_{i=m}^{n} a = (n - m + 1)a,$$

$n - m + 1$ being the number of terms of the sum.

C 2. When summing quantities which are doubly indexed, the order of summation may be reversed. For example:

$$\sum_{i=1}^{n}\sum_{j=1}^{i} a_{ij}$$

$$= \sum_{i=1}^{n}(a_{i1} + a_{i2} + \cdots + a_{ii})$$

$$= (a_{11}) + (a_{21} + a_{22}) + (a_{31} + a_{32} + a_{33})$$

$$+ \cdots + (a_{n1} + a_{n2} + \cdots + a_{nn})$$

$$= (a_{11} + a_{21} + a_{31} + \cdots + a_{n1})$$

$$+ (a_{22} + a_{32} + \cdots + a_{n2})$$

$$+ (a_{33} + \cdots a_{n3}) + \cdots + a_{nn}$$

$$= \sum_{j=1}^{n}(a_{jj} + a_{j+1,j} + \cdots + a_{nj})$$

$$= \sum_{j=1}^{n}\sum_{i=j}^{n} a_{ij}.$$

This may be visualized by considering a_{ij} as a function of the ordered pair (i, j) which represents a point in the plane. The domain of the function is the set $\{(i,j) \mid 1 \le j \le i \le n\}$, which can be pictured as in Figure 247.

$i = 1$	$\bullet$			
$i = 2$	$\bullet$	$\bullet$		
$i = 3$	$\bullet$	$\bullet$	$\bullet$	
$i = 4$	$\bullet$	$\bullet$	$\bullet$	$\bullet$
$i = 5$	$\bullet$	$\bullet$	$\bullet$	$\bullet$	$\bullet$
	$\vdots$	$\vdots$	$\vdots$	$\vdots$	$\vdots$
	$j = 1$	$j = 2$	$j = 3$	$j = 4$	$j = 5$

FIGURE 247 $\sum_{i=1}^{n}\sum_{j=1}^{i} a_{ij}$ sums a_{ij} across each row and adds the row sums.
$\sum_{j=1}^{n}\sum_{i=j}^{n} a_{ij}$ sums a_{ij} down each column and adds the column sums;

Other instances of double summation can be handled similarly.

C 3. Theory of the quadratic. Let a, b, c be real numbers with $a \ne 0$. The roots of the quadratic polynomial $ax^2 + bc + c$ are:

distinct and real if $b^2 - 4ac > 0$;
coincident and real if $b^2 - 4ac = 0$;
distinct and non real if $b^2 - 4ac < 0$.

The sum of the roots is $-\frac{b}{a}$ and the product is $\frac{c}{a}$. If $b^2 - 4ac < 0$, then for all x

$$ax^2 + bx + c$$
$$= a\left\{\left(x + \frac{b}{2a}\right)^2 + \frac{1}{4a^2}(4ac - b^2)\right\}$$

has the same sign as a. The roots are given by the expressions

$$\frac{-b + \sqrt{b^2 - 4ac}}{2a} \quad \text{and} \quad \frac{-b - \sqrt{b^2 - 4ac}}{2a}.$$

C 4. Polynomials. The Remainder Theorem asserts that, when a polynomial $p(x)$ is divided by $(x - a)$, the remainder is $p(a)$. The Factor Theorem follows immediately: a is a root of $p(x)$ if and only if $(x-a)$ divides evenly into $p(x)$. Any polynomial of degree n has at least one complex root and at most n roots. In fact, a polynomial $p(x)$ can assume any given value for at most finitely many distinct values of x.

C 5. Polynomials (roots and coefficients). Let $a_n x^n + \cdots + a_1 x + a_0$ be a polynomial with roots

$r_1, r_2, \ldots, r_n$ (which may be repeated). Then

$$a_n x^n + \cdots + a_1 x + a_0$$
$$= a_n(x - r_1)(x - r_2) \cdots (x - r_n)$$

and the coefficients are related to the roots as follows:

$$r_1 + r_2 + \cdots + r_n = -\frac{a_{n-1}}{a_n}$$

$$\sum_{i \neq j} r_i r_j = \frac{a_{n-2}}{a_n}$$

$$\sum r_{i_1} r_{i_2} \cdots r_{i_s} = (-1)^s \frac{a_{n-s}}{a_n}.$$

(the sum has $\binom{n}{s}$ terms)

$$\vdots$$

$$r_1 r_2 \cdots r_n = (-1)^n \frac{a_0}{a_n}.$$

C 6. Sums of powers of roots of a polynomial. Let $a_n x^n + a_{n-1} x^{n-1} + \cdots + a_1 x + a_0$ be a polynomial with roots $r_1, r_2, \ldots, r_n$. Let $s_k = r_1^k + r_2^k + \cdots + r_n^k$ $(k = 0, 1, 2, \ldots)$. We have

$$s_0 = n \qquad s_1 = -\frac{a_{n-1}}{a_n}$$

$$s_2 = \frac{a_{n-1}^2 - 2a_{n-2}a_n}{a_n^2}$$

In general, for $1 \leq k \leq n - 1$,

$$a_n s_k + a_{n-1} s_{k-1}$$
$$+ \cdots + a_{n-k+1} s_1 + k a_{n-k} = 0,$$

and for $n \leq k$,

$$a_n s_k + a_{n-1} s_{k-1} + \cdots a_0 s_{n-k} = 0.$$

C 7. Polynomials (repeated roots). Let $p(x)$ be a polynomial and r a root. We say that r is a root of multiplicity m if and only if for some polynomial $q(x)$,

$$p(x) = (x - r)^m q(x), \qquad q(r) \neq 0.$$

A double root is a root of multiplicity 2. A triple root is a root of multiplicity 3. A root of multiplicity 1 is said to be simple.

If $p(x) = a_n x^n + a_{n-1} x^{n-1} + \cdots + a_2 x^2 + a_1 x + a_0$, the derivative $p'(x)$ of $p(x)$ is defined by

$$p'(x) = n a_n x^{n-1} + (n-1) a_{n-1} x^{n-2}$$
$$+ \cdots + 2 a_2 x + a_1.$$

$p'(x)$ can also be written as

$$\lim_{t \to x} \frac{p(t) - p(x)}{t - x}.$$

A root r of $p(x)$ has multiplicity at least 2 if and only if $p'(r) = 0$. (One way to see this is to write

$$p(t) = (t - r)f(t).$$

Now, if r has multiplicity at least 2, then $f(r) = 0$, and hence

$$p'(r) = \lim_{t \to r} f(t) = f(r) = 0.$$

On the other hand, suppose $p'(r) = 0$. Then

$$f(r) = \lim_{t \to r} f(t) = p'(r) = 0,$$

so that $f(t) = (t - r)g(t)$ for some polynomial $g(t)$. Hence,

$$p(t) = (t - r)^2 g(t)$$

and the multiplicity of r is at least 2.)

C 8. Factorization of polynomials in several variables. If $P(x_0, x_1, \ldots, x_n)$ is a given polynomial to be factored, one way to find factors is to see whether P vanishes identically when one variable x_0 is replaced by a polynomial $Q(x_1, x_2, \ldots, x_n)$ in the remaining variables. If so, then $x_0 - Q(x_1, x_2, \ldots, x_n)$ will be a factor of P.

C 9. Factorizations

$$x^n - y^n = (x - y) \times$$
$$(x^{n-1} + x^{n-2}y + \cdots + xy^{n-2} + y^{n-1})$$

for positive integral n;

$$x^n + y^n = (x + y) \times$$
$$(x^{n-1} - x^{n-2}y + \cdots - xy^{n-2} + y^{n-1})$$

for n odd.

C 10. Polynomials (division and remainder). Let $u(x)$ and $v(x)$ be two polynomials. Then there is a unique pair of polynomials $q(x)$ (the quotient) and $r(x)$ (the remainder) for which $u(x) = q(x)v(x) + r(x)$ and degree $r(x) <$ degree $v(x)$.

C 11. Homogeneous polynomial. A polynomial in several variables is homogeneous if and only if each term is of the same degree. More technically, $p(x_1, x_2, \ldots x_n)$ is homogeneous of degree k if and only if

$$p(tx_1, tx_2, \ldots, tx_n) \equiv t^k p(x_1, x_2, \ldots, x_n).$$

For example, in three variables x, y, z,

$ax + by + cz$ is homogeneous of degree 1,

$ax^2 + by^2 + cz^2 + hxy + gxz + fyz$ is homogeneous of degree 2 (a, b, c, f, g, h are constant coefficients).

C 12. Symmetric polynomials. A polynomial in several variables is symmetric if and only if it remains unchanged under any permutation (or interchanging) of the variables.

The following are the elementary symmetric functions in 2, 3 and n variables:

$x + y, \qquad xy;$

$x + y + z, \qquad xy + xz + yz, \qquad xyz;$

$$\sum_{1 \le i \le n} x_i = x_1 + x_2 + \cdots + x_n,$$

$$\sum_{1 \le i < j \le n} x_i x_j = x_1 x_2 + x_1 x_3 + \cdots + x_1 x_n$$

$$+ x_2 x_3 + \cdots + x_{n-1} x_n,$$

$$\sum_{1 \le i < j < k \le n} x_i x_j x_k,$$

$$\vdots$$

$$x_1 x_2 \cdots x_n.$$

C 13. Every root of a polynomial with integer coefficients and leading coefficient equal to 1 which is rational is, in fact, an integer. Another way of putting this is to say that, if polynomial

with integer coefficients and leading coefficient equal to 1 has a non-integer root, then that root is irrational or nonreal.

C 14. Descartes' Rule of Signs. Let

$$a_n x^n + a_{n-1} x^{n-1} + \cdots a_1 x + a_0$$

be a polynomial with real coefficients. Write down in order the signs of the nonzero coefficients ($+$ or $-$). The number of positive roots of the polynomial does not exceed the number of changes of signs. Thus, if all nonzero coefficients are positive, there are no positive roots. For example, the number of positive roots of $8x^9 - 7x^6 - 4x^5 + 3x^3 + 1$ does not exceed 2.

C 15. Intermediate value theorem for polynomials. If a polynomial with real coefficients assumes a negative value at a real number u and a positive value at a real number v, then there is at least one root of the polynomial which lies between u and v. (This is a particular instance of a theorem holding for continuous functions such as $\sin x$, e^x or $\log x$ ($x > 0$).)

C 16. Polynomials (determined by values). Let n be a positive integer and let

$$(a_0, b_0), \ (a_1, b_1), \ldots, (a_n, b_n)$$

be $n + 1$ pairs of complex numbers. Then there is exactly one polynomial p of degree not exceeding n for which

$$p(a_i) = b_i \qquad (i = 0, \ldots, n).$$

More informally, a polynomial of degree no more than n is uniquely determined by specifying $n + 1$ of its values.

C 17. Complex numbers. Every complex number $x + yi$ can be written in the form

$$r(\cos \theta + i \sin \theta)$$

where $r = \sqrt{x^2 + y^2}$ is the absolute value and $\theta = \arctan \frac{y}{x}$ is the argument. The expression $\cos \theta + i \sin \theta$ is abbreviated to cisθ.

C 18. Roots of unity. The polynomial $x^n - 1$ has n distinct roots:

$$1, \operatorname{cis} \frac{2\pi}{n}, \operatorname{cis} \frac{4\pi}{n}, \ldots,$$
$$\operatorname{cis} \frac{2s\pi}{n}, \ldots, \operatorname{cis} \frac{2(n-1)\pi}{n}.$$

These are called the nth roots of unity. Each nth root of unity except 1 is a root of the polynomial

$$x^{n-1} + x^{n-2} + \cdots + x^2 + x + 1.$$

C 19. Conjugates. If $a + b\sqrt{d}$ is a surd with a and b rational, its conjugate is $a - b\sqrt{d}$. The product of the surd with its conjugate is the rational $a^2 - b^2 d$. Likewise, the product of a complex number $x + yi$ (x and y real) with its conjugate $x - yi$ is the positive real number $x^2 + y^2$.

C 20. Matrices. A rectangular array of numbers

$$\begin{pmatrix} a_{11} & a_{12} & a_{13} & \cdots & a_{1n} \\ a_{21} & a_{22} & a_{23} & \cdots & a_{2n} \\ & & \vdots & & \\ a_{m1} & a_{m2} & a_{m3} & \cdots & a_{mn} \end{pmatrix}$$

with m rows and n columns is called an $m \times n$ matrix. The sum and product of 2×2 and 3×3 matrices are given by

$$\begin{pmatrix} a & b \\ c & d \end{pmatrix} + \begin{pmatrix} p & q \\ r & s \end{pmatrix} = \begin{pmatrix} a+p & b+q \\ c+r & d+s \end{pmatrix}$$

$$\begin{pmatrix} a & b \\ c & d \end{pmatrix} \begin{pmatrix} p & q \\ r & s \end{pmatrix} = \begin{pmatrix} ap+br & aq+bs \\ cp+dr & cq+ds \end{pmatrix}$$

$$\begin{pmatrix} a & b & c \\ d & e & f \\ g & h & k \end{pmatrix} + \begin{pmatrix} p & q & r \\ u & v & w \\ x & y & z \end{pmatrix} = \begin{pmatrix} a+p & b+q & c+r \\ d+u & e+v & f+w \\ g+x & h+y & k+z \end{pmatrix}$$

$$\begin{pmatrix} a & b & c \\ d & e & f \\ g & h & k \end{pmatrix} \begin{pmatrix} p & q & r \\ u & v & w \\ x & y & z \end{pmatrix} = \begin{pmatrix} ap+bu+cx & aq+bv+cy & ar+bw+cz \\ dp+eu+fx & dq+ev+fy & dr+ew+fz \\ gp+hu+kx & gq+hv+ky & gr+hw+kz \end{pmatrix}$$

These definitions can be generalized to the sum of two $m \times n$ matrices and the product of an $m \times n$ and an $n \times k$ matrix. Multiplication is associative; addition is distributive, but not commutative.

C 21. Let

$$a_1 x + b_1 y + c_1 z = d_1$$
$$a_2 x + b_2 y + c_2 z = d_2$$
$$a_3 x + b_3 y + c_3 z = d_3$$

be a linear system of three linear equations in three unknowns x, y, z. This system is uniquely solvable for x, y, z if and only if the determinant

$$\det \begin{pmatrix} a_1 & b_1 & c_1 \\ a_2 & b_2 & c_2 \\ a_3 & b_3 & c_3 \end{pmatrix},$$

which is by definition equal to $a_1 b_2 c_3 - a_1 b_3 c_2 + a_2 b_3 c_1 - a_2 b_1 c_3 + a_3 b_1 c_2 - a_3 b_2 c_1$, does not vanish. If the determinant vanishes, then the system may have either no solution or infinitely many solutions.

D. Inequalities

D 1. The absolute value $|x|$ of a real number is defined by

$$|x| = \begin{cases} x & \text{if } x \geq 0, \\ -x & \text{if } x < 0. \end{cases}$$

We have that $|x| \geq 0$ and $-|x| \leq x \leq |x|$ for each real x. If x and y are real,

$$|x + y| \leq |x| + |y|,$$
$$|xy| = |x||y|,$$
$$||x| - |y|| \leq |x - y|.$$

D 2. Every square or sum of squares of real quantities is nonnegative.

D 3. The arithmetic mean of nonnegative reals is at least as great as their geometric mean, i.e., if $a_1, a_2, \ldots, a_n \geq 0$ then

$$\frac{a_1 + a_2 + \ldots + a_n}{n} \geq (a_1 a_2 \ldots a_n)^{\frac{1}{n}}.$$

In particular, if $a > 0, b > 0$, then $a + b \geq 2\sqrt{ab}$.

D 4. Cauchy-Schwarz Inequality. Given real numbers $a_1, a_2, \ldots, a_n, b_1, b_2, \ldots, b_n$,

$$(a_1 b_1 + a_2 b_2 + \cdots + a_n b_n)^2$$
$$\leq (a_1^2 + a_2^2 + \cdots + a_n^2)(b_1^2 + b_2^2 + \cdots + b_n^2).$$

D 5. Minkowski inequality. Let $p > 1$. Then, for positive real numbers $a_1, a_2, \ldots, a_n$, $b_1, b_2, \ldots, b_n$,

$$\left((a_1 + b_1)^p + (a_2 + b_2)^p \right.$$
$$\left. + \cdots + (a_n + b_n)^p\right)^{\frac{1}{p}}$$
$$\leq (a_1^p + a_2^p + \cdots + a_n^p)^{\frac{1}{p}}$$
$$+ (b_1^p + b_2^p + \cdots + b_n^p)^{\frac{1}{p}}.$$

D 6. Hölder inequality. Let p, q be real numbers exceeding 1 which satisfy

$$\frac{1}{p} + \frac{1}{q} = 1.$$

Then, for positive real numbers $a_1, a_2, \ldots, a_n$, $b_1, b_2, \ldots, b_n$,

$$(a_1 b_1 + a_2 b_2 + \cdots + a_n b_n)$$
$$\leq (a_1^p + a_2^p + \cdots + a_n^p)^{\frac{1}{p}} \times$$
$$(b_1^q + b_2^q + \cdots + b_n^q)^{\frac{1}{q}}.$$

When $p = q = 2$, we have the Cauchy-Schwarz Inequality (D 4).

D 7. Generalized means. Let $a_1, a_2, \ldots, a_n$ be any positive numbers for which $a_1 + a_2 + \cdots + a_n = 1$. For positive numbers $x_1, x_2, \ldots, x_n$ we define various means:

$$M_0 = \prod_{k=1}^{n} x_k^{a_k}$$

$$M_t = \left(\sum_{k=1}^{n} a_k x_k^t\right)^{\frac{1}{t}} \quad \text{(for } t \neq 0, t \text{ real)}$$

$$M_\infty = \max_k x_k \qquad M_{-\infty} = \min_k x_k.$$

M_1 is the weighted arithmetic mean, M_0 the weighted geometric mean and M_{-1} the weighted harmonic mean. For $s \leq t$, it is true that

$$M_{-\infty} \leq M_s \leq M_t \leq M_\infty.$$

A frequently occurring case is the one for which $a_1 = a_2 = \ldots = a_n = \frac{1}{n}$. (cf. D.3)

D 8. Jensen Inequality. Let $f(x)$ be a real-valued function of a real variable. $f(x)$ is convex if and only if for all u and v in the domain of f and all $t, 0 \leq t \leq 1$,

$$f(tu + (1 - t)v) \leq tf(u) + (1 - t)f(v).$$

The graph of a convex function looks like Figure 248.

A convex function $f(x)$ has the property: for any $u_1, u_2, \ldots, u_n$, and any $a_1, a_2, \ldots, a_n$ with sum $a_1 + \cdots + a_n = 1$ and $a_i \geq 0$,

$$f(a_1 u_1 + a_2 u_2 + \cdots + a_n u_n)$$
$$\leq a_1 f(u_1) + a_2 f(u_2) + \cdots + a_n f(u_n).$$

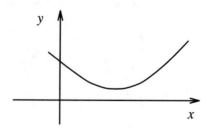

FIGURE 248

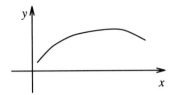

FIGURE 249

A function $f(x)$ is concave if for all u and v and all t, $0 \le t \le 1$,

$$f(tu + (1-t)v) \ge tf(u) + (1-t)f(v);$$

indeed, for a concave function, with u_i and a_i as above,

$$f(a_1u_1 + \cdots + a_nu_n)$$
$$\ge a_1f(u_1) + \cdots + a_nf(u_n).$$

The graph of a concave function looks like Figure 249.

D 9. Particular inequalities which occur frequently enough to be committed to memory are
   (a) $t(1-t) \le \frac{1}{4}$ for all real t;
   (b) $t + t^{-1} \ge 2$ for all real $t > 0$.

E. Geometry and Trigonometry

Numerous results are required from geometry. Those not stated in full can be found in any standard geometry text.
   $\overline{AB}$ denotes the length of a line segment AB.

E 1. The sum of the interior angles of convex n-gon is $2n - 4$ right angles; the sum of the exterior angles is 4 right angles. In particular the sum of the angles of a triangle is $180°$. Any exterior angle of a triangle is equal to the sum of the interior and opposite angles.

E 2. Triangle congruence theorems.

E 3. Theorems about similar triangles.

E 4. Pythagorean Theorem.

E 5. Theorems regarding the angles made by a tranversal with parallel lines. Parallel lines cut off proportional segments of two transversals.

E 6. Triangles on equal bases between the same parallels are equal in area.

E 7. The area of any triangle contained in a parallelogram is at most half the area of the parallelogram. The notation $[ABC]$ and $[ABCD]$ may be used to denote the area of triangle ABC and the area of quadrilateral $ABCD$.

E 8. The length of any side of a triangle is less than the sum of the lengths of the other two sides.

E 9. The line joining the midpoints of two sides of a triangle is parallel to the third side.

E 10. Let ABC be any triangle and D be a point in BC such that AD bisects the angle A. Then $\overline{AB} : \overline{BD} :: \overline{AC} : \overline{CD}$. Suppose the external bisector of A meets BC produced at E. Then $\overline{AB} : \overline{BE} :: \overline{AC} : \overline{CE}$.

E 11. The angles subtended at the circumference of a circle by the same or by equal arcs are equal. The angle subtended at a point outside, respectively inside, the circle by an arc is less than, respectively greater than, the angle subtended at the circumference. The angle subtended at the circumference of a circle by the diameter is right. The locus of points at which a segment subtends a given angle are two circular arcs terminating in the points. The locus of a vertex of a right-angled triangle with fixed hypotenuse is a circle with the hypotenuse as diameter.

E 12. Let ABC be a right-angled triangle, D the midpoint of the hypotenuse BC and P the foot of the perpendicular from A to BC. Then $\overline{AD} = \overline{BD} = \overline{CD}$ and $\overline{AP}^2 = \overline{BP} \cdot \overline{CP}$.

E 13. A quadrilateral is cyclic if and only if each pair of opposite angles add up to $180°$.

E 14. The angle subtended at the center of a circle by an arc is double that subtended at the circumference.

E 15. The angle between the tangent and chord to a circle through the point of tangency is equal to the angle subtended by the chord on the opposite side at the circumference. The diameter through a point on a circle is perpendicular to the tangent through the point.

E 16. Given a point and a line not containing the point, the point on the line nearest the given point is at the foot of the perpendicular from the point.

E 17. Let P be a point outside a circle TAB such that PT is tangent to the circle and PAB is a secant. Then $\overline{PT}^2 = \overline{PA} \cdot \overline{PB}$.

E 18. If APB and CPD are two intersecting chords of a circle $ABCD$, then

$$\overline{AP} \cdot \overline{PB} = \overline{CP} \cdot \overline{PD}.$$

E 19. Constructions: bisectors of angle, right bisector of segment, parallels, dropping perpendicular from point to line, circle through three given points, circle through two given points tangent to a given line.

E 20. Menelaus' Theorem. Let a transversal cut the sides BC, CA, AB (produced) of triangle ABC in F, G, H respectively. See Figure 250. Then

$$\frac{\overline{AH}}{\overline{HB}} \cdot \frac{\overline{BF}}{\overline{FC}} \cdot \frac{\overline{CG}}{\overline{GA}} = -1.$$

E 21. A cevian of a triangle is any segment joining a vertex to a point of the opposite side (possibly produced). Ceva's Theorem provides that,

if AD, BE, CF are three cevians of a triangle passing through a common point, then

$$\frac{\overline{AF}}{\overline{FB}} \cdot \frac{\overline{BD}}{\overline{DC}} \cdot \frac{\overline{CE}}{\overline{EA}} = 1.$$

E 22. Analytic geometry: distance between two points, equations of lines and conics.

E 23. Area of a triangle in the Cartesian plane. Let (x_1, y_1), (x_2, y_2), (x_3, y_3) be three points. The area of the triangle is given by

$$\frac{1}{2}|(x_2 y_3 - x_3 y_2) + (x_3 y_1 - x_1 y_3)$$
$$+ (x_1 y_2 - x_2 y_1)|.$$

E 24. Distance from a point to line in plane. Let $ax + by = c$ be a line and (x_1, y_1) a point in the Cartesian plane. The perpendicular distance from the point to the line is given by

$$\frac{|ax_1 + by_1 - c|}{\sqrt{a^2 + b^2}}.$$

E 25. Parabola. See Figure 251. Let a point F and a line D be given in the plane. The locus of those points equidistant from F and D is called a parabola. The line D is the directrix and F is the focus of the parabola. The line ℓ through F perpendicular to D is the axis of the parabola; it meets the parabola at its vertex V. Let P be a point on the parabola. The line through P parallel to the axis and the line PF make equal angles with the normal to the parabola at P. (This is called the reflection property.)

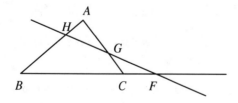

FIGURE 250

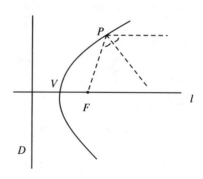

FIGURE 251

E 26. The locus of points in space equidistant from the distinct points P, Q is a plane perpendicular to and passing through the midpoint of PQ.

E 27. If A, B, C, O are points in space such that $AO \perp CO$ and $BO \perp CO$ then $PO \perp CO$ for any point P on AB or AB produced.

E 28. A plane or solid figure is said to be convex if it contains along with any two points, the line segment joining the points.

E 29. A polyhedron is a solid geometric figure whose faces are plane polygons. A tetrahedron is a polyhedron with four triangular faces. The polyhedron is regular if and only if all faces are congruent regular polygons and each vertex is incident with the same number of edges. There are only five regular polyhedra: the tetrahedron (with equilateral triangular faces), the cube, the octahedron, the dodecahedron and the icosahedron. If a polyhedron with no holes has V vertices, E edges and F faces, then Euler's formula gives the relation $E + 2 = F + V$.

E 30. A figure in space is coplanar if and only if some plane contains it; otherwise, it is skew. Two lines are skew when they neither intersect nor are parallel. Lines which all intersect in a single point are said to be concurrent.

E 31. A set in space is cospherical if and only if it is contained in the surface of some sphere. Given four non-coplanar points, there is a unique sphere which contains them. A circle drawn on the surface of a sphere whose center coincides with that of the sphere is said to be a great circle.

E 32. The angle between two intersecting planes is defined to be the angle between intersecting normals to the planes.

E 33. Trihedral angle inequality. Let the pairs of three concurrent lines make angles A, B, and C at the point of intersection. Then

$$A + B \geq C.$$

E 34. A cone in space is the smallest convex set which contains a planar set (its base) and an external point (its vertex). Examples are: tetrahedron, pyramid, right-circular cone (ice-cream cone). The volume of a cone is equal to one third of the base area multiplied by the height (perpendicular distance form the vertex to the plane of the base).

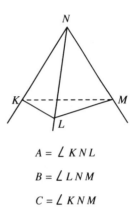

$$A = \angle KNL$$

$$B = \angle LNM$$

$$C = \angle KNM$$

FIGURE 252

E 35. A geometric figure is centrosymmetric about a point O if, whenever P is in the figure and O is the midpoint of a segment PQ, then Q is also in the figure.

E 36. A homothety (central similarity) is a transformation that fixes one point O (its center) and maps each point P to a point P' for which O, P, P' are collinear and the ratio $\overline{OP} = \overline{OP'} = k$ is constant (k can be positive or negative).

E 37. Desargue's Theorem. If two triangles have corresponding vertices joined by lines which are either parallel or concurrent, then the intersections of corresponding sides are collinear. The converse holds: if the intersections of corresponding sides are collinear, then the lines joining corresponding vertices are either parallel or concurrent.

E 38. Referred to three mutually perpendicular coordinate axes, each point in space can be coordinated (x, y, z). The distance between the points with coordinates (x, y, z) and (u, v, w) is $\sqrt{(x-u)^2 + (y-v)^2 + (z-w)^2}$.

E 39. $\overrightarrow{AB}$ denotes the vector from point A to point B. With respect to a fixed reference point O every point P has a corresponding vector $\overrightarrow{P} = \overrightarrow{OP}$. The set $\{\overrightarrow{A_iB_i} \mid i = 1, \ldots, n\}$ of vectors is linearly independent if and only if the condition

$$r_1\overrightarrow{A_1B_1} + \ldots + r_n\overrightarrow{A_nB_n} = \overrightarrow{O} = \overrightarrow{OO}$$

for real scalars $r_1, \ldots, r_n$ implies $r_1 = r_2 = \cdots = r_n = 0$. This means that none of the vectors can be written as a vector sum of multiples of the remaining vectors.

E 40. Let $\overrightarrow{OA}$ and $\overrightarrow{OB}$ be two vectors. P is a point on the segment AB if and only if $\overrightarrow{OP} = (1-t)\overrightarrow{OA} + t\overrightarrow{OB}$ for some number t with $0 \leq t \leq 1$.

E 41. Vectors in space. Let $\overrightarrow{x}$ and $\overrightarrow{y}$ be two vectors in three-dimensional space. If θ is the angle between the vectors,

$$\overrightarrow{x} \cdot \overrightarrow{y} = |\overrightarrow{x}||\overrightarrow{y}| \cos\theta.$$

Also, $\overrightarrow{x} \times \overrightarrow{y}$ is $\overrightarrow{O}$ when $\overrightarrow{x}$ and $\overrightarrow{y}$ are parallel and, otherwise, is a vector in a direction perpendicular to $\overrightarrow{x}$ and to $\overrightarrow{y}$ (the direction of the thumb if the fingers of the right hand are wrapped around through $\overrightarrow{x}$ then $\overrightarrow{y}$) with magnitude $|\overrightarrow{x}||\overrightarrow{y}| \sin\theta$ $(0 \leq \theta \leq \pi)$.

If the vectors are referred to the standard mutually perpendicular triple of unit vectors $\overrightarrow{i}, \overrightarrow{j}, \overrightarrow{k}$, i.e., $\overrightarrow{x} = x_1\overrightarrow{i} + x_2\overrightarrow{j} + x_3\overrightarrow{k}$, $\overrightarrow{y} = y_1\overrightarrow{i} + y_2\overrightarrow{j} + y_3\overrightarrow{k}$, then

$$\overrightarrow{x} \cdot \overrightarrow{y} = x_1y_1 + x_2y_2 + x_3y_3$$

$$\overrightarrow{x} \times \overrightarrow{y} = (x_2y_3 - x_3y_2)\overrightarrow{i}$$
$$+ (x_3y_1 - x_1y_3)\overrightarrow{j}$$
$$+ (x_1y_2 - x_2y_1)\overrightarrow{k}.$$

The magnitude of $\overrightarrow{x} \times \overrightarrow{y}$ can be interpreted as the area of the parallelogram with adjacent sides $\overrightarrow{x}$ and $\overrightarrow{y}$. The triple scalar product

$$\overrightarrow{x} \cdot (\overrightarrow{y} \times \overrightarrow{z}) = \overrightarrow{y} \cdot (\overrightarrow{z} \times \overrightarrow{x})$$
$$= \overrightarrow{z} \cdot (\overrightarrow{x} \times \overrightarrow{y})$$

is the volume of the parallelepiped with concurrent sides $\overrightarrow{x}, \overrightarrow{y}, \overrightarrow{z}$.

$$\overrightarrow{x} \cdot \overrightarrow{y} = 0$$

if and only if $\overrightarrow{x}$ and $\overrightarrow{y}$ are orthogonal.

E 42. Planes and lines in Space. The equation of a plane in space passing through a given point P and with normal in the direction of the given vector $\overrightarrow{u}$ is $(\overrightarrow{P} - \overrightarrow{X}) \cdot \overrightarrow{u} = 0$. A point X lies on the line through the given points P and Q if

$$\overrightarrow{X} = (1-t)\overrightarrow{P} + t\overrightarrow{Q}, \quad t \text{ a real number.}$$

The point X lies strictly between P and Q if $0 < t < 1$; the midpoint of the segment PQ is given by $\frac{1}{2}(\overrightarrow{P} + \overrightarrow{Q})$. Alternatively, if the line passes through P and its direction is $\overrightarrow{u}$, then a general point X on the line is given by $\overrightarrow{P} + t\overrightarrow{u}$, where t is real.

The distance from the point $\overrightarrow{c}$ to a plane of equation $\overrightarrow{u} \cdot \overrightarrow{x} = k$ is

$$\frac{|k - \overrightarrow{u} \cdot \overrightarrow{c}|}{|u|}$$

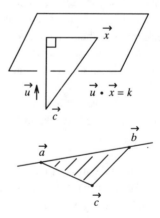

FIGURE 253

This can be seen from Figure 253, where x is any point on the plane and $\dfrac{|(\overrightarrow{x}-\overrightarrow{c})\cdot\overrightarrow{u}|}{|\overrightarrow{u}|}$ is the length of the projection of $\overrightarrow{x}-\overrightarrow{c}$ onto the direction $\overrightarrow{u}$.

The distance from the point $\overrightarrow{c}$ to the line through $\overrightarrow{a}$ and $\overrightarrow{b}$ is

$$\frac{|(\overrightarrow{a}-\overrightarrow{c})\times(\overrightarrow{b}-\overrightarrow{c})|}{|\overrightarrow{a}-\overrightarrow{b}|}$$

(The numerator is twice the area of the shaded region, the denominator the length of the base from a to b. See Figure 253.)

The distance from the point c to the line through point a in the direction $\overrightarrow{u}$ is

$$\frac{|(\overrightarrow{a}-\overrightarrow{c})\times\overrightarrow{u}|}{|u|}.$$

The perpendicular distance from the line through point a in direction $\overrightarrow{u}$ and the line through point b in the direction $\overrightarrow{v}$ is

$$\frac{|(\overrightarrow{a}-\overrightarrow{b})\cdot(\overrightarrow{u}\times\overrightarrow{v})|}{|\overrightarrow{u}\times\overrightarrow{v}|}.$$

The lines are assumed to be skew. See Figure 254.

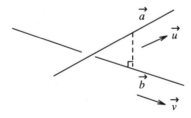

FIGURE 254

(This amounts to finding the length of the projection of $\overrightarrow{a}-\overrightarrow{b}$ onto the direction $\overrightarrow{u}\times\overrightarrow{v}$ of the common perpendicular of the lines.)

E 43. The Ptolemaic inequality. Let a, b, c, d be the lengths of the sides (taken in order) of a quadrilateral and let u,v be the lengths of the diagonals. Then $ac+bd\geq uv$, with equality if and only if the quadrilateral is concyclic.

E 44. The area of a circle is πr^2 where r is the radius. The area of a triangle is given by

$$\frac{1}{2}bh = \left[s(s-a)(s-b)(s-c)\right]^{\frac{1}{2}}$$
$$= \frac{1}{2}ab\sin C,$$

where a,b,c are the lengths of the sides, C is the angle opposite c, h is the length of the altitude perpendicular to the side of length b and s is half the perimeter.

E 45. Let ABC be a triangle with sides opposite A, B, C of length a,b,c respectively. Let $\triangle$ be its area and R its circumradius—the radius of the circumcircle, which passes through A, B, C and has its center at the intersection of the right bisectors of the sides of the triangle. Then

$$a = 2R\sin A, \quad b = 2R\sin B, \quad c = 2R\sin C,$$

$$\triangle = \frac{1}{2}ab\sin C = \frac{1}{2}ca\sin B$$
$$= \frac{1}{2}bc\sin A = \frac{abc}{4R}.$$

From these relations we have the Law of Sines:

$$\frac{\sin A}{a} = \frac{\sin B}{b} = \frac{\sin C}{c}.$$

The incircle which lies within the triangle and is tangent to the sides has its center at the intersection of the bisectors of the angles of the triangle; the radius r of the incircle is given by $2\triangle = r(a+b+c)$, where $\triangle$ is the area of the triangle.

E 46. Let r be the inradius and r_a, r_b, r_c be the three radii of the escribed circles of a triangle ABC (the circle with radius r_1 touches the side BC externally and sides AB and AC produced, for example). Then

$$r = \frac{\triangle}{s}, \quad r_a = \frac{\triangle}{s-a},$$
$$r_b = \frac{\triangle}{s-b}, \quad r_c = \frac{\triangle}{s-c},$$

$$r = (s-a)\tan\frac{A}{2} = a\sec\frac{A}{2}\sin\frac{B}{2}\sin\frac{C}{2}$$
$$= 4R\sin\frac{A}{2}\sin\frac{B}{2}\sin\frac{C}{2}$$
$$r_a = 4R\sin\frac{A}{2}\cos\frac{B}{2}\cos\frac{C}{2}$$
$$= a\sec\frac{A}{2}\cos\frac{B}{2}\cos\frac{C}{2}$$

There are analogous equations involving b and c.

E 47.   For any angles α and β

$$\tan(\alpha+\beta) = \frac{\tan\alpha + \tan\beta}{1 - \tan\alpha\tan\beta}.$$

E 48.   Pythagorean triples. A set (a,b,c) of three numbers is a Pythagorean triple if there is a right-angled triangle with sides of lengths a, b, c. If c is the length of the hypotenuse, this is equivalent to the assertion : $c^2 = a^2 + b^2$.

If a, b, and c are integers, the triple is primitive if the greatest common divisor of a, b, c is 1. All primitive (integer) Pythagorean triples (a, b, c) are given parametrically by

$$a = 2uv, \qquad b = u^2 - v^2, \qquad c = u^2 + v^2,$$

where u and v are relatively prime, are of different parity and satisfy $u > v$.

F. Analysis

F 1.   Logarithms. Let b, a be positive real numbers with $b \neq 1$. Then $x = \log_b a$ (logarithm to base b of a) if and only if $a = b^x$. The following hold:

$$\log_b(uv) = \log_b u + \log_b v,$$
$$\log_b 1 = 0, \qquad \log_b b = 1.$$

Chain rule:

$$\log_a b \log_b c = \log_a c \qquad (a, b, c > 0, a, b \neq 1),$$
$$\log_a b = \frac{1}{\log_b a} \qquad (a, b > 0, a, b \neq 1).$$

F 2.   Let a and b be real numbers. Then

$$\max(a,b) = \begin{cases} b & \text{if } a \leq b, \\ a & \text{if } b < a; \end{cases}$$
$$\min(a,b) = \begin{cases} a & \text{if } a \leq b, \\ b & \text{if } b < a. \end{cases}$$

F 3.   Difference equation. Let a, b be real numbers. Suppose $(x_1, x_2, x_3, \ldots, x_n, \ldots)$ is a sequence for which

$$x_n = ax_{n-1} + bx_{n-2} \quad (n = 3, 4, 5, 6, \ldots). \quad (*)$$

The equation $t^2 = at + b$ is the characteristic equation for $(*)$. If it has two distinct solutions t_1 and t_2, then

$$x_n = c_1 t_1^n + c_2 t_2^n \quad (n = 1, 2, \ldots),$$

where c_1 and c_2 are determined by

$$x_1 = c_1 t_1 + c_2 t_2, \qquad x_2 = c_1 t_1^2 + c_2 t_2^2.$$

If, on the other hand, the characteristic equation has only one solution, m (a double root of $t^2 - at - b$), then $(*)$ has the form

$$x_n = 2mx_{n-1} - m^2 x_{n-2} \quad (n = 3, 4, 5, 6, \ldots).$$

In this case

$$x_n = (c_1 + c_2 n)m^n \quad (n = 1, 2, \ldots)$$

where c_1 and c_2 are chosen to satisfy

$$x_1 = (c_1 + c_2)m, \qquad x_2 = (c_1 + 2c_2)m^2.$$

For example, the Fibonacci sequence $(1, 1, 2, 3, 5, \ldots)$ satisfies

$$f_1 = f_2 = 1, \qquad f_n = f_{n-1} + f_{n-2},$$

so that the characteristic equation is $t^2 = t + 1$, with solutions

$$t_1 = \frac{1}{2}(1 + \sqrt{5}), \qquad t_2 = \frac{1}{2}(1 - \sqrt{5}).$$

Thus,

$$f_n = \frac{1}{\sqrt{5}}\left\{\left(\frac{1+\sqrt{5}}{2}\right)^n - \left(\frac{1-\sqrt{5}}{2}\right)^n\right\}.$$

F 4.   Intervals:

$$(a,b) = \{x \mid a < x < b\},$$

$$[a,b] = \{x \mid a \le x \le b\},$$

$$[a,b) = \{x \mid a \le x < b\},$$

$$(a,b] = \{x \mid a < x \le b\}.$$

F 5.   Euler's equation: $e^{i\theta} = \cos\theta + i\sin\theta$ where θ is real and e is the base of natural logarithms.

Every complex number $z = x + yi$ can be written in the form $re^{i\theta}$ where $r = (x^2 + y^2)^{1/2}$ and $\theta = \arctan\frac{y}{x}$.

F 6.   Let z and w be two complex numbers. Then the set $\{(1-t)z + tw \mid 0 \le t \le 1\}$ is the line segment joining z and w in the complex plane. The point v which divides the segment joining z and w in the ratio $v - z : w - v = r : s$ is

$$v = \frac{s}{r+s}z + \frac{r}{r+s}w.$$

An ex-student of G. A. Miller, returning to the Illinois campus, noted that Miller was dressed extremely shabbily. The student had always liked Miller and now he was a successful business man. He therefore sent Miller a gift—a check for $100. He was rather puzzled that this cheque was never cashed. The reason became clear after Miller died and left $1,000,000 to the university.

Index of Problems